Books are to be returned on or before
the last date below.

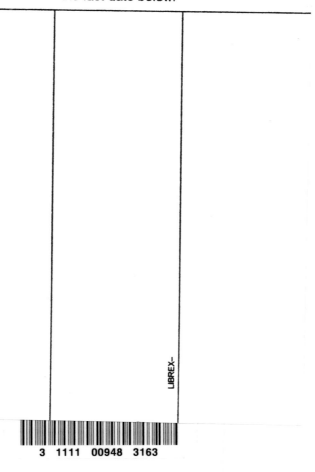

LIBREX–

EXS 93

Modern Methods of Drug Discovery

Edited by A. Hillisch and R. Hilgenfeld

Birkhäuser Verlag
Basel · Boston · Berlin

Editors

Dr. Alexander Hillisch
EnTec GmbH
Structural Bioinformatics and Drug Design
Adolf-Reichwein-Straße 20
D-07745 Jena, Germany
alexander.hillisch@entec-jena.de
www.entec-jena.de

Prof. Rolf Hilgenfeld
Institut für Molekulare Biotechnologie e.V.
Abt. Strukturbiologie-Kristallographie
Beutenbergstraße 11
D-07745 Jena, Germany
hilgenfeld@imb-jena.de
www.imb-jena.de/www_sbx

Library of Congress Cataloging-in-Publication Data
Modern methods of drug discovery / Alexander Hillisch, Rolf Hilgenfeld, editors.
    p. cm
  Includes bibliographical references and index.
  ISBN 376436081X (alk. paper)
  1. Drugs--Research. 2. Pharmaceutical technology. 3. Pharmaceutical chemistry--Data
  processing. 4. QSAR (Biochemistry) 5. Combinatorial chemistry. I. Hillisch, Alexander,
  1971- II. Hilgenfeld, R. (Rolf)

  RM301.25 .M63 2002
  615'.19--dc21
                            2002028141

Bibliographic information published by Die Deutsche Bibliothek
Die Deutsche Bibliothek lists this publication in the Deutsche Nationalbibliografie; detailed
bibliographic data is available in the internet at http://dnb.ddb.de.

ISBN 3-7643-6081-X Birkhäuser Verlag, Basel - Boston - Berlin

© 2003 Birkhäuser Verlag, P.O. Box 133, CH-4010 Basel, Switzerland
Member of the BertelsmannSpringer Publishing Group
Printed on acid-free paper produced from chlorine-free pulp. TCF ∞
Cover design: Micha Lotrovsky, Therwil, Switzerland
Cover illustration: Background picture: x-ray structure of influenza virus B neuraminidase complexed with Zanamivir™. This compound was designed on the basis of the 3D-structure of the enzyme. The surface of the drug is color coded according to the electrostatic potential (blue negatively and red positively charged). The picture was generated using SYBYL 6.8. Icons (from left to right): tips of a pipetting robot used for high-throughput screening of compound libraries (provided by CyBio AG, Jena, Germany); Yew tree (*Taxus baccata*) needles and fruits containing taxol, a potent anticancer drug; sequence alignment of the nuclear receptor protein family (provided by EnTec GmbH, Hamburg/Jena, Germany); DNA-microarray used to determine gene expression profiles (provided by Clondiag Chip Technologies GmbH, Jena, Germany).

Printed in Germany
ISBN 3-7643-6081-X
9 8 7 6 5 4 3 2 1
                                          http://www.birkhauser.ch

Contents

List of contributors

James R. Blinn, Pharmacia, Discovery Research, 7000 Portage Road, Kalamazoo, MI 49008, USA

Mark T.D. Cronin, Liverpool John Moores University, School of Pharmacy and Chemistry, Byrom Street, Liverpool, L3 3AF, UK; e-mail: m.t.cronin@livjm.ac.uk

Mary J. Cunningham, Incyte Pharmaceuticals, 3160 Porter Drive, Palo Alto, CA 94304, USA

Klaus Fiebig, Institut für Organische Chemie, Johann-Wolfgang von Goethe-Universität Frankfurt/Main, Marie-Curie Straße 11, D-60431 Frankfurt a.M., Germany; e-mail: kf@org.chemie.uni-frankfurt.de

Hua Gao, Pharmacia, Discovery Research, 7000 Portage Road, Kalamazoo, MI 49008, USA

Helmut Giersiefen, Curacyte AG, Gollierstraße 70, 80339 München, Germany

Susanne Grabley, Hans-Knöll-Institut für Naturstoff-Forschung e.V., Beutenbergstraße 11a, D-07745 Jena, Germany; e-mail: sgrabley@pmail.hki-jena.de

Rolf Hilgenfeld, Institut für Molekulare Biotechnologie e.V., Beutenbergstraße 11, D-07745 Jena, Germany

Alexander Hillisch, EnTec GmbH, Adolf-Reichwein-Straße 20, D-07745 Jena, Germany; e-mail: alexander.hillisch@entec-jena.de

Gordon Holt, Oxford GlycoSciences (UK) Ltd, 10 The Quadrant, Abingdon Science Park, Abingdon OX14 3YS, UK

Richard A. Houghten, Torrey Pines Institute for Molecular Studies, 3550 General Atomics Court, San Diego, CA 92121, USA; e-mail: rhoughten@tpims.org

David B. Jackson, Cellzome, Meyerhofstraße 1, 69117 Heidelberg, Germany; e-mail: david.jackson@cellzome.de

Glen E. Kellogg, Virginia Commonwealth University, Department of Medicinal Chemistry, School of Pharmacy, Richmond, VA 23298-0540, USA; e-mail: glen.kellogg@vcu.edu

Alastair Matheson, 5 Highcroft Road, London N19 3AQ, UK

Hans Matter, Aventis Pharma Deutschland GmbH, DI & A Chemistry, Molecular Modelling, Building G878, D-65926 Frankfurt a.M., Germany; e-mail: hans.matter@aventis.com

Eric Minch, LION bioscience AG, Waldhof Straße 98, Wieblingen, D-69123 Heidelberg, Germany

Chris Moyses, Oxford GlycoSciences (UK) Ltd, 10 The Quadrant, Abingdon Science Park, Abingdon OX14 3YS, UK

Robin E. Munro, LION bioscience AG, Waldhof Straße 98, Wieblingen, D-69123 Heidelberg, Germany

Adel Nefzi, Torrey Pines Institute for Molecular Studies, 3550 General Atomics Court, San Diego, CA 92121, USA

John M. Ostresh, Torrey Pines Institute for Molecular Studies, 3550 General Atomics Court, San Diego, CA 92121, USA

Martin J. Page, OSI Pharmaceuticals, Cancer Biology, Watlington Road, Oxford OX4 6LT, UK; e-mail: mpage@osip.com

Douglas C. Rohrer, Pharmacia, Discovery Research, 7000 Portage Road, Kalamazoo, MI 49008, USA

Isabel Sattler, Hans-Knöll-Institut für Naturstoff-Forschung e.V., Beutenberg-straße 11a, D-07745 Jena, Germany

Simon F. Semus, Wyeth-Ayerst Research, Department of Biological Chemistry, CN8000, Princeton, NJ 08543, USA; present address: GlaxoSmithKline, 709 Swedeland Road, King of Prussia, PA 19406, USA; e-mail: Simon.F.Semus@gsk.com

Ralf Thiericke, CyBio Screening GmbH, Winzerlaer Strasse 2a, D-07745 Jena, Germany; e-mail: ralf.thiericke@cybio-ag.com

Han van de Waterbeemd, Global Research and Development, PDM, Sandwich, Kent CT13 9NJ, UK; e-mail: han_waterbeemd@sandwich.pfizer.com

John H. van Drie, Vertex Pharmaceuticals, 130 Wavely St, Cambridge, MA 02139; e-mail: vandrie@mindspring.com

Martin Vogtherr, Institut für Organische Chemie, Johann-Wolfgang von Goethe-Universität Frankfurt/Main, Marie-Curie Straße 11, D-60431 Frankfurt a.M., Germany

Preface

Research in the pharmaceutical industry today is in many respects quite different from what it used to be only fifteen years ago. There have been dramatic changes in approaches for identifying new chemical entities with a desired biological activity. While chemical modification of existing leads was the most important approach in the 1970s and 1980s, high-throughput screening and structure-based design are now major players among a multitude of methods used in drug discovery. Quite often, companies favor one of these relatively new approaches over the other, e.g., screening over rational design, or *vice versa*, but we believe that an intelligent and concerted use of several or all methods currently available to drug discovery will be more successful in the medium term.

What has changed most significantly in the past few years is the time available for identifying new chemical entities. Because of the high costs of drug discovery projects, pressure for maximum success in the shortest possible time is higher than ever. In addition, the multidisciplinary character of the field is much more pronounced today than it used to be. As a consequence, researchers and project managers in the pharmaceutical industry should have a solid knowledge of the more important methods available to drug discovery, because it is the rapidly and intelligently combined use of these which will determine the success or failure of preclinical projects.

In spite of this, few publications are available that provide an overview of the rich spectrum of methods available. To fill that need, we conceived this book, which we hope will be useful for both the active researcher and the manager in the pharmaceutical industry and in academia.

We are most grateful to the authors of the individual chapters who contributed excellent texts despite the demands of all actively being involved in drug discovery research. We are indebted to Birkhäuser Publishing Ltd., in particular Dr. Klüber, for constant encouragement of the project, even though it did have its difficulties with timelines. A.H. also thanks his employers, Entec GmbH, for supporting this project.

Jena, August 2002
Rolf Hilgenfeld
Alexander Hillisch

Modern Methods of Drug Discovery
ed. by A. Hillisch and R. Hilgenfeld
© 2003 Birkhäuser Verlag/Switzerland

1   Modern methods of drug discovery: An introduction

Helmut Giersiefen[1], Rolf Hilgenfeld[2] and Alexander Hillisch[3]

[1] *Curacyte AG, Gollierstr. 70, D-80339 München, Germany*
[2] *Institut für Molekulare Biotechnologie e.V., Beutenbergstr. 11, D-07745 Jena, Germany*
[3] *EnTec GmbH, Adolf-Reichwein-Str. 20, D-07745 Jena, Germany*

1.1   Introduction

The pharmaceutical industry is continuing to attempt double-digit growth rates driven by high market capitalization. Standard responses to this challenge have only provided limited impact. Besides scaling-up businesses through mergers or selective acquisitions of platform technologies or drug candidates, an increase of Research and Development (R&D) productivity still represents a sure approach to address this challenge.

An increase of R&D productivity can either be accomplished by increasing the efficiency of R&D (lowering cost or decreasing time-to-market), or by reducing the failure rates throughout the pharmaceutical value chain. Over the past decade pharmaceutical companies successfully increased R&D productivity, specifically by re-engineering their development processes. However, these optimization approaches may be reaching their limits.

In this chapter, we give an overview of the economics of drug discovery and development, and discuss some of the most important methods that have potential to accelerate the drug discovery process.

1.2   Economics of drug discovery and development

Research-based pharmaceutical companies, on average, spend about 20% of their sales for R&D [1]. This percentage is significantly higher than in virtually any other industry, including electronics, aerospace, automobiles and computers [1]. Since 1980, U.S. pharmaceutical companies have practically doubled spending on R&D every five years [2].

Despite these enormous expenditures and efforts of pharmaceutical companies, there has been a steady decline in the number of drugs introduced each year into human therapy, from 70–100 in the 1960s, 60–70 in the 1970s, to about 50 in the 1980s and below 40 in the 1990s.

In 1996, the term "innovation deficit" was introduced by Jürgen Drews, president of international research at Hoffmann-la Roche. "Innovation deficit" defines the gap between the number of new chemical entities (NCEs) required to be launched in order to accomplish an annual 10% revenue increase and the actual number of NCEs introduced in the market by the top 10 pharmaceutical companies. While Drews predicted a deficit of 1.3 NCEs per company for 1999, a world-leading management consulting firm recently published a real lack of 1.5 NCEs in 2000 and expected this trend to continue, resulting in a deficit of 2.3 NCEs by 2005 [3] (see Fig. 1).

A new drug today requires an average investment of $880 million [4] and 15 years of development (see Fig. 2), including the cost and time to discover potential biological targets. About 75% of this cost (~$660 million) is attributable to failure along the pharmaceutical value chain [4, 5]. For example, 90% of all drug development candidates fail to make it to the market. Out of the ~15 years in development time of a successful compound, about 6 years are devoted to the drug discovery and the preclinical phase, 6.7 years to clinical trials and 2.2 years to the approval phase [6]. Figure 3 shows that nearly one-third of company financed

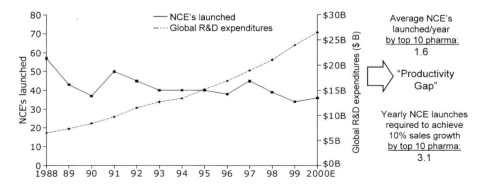

Figure 1. The innovation deficit in the pharmaceutical industry: The gap between the number of new chemical entities (NCEs) required to be launched in order to accomplish an annual 10% revenue increase and the actual number of NCEs introduced in the market by the top 10 pharmaceutical companies. In 2000 this gap was 1.5 NCEs.

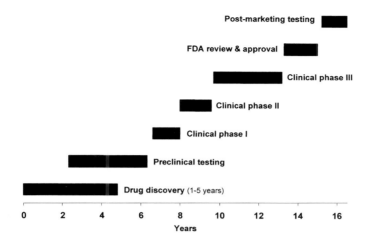

Source: **PhRMA,** based on data from Center for the Study of Drug Development, Tufts University, 1995

Figure 2. Average periods for steps in the drug discovery and developement procedure (FDA: Food and Drug Administration).

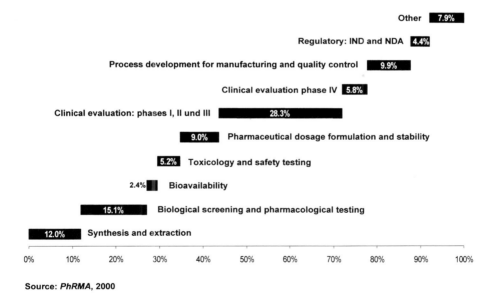

Source: **PhRMA,** 2000

Figure 3. Allocation of R&D by function in percentage terms (IND: Investigational New Drug, NDA: New Drug Application).

R&D is devoted to the drug discovery phase and to preclinical development, whereas about half of all activities are spent on clinical and chemical/pharmaceutical development (see Fig. 3).

One reason for "innovation deficit" is the increased demand on safety of drugs. The average number of clinical trials per new drug application (NDA) steadily increased from 30 in the 1970s, about 40 in the 1980s to 70 in the 1990s [7]. If 1,500 patients were required for an NDA in the early 1980s, today the average number of patients is around 4,000. The increased demand on safety is consequently reflected in a prolonged duration of the drug development process. In the 1960s, the total development time was 8.1 years. This rose to 11.8 years in the 1970s, 14.2 years in the 1980s to finally 14.9 years in the 1990s [6].

Since drug safety cannot be compromised, methods that enhance R&D productivity, namely accelerate drug discovery and reduce failure rates during later drug development, are desirable.

Can the new technologies pertaining to target and drug discovery fulfill these needs? Recent advances, for example, in functional genomics, proteomics, assay development, combinatorial chemistry, high-throughput screening, structure-based ligand design and pharmcogenomics will definitively enrich the basis for discovering and developing innovative therapeutic products, but can these methods also improve the economics of drug research, development and commercialization *per se*?

1.3   The drug discovery process: methods and strategies

In the following current and future drug discovery methods and strategies are reviewed in chronological order throughout the drug discovery process. Many of these techniques are described in more detail in the other chapters of this book. We refer solely to small organic molecules and not therapeutic proteins ("biologicals"), for which the situation may be quite different.

The drug discovery process can be divided into four steps: drug target identification, target validation, lead compound identification and optimization. In Figure 4 some of the key methods used to discover new drugs are depicted.

1.3.1   Target identification

The identification of new and clinically relevant molecular targets for drug intervention is of outstanding importance to the discovery of innovative drugs.

It is estimated that up to 10 genes contribute to multifactorial diseases [8]. Typically, these "disease genes" are linked to another five to 10 gene products in physiological or pathophysiological circuits which are also suitable for intervention with drugs. If these numbers are multiplied with the number of diseases that pose a major medical problem in the industrial world, which is between 100 and 150, we end up with about 5,000 to 10,000 potential drug targets. A recent analysis shows that current drug therapy is based on less than 500 molecular targets [8], with 45% of these being G-protein coupled receptors, 28% enzymes, 11% hormones and factors, 5% ion channels and 2% nuclear receptors. Thus, the number

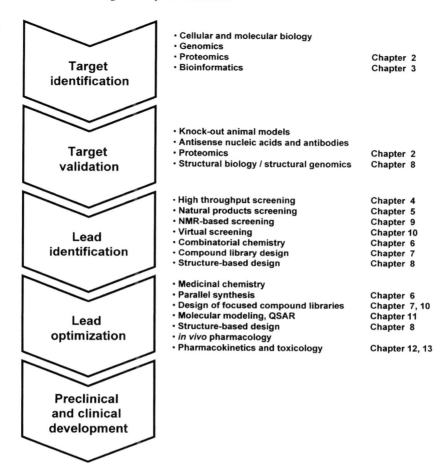

Figure 4. Phases of the drug discovery process. After target identification and validation chemical substances are discovered and optimized with regard to efficacy, potency, selectivity as well as pharmacokinetic, toxicological and metabolic aspects. This process is followed by preclinical and clinical development. Methods reviewed in this book are indicated by references to the respective chapters.

of drug targets that can be exploited in the future is at least 10 times the number of targets currently implemented in drug therapy.

Besides classical methods of cellular and molecular biology, new techniques of target identification are becoming increasingly important. These methods aim at i) discovering new genes and proteins and ii) quantifying and analyzing gene and protein expression between diseased and normal cells.

Methods routinely used to identify new drug targets are genomics [9] and proteomics [10, 11] (see Chapter 2).

The term genomics was coined in the mid 1980s. This new discipline has evolved from two independent advances:

- Automation—resulting in a significant increase in the number of experiments that could be conducted in a given time and thereby generating vast amounts of scientific data.
- Informatics—the ability to transform raw data into meaningful information and knowledge by applying computerized techniques for managing, analyzing and interpreting data obtained in experiments (for bioinformatics see Chapter 3) [12]. Without informatics, data would remain raw material. This latter trend continues due to the ever-accelerating power of computers, enhanced algorithms, integration of data and technology platforms and the versatility of the internet that enables global sharing and exchange of scientific knowledge and experience [13].

Clearly, the identification of new and clinically relevant biological targets has benefited from the genomics approach. The huge genomics initiatives which started in the early 1990s have lead to an enormous amount of DNA sequence information. The entire genomes of 695 viruses, 22 viroids, 13 archaeae, 57 bacteriae and 4 eukaryotae including one fungus, one plant, two animals have been sequenced so far. Two independent drafts of the human genome, each approximately 90% complete have recently been published by the "Human Genome Consortium" [14] and Celera Genomics Inc. [15]. The sequence information of the 98% complete version (August 2002) of the human genome is freely available to the public on the internet [16]. Many more genomes, such as mouse, rat, zebrafish, corn, wheat and rice are currently being sequenced. Bioinformatics methods (see Chapter 3) are used to transform the raw sequence data into meaningful information (e.g., genes and their encoded proteins) [17] and to compare whole genomes. However, gene prediction using bioinformatics is a particularly sore point in deducing meaningful information from sequence data. Currently, the correlation between predicted and actual genes is around 70% with just 40–50% exons predicted correctly (see Chapter 3). This is exemplified even more drastically in the case of the human genome. The Human Genome Consortium found evidence for 29,691 transcripts [14], whereas the commercial genome project of Celera Genomics Inc. found 39,114 genes [15]. Both data sets agree well for 9,300 known genes. Although the consortium and Celera together predicted 31,100 novel genes, only 21% appear on both lists [18]. Comparing experimental data on gene expression with the available sequence information now yields an estimated 65,000 to 75,000 human genes [19].

The *in silico* identification of novel drug targets is now feasible by systematically searching for paralogues (paralogues are evolutionary related proteins that perform different but related functions within one organism) of known drug targets in, for example, the human genome [14, 20, 21].

The rapid progress in sequencing whole genomes will enable the identification of novel drug targets for antimicrobial and antiparasitic chemotherapy [22].

Bioinformatics will allow the subtractive comparative analysis of complete genome sequences of closely related microbial species as well as of pairs of isolates from the same species with different features, such as a pathogenic and an apathogenic representative and the subsequent identification of target proteins associated with pathogenism [23].

Considering the gene expression microarrays and especially gene chip technology, scientists today can use a single micro-device to evaluate and compare the expression of up to 20,000 genes of healthy and diseased individuals at once [24]. Thirty-three thousand genes of the human genome can thus be analyzed simultaneously using two high-density arrays of oligonucleotide probes, each at a size of about a square centimeter (Affymetrix-technology) [25].

However, it is becoming increasingly evident that the complexity of biological systems lies at the level of the proteins, and that genomics alone will definitely not suffice to understand these systems. It is also at the protein level that disease processes become manifest and at which at least 91% of drugs act [8]. Thus, the analysis of proteins (including protein-protein-, protein-nucleic acid- and protein-ligand-interactions) will be of utmost importance to target discovery. Proteomics (see Chapter 2), the systematic high-throughput separation and characterization of proteins within biological systems, help to fulfill some of these needs. Concerning expression analysis, it has been shown that the correlation between RNA and protein expression is rather weak and ranges in yeast from 10% to 40% for lower-abundance proteins up to 94% for higher-abundance proteins [26]. Some of these differences can be explained with posttranslational modifications, accessible only to proteomics analysis in this context. Target identification with proteomics is performed by comparing the protein expression levels in normal and diseased tissues. Two-dimensional polyacrylamide gel electrophoresis (2D-PAGE) is used to separate the proteins which are subsequently identified and fully characterized with matrix-assisted laser desorption/ionization mass spectrometry (MALDI-MS) [27]. Differential protein expression between normal and diseased samples probably indicates that the protein of interest is involved in the pathophysiology and could therefore be a potential drug target. This assumption has to be confirmed during the target validation process.

1.3.2   Target validation

The validation of a drug target involves demonstrating the relevance of the target protein in a disease process and ideally requires both gain and loss of function studies. This is accomplished primarily with highly specialized knock-out (loss of function) and knock-in (gain of function) animal models that are capable of mimiking the targeted disease state [28–30], or with the use of small molecules (inhibitors, agonists, antagonists), antisense nucleic acid constructs [31], ribozymes and neutralizing antibodies [32] (see also Chapter 2). The development of *in vivo* models is always costly and time consuming. Provided that genomics and proteomics will uncover novel biological targets focusing on new pathomechanisms, many of the required models have yet to be designed. Nevertheless, they will always be an approximation of the real clinical situation in humans and can thus not necessarily contribute to the predictability of drug development. Methods used for target identification at the *in vitro* level are RNA and protein expression analysis and cell based assays.

By far, not all target proteins that are relevant for a certain disease process can be affected with preferably orally bioavailable small molecules. Since strong interactions between a protein and its ligand are characterized by a high degree of complementarity, knowledge of the protein three-dimensional structure will enable the prediction of "druggability" of the protein (see Chapter 8). For example, a protein with a large hydrophobic ligand-binding pocket will also interact preferably with large apolar ligands that are certainly problematic cases with respect to physicochemical (e.g., aqueous solubility, logP) and pharmacokinetic properties. In the future, protein structure information (x-ray, NMR and comparative models) may be applied to rule out such difficult drug targets. To cope with the thousands of potential drug targets, structural genomics (systematic, large-scale efforts towards the structural characterization of all proteins) [33, 34] will play an important role. It is conceivable that tool compounds, designed on the basis of protein three-dimensional structure information, will help to unravel the physiological roles of target proteins and contribute to "chemical" target validation [35-37] (see Chapter 8). Probably such tool compounds can later be quickly turned into drugs if the target shows disease relevance.

In summary, target validation is one of the bottlenecks in the drug discovery process, since this phase is less adaptable to automation. The careful validation of target proteins, not only with respect to disease relevance, but also "druggablity," will reduce the failure rate and increase the overall efficiency of the drug discovery process.

1.3.3   Lead identification

In the lead generation phase, compounds are identified which interact with the target protein and modulate its activity. Such compounds are mainly identified using random (screening) or rational (design) approaches.

High-throughput screening (HTS, see Chapter 4) is used to test large numbers of compounds for their ability to affect the activity of target proteins. Today, entire in-house compound libraries with millions of compounds can be screened with a throughput of 10,000 (HTS) up to 100,000 compounds per day (uHTS, ultra high-throughput screening) [38, 39] using very robust test assays (see Chapter 4). Combinatorial chemistry and parallel synthesis (Chapter 6) are employed to generate such huge numbers of compounds. However, there are some concerns regarding the pure "number"-approach [40]. In theory, generating the entire or at least significant parts of chemical space for drug molecules and testing them would be an elegant approach to drug discovery. In practice, this is not feasible, since we are faced with a chemical space of 10^{62}–10^{63} compounds that comprise carbon, oxygen, nitrogen, hydrogen, sulfur, phosphorous, fluorine, chlorine and bromine, having a molecular weight of less than 500 Da [41]. Unfortunately, even the largest chemical libraries available today, including up to 10 million compounds (10^7), comprise only a tiny fraction of the entire chemical space. With existing chemistry technologies we will not exceed library sizes of

10^8 compounds in the near future. Since it seems that we have to live with that problem, concepts are needed to synthesize and select biologically relevant compounds. One solution, according to Wess et al., could be to accumulate as much knowledge as possible on biological targets (e.g., structure, function, interactions, ligands) and fill the limited chemical space with targeted approaches of chemical synthesis [40]. Natural compounds and synthetic derivatives thereof can contribute essentially to such approaches [42] (Chapter 5). This is simply because nature has provided clues as to the structural requirements for biological activity. In other words, if lead identification is compared to the proverbial "search for the needle in the haystack," then simply adding more "hay" will not yield the solution. Chemical synthesis, be it traditional solution phase chemistry, natural products derivatization, combinatorial or parallel synthesis, has to work closely together with design approaches like structural biology, molecular modeling, library design and physicochemical approaches, in order to select smart "haystacks" with more "needles."

Concluding this, the design of compound libraries will gain further importance for high- or medium throughput screening, since besides molecular similarity and diversity issues, physicochemical (Chapter 12) and toxicological aspects (Chapter 13) can now be considered to select "drug-like" compounds for screening.

Another crucial point for reliable results with high-throughput screening is the robustness and quality of the biological test assays. Homogenous "mix and measure" assays are preferred for HTS as they avoid filtration, separation, and wash steps that can be time-consuming and difficult to automate. Assays for HTS can be grouped into two categories: so-called solution-based biochemical assays and cell-based assays [43, 44]. The former are based on radioactive (scintillation proximity assay, SPA) and fluorescence (fluorescence resonance energy transfer FRET, fluorescence polarization FP, homogenous time resolved fluorescence HTRF, and fluorescence correlation spectroscopy FCS) detection methods to quantify the interaction of test compounds with biological target molecules. Scintillation proximity assays in HTS have largely replaced heterogenous assays that make use of radio-labeled ligands with subsequent filtration steps to measure high-affinity binding to receptors. Cell-based assays for HTS can be categorized as follows: i) second messenger assays that monitor signal transduction, ii) reporter gene assays that monitor cellular responses at the transcriptional/translational level, and iii) cell proliferation assays that detect induction or inhibition of cell growth.

A different approach to lead compound discovery is *in silico* or virtual screening [45, 46] (see also Chapter 10). With this computer method, three-dimensional (3D) structures of compounds from virtual or physically existing libraries are docked into binding sites of target proteins with known or predicted structure. Empirical scoring functions are used to evaluate the steric and electrostatic complementarity (the fit) between the compounds and the target protein. The highest ranked compounds are then suggested for biological testing. These software tools are attractive and cost-effective approaches to generate chemical lead structures virtually and before using expensive synthetic chemistry. Furthermore, they allow rapid and thorough understanding of the relationship between chemical structure and biological function. The virtual screening of small molecules takes less than a minute per chemical structure per computer processor (CPU) when using the more sophisticated computer algorithms that are available today [45]. Utilizing clusters of CPUs results in a high degree of parallelization. The throughput with 100 parallel CPUs is even higher compared to current uHTS technologies. The main advantage is that the method does not depend on the availability of compounds, meaning that not only in-house libraries can be searched, but also external or virtual libraries. The application of scoring functions on the resulting data sets facilitate smart decisions about what chemical structures bear the potential to exhibit the desired biological activity.

However, one important prerequisite for these technologies is the availability of structural data of the target and—if possible—in complex with the biological ligand or an effector molecule. Currently, there exists experimental high resolution structure information on only about 1% of all highly annotated protein sequences (see Chapter 8). Comparative models for more than 40% of these proteins are available (see Chapter 3 and 8). So, in silico screening can be applied on average to each third and probably in the future to each second drug discovery project. Structural biology will therefore become an increasingly important discipline in drug discovery. This is grounded in three factors that determine the ability of in silico screening tools for identifying lead compounds:

i) quality of the structural models determined by nuclear magnetic resonance spectroscopy (NMR) or x-ray diffraction, ii) current understanding and assumptions about protein (target) and ligand (drug) binding interactions and the transposition of this understanding into suitable models, and iii) ability to apply computational chemistry as an enabler of a cross-functional and reiterative process whereby laboratory data demonstrating the desired effect of a ligand can provide insights for refining and optimizing chemical structures. The power of in silico screening is expected to increase in the future dramatically due to the availability of protein structures from structural genomics initiatives (Chapter 8), larger structure activity data sets (Chapter 11) and experience, necessary to refine today's computer algorithms to increase their predictability (Chapter 10).

Another screening approach, namely NMR-based screening (Chapter 9) fills the gap between HTS and virtual screening [47]. This method combines the random screening approach (currently up to 1000 compounds/day) with the rational structure-based approach to lead discovery. Small organic molecules that bind to proximal subsites of a protein are identified through screening, and linked together in a rational approach, to produce high-affinity ligands (Chapter 9).

Once hits (compounds that elicit a positive response in a particular biologically assay) have been identified by applying the different screening approaches, these are validated by re-testing them and checking the purity and structure of the compounds. This is to avoid spending too much time with further characterizations of false positives from the screening. Only if the hits fulfill certain criteria are these regarded as leads. The criteria can originate from different directions such as:
• pharmacodynamic properties: efficacy, potency and selectivity in vitro and in vivo;

- physicochemical properties: e.g., Lipinski's "rule-of-five," water-solubility and chemical stability (see Chapters 7 and 12);
- pharmacokinetic properties: e.g., permeability in the Caco-2 assay (see Chapter 7 and 12), metabolic stability and toxicological aspects (see Chapter 13);
- chemical optimization potential: ease of chemical synthesis can be crucial, "dead-end-leads" which are synthetically not easily amendable to many variations should be avoided;
- patentability: compounds that are to some extent protected by competitor's patents are certainly less interesting than entirely new lead structures.

Clearly, the early consideration of pharmacokinetic and toxicological aspects seems crucial, since in the preclinical drug development phase on average 40% of the compounds fail due to poor pharmacokinetics and 11% due to toxicity [48].

1.3.4   Lead optimization

During the lead optimization phase, small organic molecules are chemically modified and subsequently pharmacologically characterized in order to obtain compounds with suitable pharmacodynamic and pharmacokinetic properties to become a drug. This process ideally requires the simultaneous optimization of multiple parameters and is thus a time-consuming and costly step, which probably constitutes the "tightest" bottleneck in drug discovery. However, by turning a biologically active chemical into an effective and safe drug, lead optimization contributes essentially towards added value in the drug discovery process.

Lead optimization is mainly a cross-talk between the traditional disciplines in drug discovery: medicinal chemistry and *in vitro/in vivo* pharmacology. Leads are characterized with respect to pharmacodynamic properties such as efficacy, potency and selectivity *in vitro* and *in vivo*, physicochemical properties (see Chapters 7 and 12) and pharmacokinetic properties (absorption, distribution, metabolism and elimination, ADME, see Chapter 7 and 12), and toxicological aspects (see Chapter 13). *In vitro* test assays allow a larger number of compounds to be characterized. The assays used to test for potency and selectivity are similar to those described for HTS (see Chapter 4) with the exception that more time-consuming detection methods (e.g. calorimetric, chromatographic, surface plasmon resonance methods) are used [50].

In parallel to compound characterization with respect to potency and selectivity, *in vitro* assays for the prediction of pharmacokinetic properties should be performed. Such *in vitro* systems utilize either Caco-2 or Madin-Darby Canine Kidney cells, human hepatic microsomes, heterologously expressed cytochrome P450 enzymes or human hepatocytes [49]. Once compounds with desirable *in vitro* profiles have been identified, these are characterized using *in vivo* models [50]. These tests usually require larger quantities of compounds and are conducted with selected molecules only.

Compounds with properties which do not fulfill the requirements for a successful drug development candidate have to be optimized through the synthesis of "better suited" derivatives. Hints and ideas on how to modify a lead compound with respect to pharmacodynamic properties (e.g., binding to a receptor or enzyme) can originate from molecular modeling, quantitative structure activity relationship (QSAR)-studies (see Chapter 11) and from structural biology (see "structure-based drug design cycle" in Chapter 8). If the three-dimensional structure of the target protein is not available, the ligand-based approach of molecular modeling can lead to new insights into structure-activity-relationships (see Chapter 11). If, for example, the steric and electrostatic properties of two different series of lead compounds that bind to the same site at the target protein are compared on the computer screen in three-dimensions, one can infer important substituents from one series to another. QSAR studies can yield mathematical equations which correlate biological activity with chemical structure (see Chapter 11). Biological activity for compounds that may or may not exist can be predicted from such equations. In cases where the three-dimensional structure of the target protein is known, docking calculations are applied to predict the relative orientation of ligands with respect to the target protein. The thorough analysis of the complexes allows one to predict how to increase the binding affinity of the ligands. These rational approaches are valuable tools in the lead optimization phase, since they connect biology with chemistry and allow a thorough understanding of the relationship between chemical structure and biological function [35]. This has been shown to accelerate the lead optimization phase (see Chapter 8).

Concerning the generation of ideas on how to modify compounds with respect to physicochemical and pharmacokinetic properties, *in silico* methods play an important role [51]. Probably the most widely used ADME-model is Lipinski's "rule-of-five," which is used to filter out compounds likely to be purely absorbed through the human intestine, based on four simple rules related to molecular properties (see Chapter 7 and 12). Other models for human intestinal absorption, blood-brain barrier (BBB) permeation or human bioavailability are based on empirical approaches such as QSAR, and require a significant quantity of high quality data from which to deduce a relationship between structure and activity [52] (see Chapter 11). Quantum mechanical simulations, structure-based design methods and expert systems can be applied to predict metabolism and toxicity (see Chapter 13).

The chemical synthesis of the proposed, better suited compounds is accomplished using traditional medicinal chemistry. Serendipity (chance findings) at this stage of drug discovery also plays an important role and should not be overlooked. Synthesizing compounds simply because they are chemically easily accessible is a common approach and has lead to important discoveries. The preparation of focused combinatorial compound libraries is also a powerful approach for the derivatization of leads (see Chapter 6 and 10), but requires meaningful HTS-assays for subsequent biological testing. However, in order to supply the most advanced and complicated test systems, the animal models [50], with sufficient amounts of pure test compounds, "traditional" chemical synthesis plays a dominant role in the optimization phase.

In conclusion, it is vital to conceive lead optimization as a simultaneous multi-dimensional optimization process rather than a sequential one. Optimizing leads first with respect to pharmacodynamic properties (e.g., potency, selectivity) and looking at pharmacokinetic parameters of these optimized compounds later, guarantees frustration in the late optimization phase. With current methods this should be partly avoidable.

1.4 Trends and visions in drug development

1.4.1 Genomic approaches

The recent regulatory approval of the Novartis drug Gleevec®, formely known as STI-571, an inhibitor to Bcr-Abl kinase and a treatment of chronic myeloid leukemia (CML) set a new benchmark in terms of development and approval time. Given the specificity challenges associated with kinase inhibition, the rigorous and selective deployment of modern drug discovery tools translated in clear customer value in terms of unsurpassed efficacy, fewer side-effects and a pharmacokinetic profile that supports once-daily oral administration.

The most important lever in terms of improving the economics of R&D is the ability to distinguish failures from successful candidates early in the process. This lever exists in every segment of the pharmaceutical value chain and has a dramatic impact on success rates of R&D portfolios. Considering, for example, all biological targets and corresponding lead drug candidates of a given pharmaceutical portfolio that fail prior to entering clinical development—if that company were able to avoid just one out of 10 earlier failures it could save an additional $100 million per drug.

Genomic-based information combined with the ability to analyze it productively will provide companies an enormous advantage. Such companies can now make more informed decisions on targets and leads to be pursued. However, improving portfolio management and decision-making will take more than just acquiring and implementing genomics technologies. It will also require strategic thinking, for example, the definition and application of rigorous selection criteria that will ultimately shape the company's product portfolio and can reduce development risk, decisions on building in-house competence or partnering, acquisition and divestiture of targets and leads, etc.

The cascade of portfolio decisions to be made again and again extends into drug discovery and drug development. Historically, pharmaceutical companies tended to develop new drugs inductively, whereby the data generated in each segment of the value chain were driving subsequent experiments and studies. In other words, experiments were designed based on what one knew about the target and the drug candidate. Such an approach bears significant risks in terms of identifying candidates that are likely to fail during drug development later than earlier, resulting in increased development costs and time-to-market. Furthermore, the generated data may not always support the ultimate value proposition that a drug has to deliver to its customers. The generation of "scientific noise" may even have posed questions that cannot not be answered easily.

Especially when pursuing new targets and pathomechanisms that may arise from the genomics approach, it is crucial to envision the desired clinical benefit early and prior to committing funds for expensive formal drug development. The ability of a company to translate clinical relevance to customers into meaningful experiments and studies in drug discovery and development will influence portfolio success rates. The discovery and development of new drugs should be guided by an understanding of the anticipated product benefit. Drug product prototypes that are structured around the key questions that customers always ask may serve as a portfolio management tool by providing a basis for stringent decision-making throughout the process. Customer questions and anticipated answers may cover essential value criteria such as efficacy, safety, dosing, quality, risk/benefit and cost/performance ratios. Experiments and studies will not only serve the generation of scientific data which is ultimately required to obtain market approval, but more importantly, the continuous analysis and interpretation of this data relative to the defined product value proposition will serve rational portfolio decisions. In case it becomes obvious from the data analysis that the envisioned value proposition cannot be accomplished, early discontinuation or mid-term corrections may be necessary. Drug discovery and development become deductive, whereby scientists become more interested in what they should know about their targets and drugs rather than what they already know.

1.4.2   Genetic approaches

While pharmaceutical companies are attempting to adopt and cope with the technological advances described in this book, the race for new technologies continues.

Having discussed the genomics wave and its potential to enhance R&D productivity, it will be genetics that will shape the future of how we discover and develop new therapies. Moreover, it can ultimately impact the entire health care system.

Two genetic approaches can impact R&D at different stages of the value chain: disease genetics and pharmacogenetics [53-55].

Disease genetics is implemented earlier in drug discovery and involves the search for genes that determine the human susceptibility for certain diseases. Once more, the discovery of new targets is the objective. However, this time new targets may not only enrich the physician's future armamentarium, they also specifically hold the prospect of transforming R&D success rates.

Pharmacogenetics will influence R&D later, in particular during clinical development. It entails the genetic-based study of genetic variations between individuals and how these variations influence drug response. Inherited differences in DNA sequence such as single-nucleotide polymorphism (SNP) contribute to phenotypic variation, influencing an individual's anthropometric characteristics, risk of disease and response to the environment (e.g., a drug) [56]. SNPs occur (on average) every 1,000–2,000 bases when two human chromosomes are compared.

With the availability of whole-genome SNP maps, it will soon be possible to create an SNP profile for patients who experience adverse events or who respond clinically to the medicine. Pharmacogenetics can thereby help to understand and predict the efficacy and safety of a potential drug candidate. There are three relevant approaches available today:

- *Pharmacogenetics* predicts drug responses by analyzing genetic variations on a DNA level;
- *Expression pharmacogenetics* accomplishes the same by specifically comparing RNA levels in different samples to determine which genes are expressed in certain disease states;
- *Proteomic pharmacogenetics* compares protein readings in different tissue samples to identify proteins that differ in structure or in expression levels (see Chapter 2).

Short-term disease genetics and pharmacogenetics will enable drug developers to predict drug responses by studying underlying genetic variants or polymorphisms. By comparing the individual's genetic variations against the "standard genome" scientists should be able to determine the risks for developing specific diseases or how well suited a patient is to a particular therapeutic intervention. The economic impact is twofold: The approach described will help to refine inclusion criteria for clinical trials and can thereby reduce trial cost and the probability of study failure in the most expensive segment of the value chain. This is especially important for disease areas that often suffered from non-responders, for example, cancer or inflammatory diseases. These diseases can have multiple underlying pathomechanisms. Therapeutic success in such cases may be limited when attacking mono-specific biological targets. In addition, the predictive knowledge regarding drug response can positively influence the prescription habit of physicians. Savings to the health care system could be paramount considering that today's drugs work only for 60% of patients at most.

Long-term, pharmacogenetics can deliver individualized therapeutic options. Customized drugs for patient sub-populations have the potential to provide predictable and improved clinical outcomes in terms of efficacy and safety. This opportunity includes the treatment of patients for whom current treatment options are ineffective or unsafe; other important drugs that have been abandoned due to a lack of safety in only a few patients could be resuscitated and be made available to those who tolerate them.

Pharmacogenetics will revise the financial basis for pharmaceutical R&D. Economic savings are likely to come from improved efficiency in discovery, improved target validation and clinical trial outcomes. The improved efficiency will be based on a consolidation of target discovery into a single step; the improved success rates are based on the ability to pinpoint targets associated with disease susceptibility. Unlike traditional targets that are usually identified through the use of cell-based or animal models and whose clinical relevance is initially speculative, pharmacogenetics begins with target validation using human disease phenotypes.

Despite the potential cost savings, pharmacogenetics will not support the industry's beloved blockbuster model. Excluding patients from clinical trials due to increasingly stringent inclusion criteria can eventually result in restricted labeling. Therefore, the pharmaceutical industry will have to carefully analyze compensatory effects including potential market upsides, price premiums that are justifiable from a pharmacoeconomic standpoint, shifts in market share and new patients eligible for individualized therapies.

The pharmaceutical industry is changing its approach towards R&D by applying knowledge and new techniques to understand the root causes rather than the symptoms of diseases. This has fundamentally changed the way new drugs are discovered and developed. The advances in drug discovery and development evolved from a detailed understanding of the human genome forming a rich basis for the discovery of many potential drug therapies. Furthermore, disease genetics and pharmacogenetics can add to the understanding of how genes and proteins work collectively in a regular and a diseased living system. New techniques for screening, designing and synthesizing chemicals using robots and computers, and most importantly, advances in information technology have facilitated this development. This new way of discovering and developing drugs is information-based. The information generated and how we use it will further accelerate and change the processes leading to innovative therapies. Recently, new software tools such as the E-CELL system were presented [57]. These tools may be able to further utilize available genetic information to ultimately simulate molecular processes in cellular systems. Naturally, the resulting independency from living systems to study drug responses would turn pharmaceutical R&D upside down. However, at present the replacement of *in vivo* models with E-CELL approaches is not realistic.

1.5   Conclusions

New technologies in drug discovery and development reviewed in this book can only provide economic impact if they are accompanied with strategic thinking. Only if pharma-managers envision new technologies in terms of their ability to deliver innovative medicines at lower cost, reduced time and with superior quality, will they be able to implement and utilize them in an economically profound way. None of the reviewed methods present all-in-one strategies for pharmaceutical R&D. For any known disease they need to be well orchestrated in order to deliver the desired outcomes.

1.6   Acknowledgements

The authors thank Peter Kengel, Frank Holfter (EnTec GmbH), Tanis Hogg (JenaDrugDiscovery GmbH) and Johannes von Langen (IMB-Jena) for their help in preparing the manuscript and Walter Elger (EnTec GmbH) for stimulating discussions.

1.7 References

1 PhRMA (2000) Based on data from PhRMA Annual Survey and Standard & Poor's Compustat, a division of McGraw-Hill
2 PhRMA Annual Survey (2000)
3 Report issued by Bain & Company Germany Inc. (2001)
4 Boston Consulting Group. A revolution in R&D. (2001)
5 DiMasi JA (2001) Risks in new drug development: approval success rates for investigational drugs. *Clin Pharmacol Ther* 69: 297–307
6 Report issued by the Tufts Center for the Study of Drug Development, Tufts University, Boston, MA, USA (1998)
7 Peck CC (1997) Drug development: improving the process. *Food Drug Law J* 52: 163–167
8 Drews J (2000) Drug discovery: a historical perspective. *Science* 287: 1960–1964
9 Ward SJ (2001) Impact of genomics in drug discovery. *Biotechniques* 31: 626–630
10 Cunningham MJ (2000) Genomics and proteomics: the new millennium of drug discovery and development. *J Pharmacol Toxicol Methods* 44: 291–300
11 Dongre AR, Opiteck G, Cosand WL et al (2001) Proteomics in the post-genome age. *Biopolymers* 60: 206–211
12 Augen J (2002) The evolving role of information technology in the drug discovery process. *Drug Discov Today* 7: 315–323
13 Davies EK, Richards WG (2002) The potential of Internet computing for drug discovery. *Drug Discov Today* 7: S99-S103
14 Lander ES, Linton LM, Birren B et al (2001) Initial sequencing and analysis of the human genome. *Nature* 409: 860–921
15 Venter JC, Adams MD, Myers EW et al (2001) The sequence of the human genome. *Science* 291: 1304–1351
16 National Center for Biotechnology Information, National Library of Medicine, Bethesda, MD 20894, http://www.ncbi.nlm.nih.gov. (2002)
17 Searls DB (2000) Using bioinformatics in gene and drug discovery. *Drug Discov Today* 5: 135–143
18 Hogenesch JB, Ching KA, Batalov S et al (2001) A comparison of the Celera and Ensembl predicted gene sets reveals little overlap in novel genes. *Cell* 106: 413–415
19 Wright FA, Lemon WJ, Zhao WD et al (2001) A draft annotation and overview of the human genome. *Genome Biol* 2: RESEARCH0025.1–0025.18
20 Duckworth DM, Sanseau P (2002) In silico identification of novel therapeutic targets. *Drug Discov Today* 7: S64-S69
21 Sanseau P (2001) Impact of human genome sequencing for in silico target discovery. *Drug Discov Today* 6: 316–323
22 Tang CM, Moxon ER (2001) The impact of microbial genomics on antimicrobial drug development. *Annu Rev Genomics Hum Genet* 2: 259–269
23 Herrmann R, Reiner B (1998) Mycoplasma pneumoniae and Mycoplasma genitalium: a comparison of two closely related bacterial species. *Curr Opin Microbiol* 1: 572–579
24 Noordewier MO, Warren PV (2001) Gene expression microarrays and the integration of biological knowledge. *Trends Biotechnol* 19: 412–415
25 Kennedy GC (2000) The impact of genomics on therapeutic drug development. *EXS* 89: 1–10
26 Gygi SP, Rochon Y, Franza BR et al (1999) Correlation between protein and mRNA abundance in yeast. *Mol Cell Biol* 19: 1720–1730
27 Loo JA, DeJohn DE, Du P et al (1999) Application of mass spectrometry for target identification and characterization. *Med Res Rev* 19: 307–319
28 Abuin A, Holt KH, Platt KA et al (2002) Full-speed mammalian genetics: *in vivo* target validation in the drug discovery process. *Trends Biotechnol* 20: 36–42
29 Tornell J, Snaith M (2002) Transgenic systems in drug discovery: from target identification to humanized mice. *Drug Discov Today* 7: 461–470
30 Sanseau P (2001) Transgenic gene knockouts: a functional platform for the industry. *Drug Discov Today* 6: 770–771
31 Dean NM (2001) Functional genomics and target validation approaches using antisense oligonucleotide technology. *Curr Opin Biotechnol* 12: 622–625
32 Tse E, Lobato MN, Forster A et al (2002) Intracellular antibody capture technology: application to

selection of intracellular antibodies recognising the BCR-ABL oncogenic protein. *J Mol Biol* 317: 85–94

33 Burley SK (2000) An overview of structural genomics. *Nat Struct Biol* 7 Suppl: 932–934

34 Service RF (2000) Structural genomics offers high-speed look at proteins. *Science* 287: 1954–1956

35 Dean PM, Zanders ED (2002) The use of chemical design tools to transform proteomics data into drug candidates. *Biotechniques* Suppl: 28–33

36 Willson TM, Jones SA, Moore JT et al (2001) Chemical genomics: functional analysis of orphan nuclear receptors in the regulation of bile acid metabolism. *Med Res Rev* 21: 513–522

37 Lenz GR, Nash HM, Jindal S (2000) Chemical ligands, genomics and drug discovery. *Drug Discov Today* 5: 145–156

38 Hertzberg RP, Pope AJ (2000) High-throughput screening: new technology for the 21st century. *Curr Opin Chem Biol* 4: 445–451

39 Wolcke J, Ullmann D (2001) Miniaturized HTS technologies – uHTS. *Drug Discov Today* 6: 637–646

40 Wess G, Urmann M, Sickenberger B (2001) Medicinal Chemistry: Challenges and Opportunities. *Angew Chem Int Ed Engl* 40: 3341–3350

41 Drews J (2000) Drug discovery today – and tomorrow. *Drug Discov Today* 5: 2–4

42 Harvey AL (1999) Medicines from nature: are natural products still relevant to drug discovery? *Trends Pharmacol Sci* 20: 196–198

43 Sundberg SA (2000) High-throughput and ultra-high-throughput screening: solution- and cell-based approaches. *Curr Opin Biotechnol* 11: 47–53

44 Silverman L, Campbell R, Broach JR (1998) New assay technologies for high-throughput screening. *Curr Opin Chem Biol* 2: 397–403

45 Schneider G, Böhm HJ (2002) Virtual screening and fast automated docking methods. *Drug Discov Today* 7: 64–70

46 Toledo-Sherman LM, Chen D (2002) High-throughput virtual screening for drug discovery in parallel. *Curr Opin Drug Discov Devel* 5: 414–421

47 Shuker SB, Hajduk PJ, Meadows RP et al (1996) Discovering high-affinity ligands for proteins: SAR by NMR. *Science* 274: 1531–1534

48 Brennan MB (2000) Drug Discovery: filtering out failures early in the game. *Chemical & Engineering News* 78: 63–73

49 Bachmann KA, Ghosh R (2001) The use of *in vitro* methods to predict *in vivo* pharmacokinetics and drug interactions. *Curr Drug Metab* 2: 299–314

50 Vogel HG (2001) *Drug Discovery and Evaluation. Pharmacological Assays*. Springer-Verlag, Berlin Heidelberg

51 Beresford AP, Selick HE, Tarbit MH (2002) The emerging importance of predictive ADME simulation in drug discovery. *Drug Discov Today* 7: 109–116

52 Butina D, Segall MD, Frankcombe K (2002) Predicting ADME properties in silico: methods and models. *Drug Discov Today* 7: S83–S88

53 Roses AD (2001) Pharmacogenetics. *Hum Mol Genet* 10: 2261–2267

54 Roses AD (2000) Pharmacogenetics and the practice of medicine. *Nature* 405: 857–865

55 Vesell ES (2000) Advances in pharmacogenetics and pharmacogenomics. *J Clin Pharmacol* 40: 930–938

56 McCarthy JJ, Hilfiker R (2000) The use of single-nucleotide polymorphism maps in pharmacogenomics. *Nat Biotechnol* 18: 505–508

57 Tomita M (2001) Whole-cell simulation: a grand challenge of the 21st century. *Trends Biotechnol* 19: 205–210

Modern Methods of Drug Discovery
ed. by A. Hillisch and R. Hilgenfeld
© 2003 Birkhäuser Verlag/Switzerland

2   Proteomics

Martin J. Page[1], Chris Moyses[2], Mary J. Cunningham[3], Gordon Holt[2] and Alastair Matheson[4]

[1] OSI Pharmaceuticals, Cancer Biology, Watlington Road, Oxford OX4 6LT, UK
[2] Oxford GlycoSciences (UK) Ltd, 10 The Quadrant, Abingdon Science Park, Abingdon OX14 3YS, UK
[3] Incyte Pharmaceuticals, 3160 Porter Drive, Palo Alto, CA 94304, USA
[4] 5 Highcroft Road, London N19 3AQ, UK

2.1   Introduction

Several disciplines have recently emerged within biotechnology which for the first time enable biological systems to be studied on a scale commensurate with their inherent complexity. From the point of view of drug discovery, a critical role is fulfilled by proteomics, the systematic high-throughput separation and characterization of proteins within biological systems. Importantly, it is at the protein level that disease processes become manifest and at which most drugs act. Proteomics is consequently assuming a central place in modern drug development, with a wide spectrum of practical applications embracing diagnostics, target discovery, target validation, lead compound selection, investigation of drug modes of action, toxicology and clinical development.

The central technologies of proteomics as it is currently practiced can be grouped into two stages (Fig. 1):

- Technologies for protein mapping, i.e., separating, distinguishing and quantifying the proteins present in individual samples.
- Technologies for identifying proteins and characterizing their full structural properties and functional roles.

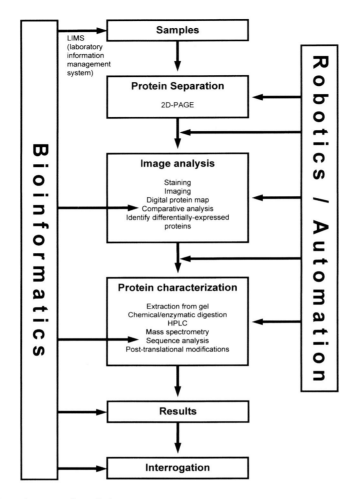

Figure 1. Stages in proteomic analysis.

At the major proteomics centers, these stages, each of which involves many individual steps, are integrated into high-throughput, automated systems, supported by sophisticated robotics and dependent at every step on powerful bioinformatics (see Chapter 3).

2.2 Protein mapping

At present, the foremost technology for protein separation is two-dimensional polyacrylamide gel electrophoresis (2D-PAGE). Typically, broad range gels with a pH range of approximately 3–10 are used to generate an initial protein expression profile for a given cell/sample type, after which greater resolution can be achieved by the use of a series of gels with narrower but overlapping pH ranges. For highly basic proteins with pH's between 10 and 12, the older 2D-PAGE technology using carrier ampholytes is still useful to achieve maximum resolution.

Sample preparation is critical to successful 2D-PAGE. Proteins must be effectively disaggregated and solubilized without degrading or modifying their constituent polypeptides. Subfractionation into membrane, cytosol, nuclear, cytoskeletal and other subcellular fractions has become a key technique, bolstering the overall resolving power of 2D-PAGE by increasing the copy number of rare proteins and enabling buffers to be customized for different cellular factions. It also yields information on the subcellular localization of proteins that may prove vital in assessing their function and pharmaceutical potential. Innovations in sample preparation are also addressing the problem of rare proteins becoming obscured on gels by more abundant proteins that migrate to similar gel locations. A number of groups have developed techniques based on immuno-affinity chromatography for removing abundant proteins from samples prior to the 2D-PAGE step. These include immuno-depletion methodology developed and extensively used by Oxford GlycoSciences (OGS; Abingdon, UK) and the ProteoClear™ technology developed by Proteomix Inc (San Diego, CA, USA).

Most image-analysis systems work by first detecting the center of gravity or peak density of a feature, then recording other parameters such as the size of the feature and its overall abundance. An alternative is edge-detection software, which is useful for complex feature outlines. Additional bioinformatics procedures come into play during the curation of feature analysis data obtained from individual gels. This involves landmarking and warping of images such that they all align into a common geometry. Usually several replicates of the test sample are run to derive multiple images, from which an integrated, composite image or mastergroup is derived. This is an important part of the experimental design, since it captures the natural biological variation present, and assists in reducing minor process variations.

The result of 2D-PAGE and subsequent image analysis is a two-dimensional map, or protein expression map (PEM™) [1], of the proteins present in a particular cell type under defined conditions. Figure 2 shows composite PEMs™ derived from 10 matched clinical preparations of purified ductal and myoepithelial cells obtained from the human breast, constructed by Oxford GlycoSciences in collaboration with the Ludwig Institute for Cancer Research (LICR; London, UK) [2]. These are currently being compared to another PEM™ derived from a large series of purified human breast cancer samples in order to identify disease-specific protein changes and ultimately novel pharmacological targets against breast cancer.

2.2.1 Challenges for 2D-PAGE

2D-PAGE continues to evolve and faces a number of significant challenges. Currently, a single gel is capable of resolving over 2000 proteins, which although

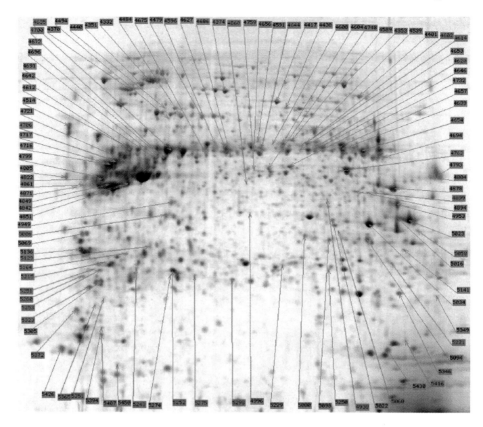

Figure 2. Protein expression map (PEM™) derived from normal human breast luminal and myoepithelial cells.

highly impressive compared with other separation technologies, is still below the approximately 6000 proteins, or more, believed to be expressed by any given tissue type [3]. The resolving power of 2D-PAGE is limited for several categories of protein. These include very basic proteins, which remain difficult to separate in reproducible patterns, and large proteins, which do not load or migrate well on current gels. Most importantly, hydrophobic proteins do not readily solubilize in the aqueous media on which current 2D-PAGE techniques are based. This is a significant problem because membrane proteins with hydrophobic domains include many cellular receptors and transporters which may provide valuable drug targets. One approach to this problem is the development of new zwitterionic detergents which increase the aqueous solubility of these proteins [4], although it is possible that hydrophobic media capable of supporting electrophoretic separation may be developed in the future.

2.3   Protein characterization

Once proteins have been separated, imaged and incorporated into a mastergroup, selected proteins of interest are identified and fully characterized using mass spectrometry (MS). In contemporary proteomics there has been particular emphasis on two ionization techniques, electrospray ionization (ESI) and matrix assisted laser desorption and ionization (MALDI). Tandem MS (MS/MS), involving successive phases of fragmentation and mass analysis, is also frequently used. It is highly accurate, very useful in the analysis of post-translational modifications and N-terminal sequencing, and applicable to mixtures of peptides.

The use of MS to characterize proteins is highly dependent on bioinformatics programs that seek to match the MS-derived data with data from existing EST, gene and protein sequence databases (see Chapter 3). In the technique known as peptide mass finger-printing, the spectrum for a given protein digest is compared against a database of predicted spectra generated for each of the peptides within a primary sequence database, assuming these peptides had been subjected to the same pattern of fragmentation.

The use of genomics-derived sequence databases in MS-based analysis constitutes a key crossover point between genomics and proteomics. On the one hand, the identification of proteins by proteomic analysis benefits from matches obtained from genomics databases (see Chapter 3). On the other, finding a match within such a database helps to annotate the genomic information by specifying a specific cell type and physiological situation in which the gene's protein product is expressed. Post-translational modifications, however, cannot be identified with reference to genomic data, and are typically identified following extensive fragmentation using MS/MS by reference to a database of spectra generated by specific side chain variants. MS/MS can also be deployed in place of traditional Edman degradation to fully sequence novel proteins. These are discovered when no match is found between the protein being subjected to MS and data from the genomics databases. Such *de novo* data represents an important level of real discovery by virtue of the proteomic approach.

2.4   New directions in proteomics technologies

The main 2D-PAGE/MS approach to proteomics is versatile and powerful, constituting the most sensitive, accurate, rapid and highly parallel methodology yet devised for the assessment of protein structure, distribution and dynamics. However, a number of further technologies, such as capillary electrophoresis and HPLC, continue to grow in importance as alternative separation methods. The use of peptide ligands to isolate and identify proteins, including mass antibody-based techniques, is also finding a growing niche. Interestingly, techniques of this kind are developing at the same time that protein microarray technology is emerging, so that in the near future we are likely to see the production of ligand-based protein chips capable of isolating and identifying a wide array of proteins. The two-hybrid and related assays have proved a useful way of analyzing protein-protein interactions and when used to analyze the full genomes of simple organisms are a powerful approach to the elucidation of protein interaction networks [5].

2.5   Applications of proteomics to drug discovery

During the next few years, proteomics is widely expected to supersede genomics as the single most productive application of biotechnology in pharmaceutical development. Its principal applications include diagnostics, discovery of targets and naturally occurring protein therapeutics, target validation, drug candidate selection and mode of action studies and toxicology (Tab. 1). There are also applications within clinical development.

Table 1. Applications of proteomics to drug discovery and development.

Preclinical	Clinical
• Target identification	• Diagnostics
• Target validation	• Markers of response
• Lead candidate selection	• Inclusion & exclusion criteria
• Mode of action studies	• Subgroups of responders/adverse reactions
• Toxicology	• Post-launch differentiation of competitors
– no observed effect level	
– screening	
– mechanism of action	

2.5.1   Databases

One of the most important applications of proteomics has been the establishment of subscription databases documenting the protein profiles of specific cell types under different conditions. For instance, a series of integrated proteomics/genomics databases has been developed jointly by OGS and Incyte (Palo Alto, CA, USA) using a range of clinically important human and microbial cell types. This combined approach is proving extremely useful since for each sample type, it uniquely brings together information linking the genotype and phenotype, via genomic (genes), mRNA (transcriptional) and proteomic (translational) data.

2.5.2   Diagnostics

Proteomics is ideally placed for the development of novel diagnostics based on the analysis of human body fluids. Secreted proteins are a more abundant and specif-

ic source of markers for diseased tissues than nucleic acids, so that whereas genomics-based diagnostic approaches may require a sample of the diseased tissue, proteomics-based approaches typically do not. By comparing protein expression profiles obtained from the body fluids of healthy and diseased individuals, changes in the abundance of proteins can readily be detected, supporting the development of markers for the diagnosis and monitoring of diseases. Proteomics projects are researching a range of diagnostic markers for conditions as diverse as bladder cancer [6], endometrial hyperplasia and carcinoma [7], ovarian cancer [8] chronic transplant rejection [9], male infertility [10], cardiac hypertrophy [11], schizophrenia [12] and others.

2.5.3   Target discovery

Target discovery, in which samples from normal and diseased tissues are compared directly, is one of the most important pharmaceutical applications of proteomics (Tab. 2). Taking oncology as an example, proteomic studies have been reported in the analysis of neuroblastomas [13], breast cancer [2, 14], lung [15], colon [16], renal cell carcinoma [17] and others.

Table 2. Diagnostics/target discovery: selected programs screening for disease-specific proteins.

Company	Therapeutic areas
Oxford GlycoSciences (Abingdon, UK)	Prostate cancer, liver cancer, Alzheimer's disease, CNS diseases, breast cancer, angiogenesis, diabetes, obesity, inflammation, anti-fungals & cardiovascular
Proteome Sciences (Cobham, UK)	Oncology, neurological disease, cardiovascular disease, transplant rejection, diabetes.
Biovation (Aberdeen, UK)	CNS diseases
Genome Pharmaceuticals Corporation (Martinsried, Germany)	Rheumatoid arthritis, multiple sclerosis, transplant rejection
MitoKor (San Diego, CA, USA)	Alzheimer's disease, diabetes
Genetics Institute (Cambridge, MA, USA)	Hematopoiesis, inflammation, autoimmune diseases

Obtaining high-quality clinical samples is a prerequisite for effective proteomics-based target discovery, and even when good samples are available care must be taken to analyze the different cell types in a purified form in order to obtain reliable controls. In the breast cancer target discovery program described above [2], OGS has used a high-quality set of samples obtained from the LICR, which crucially are separated and purified prior to analysis, using specialist techniques, into luminal and myoepithelial cell types. This is vital to obtain accurate PEMs™ from each cell type, and to eliminate confusion in their interrogation by virtue of mixed cell heterogeneity. It is noteworthy that approximately 95% of breast tumors arise specifically from the luminal cell population. The derivation of a specific PEM™ for this cell type allows comparison with a PEM™ of similarly purified breast tumor cells. OGS and LICR believe this approach will represent a major advance towards finding tumour-specific proteins of therapeutic relevance.

Generating PEMs™ for normal and diseased cell types and determining how they differ is only a first step towards identifying molecules that play a causal role in the disease process. It is typical for the expression of some 50–300 proteins to differ significantly between the PEMs™ of normal and diseased samples. PEMs™ may not only differ with regard to the absolute abundance of specific proteins, but also with regard to post-translational modifications such as phosphorylation, for which huge differences may exist between normal and disease states even in cases where the level of protein abundance does not differ significantly. To identify the causally important changes within this wider data set, the biological roles of the proteins must be elucidated. Since the process of protein characterization is linked to protein databases, biological information on many proteins can now be very rapidly obtained. Annotation on this basis coupled to comparative PEM™ analysis enables the researcher to gain a broad picture of the biology of the system under study and to identify those molecular pathways which are perturbed and in which drug targets may be found.

2.5.4 Target validation

More detailed target validation ideally requires both gain and loss of function studies to determine whether a change in the expression of a given protein leads to a disease phenotype and whether correction of the change reverses it. To validate function, proteomic techniques are combined with established molecular biology approaches, including the use of small molecule inhibitors and agonists, antisense nucleic acid constructs, dominant-negative proteins, and neutralizing antibodies microinjected into cells. It is to be expected that the continuing development of functional proteomics will lead to more rapid screens for function than are presently available.

Once the relevant components of a signal transduction pathway causative of disease have been identified, immunological techniques can significantly increase the resolving power of proteomics to analyze their behavior during target validation studies. Proteomes obtained from normal and diseased tissues can, for instance, be blotted onto membranes and probed using antibodies against selected proteins. Since proteomics has the power to resolve over 2000 proteins on a single gel, data on, for example, specific isoforms of signaling molecules such as glycosylated and phosphorylated variants can readily be derived using this approach. Frequently such subtle changes constitute the primary difference between a normal and disease phenotype.

Similarly, in immunoprecipitation studies large quantities of protein (i.e., several milligrams) can be incubated with high affinity antibodies and the proteins captured eluted and electrophorezed to provide a high-resolution proteome of a specific subset of proteins. This approach is important because it also allows the identification of multiprotein complexes or other proteins that co-precipitate with the target protein. Such proteins frequently represent signaling partners for the target protein and their identification can greatly facilitate exploration of the target's biological role and pharmaceutical potential.

Western immunoblotting and immunoprecipitation using high affinity antibodies can resolve proteins when as few as 10 copies per cell are present, a significant advance over the best fluorescent dyes available. Thus very high levels of sensitivity can be achieved when a protein is specifically targeted, although it is notable that for this level of resolution, proteomics currently requires prior knowledge of the proteins subjected to analysis. With appropriate fractionation techniques and the most sensitive dyes, even lower resolutions of 1–10 copies per cell are now becoming realistic, putting the sensitivity of proteomics on a par with that of genomics.

2.5.5   Candidate selection

Once drug candidates have been raised against validated targets, proteomics can play an integral role in the selection and optimization of lead candidates for further development. In particular, it is important to confirm that their efficacy is achieved through the expected mechanism of action and to compare the effects of candidates against a range of markers of efficacy and toxicity.

One example of the power of proteomics to characterize mechanisms of action is provided by a recent study of clofibrate's effects on rat liver *in vivo*. Clofibrate is a peroxisome proliferator activation receptor (PPAR) inducer which stimulates fatty acid oxidation by a complex multi-pathway modulation [18] and also induces hepatocellular carcinomas in rodents via a non-genotoxic transformation that is largely uncharacterized. In a proteomic study comparing liver cells from four clofibrate-treated rats with four controls, a total of 304 out of 92,000 gel features examined differed significantly between treated and control animals. Sequence annotations of these features demonstrated that many were proteins known to be affected by clofibrate treatment such as catalase, glutathione sulfotransferase, enol-CoA hydratase, and estrogen metabolic enzymes. In effect, all of the key findings in hundreds of clofibrate-focused publications were confirmed in one simple eight-animal proteomics study. Moreover, many new discoveries were made. For example, clofibrate was found to modulate the levels of over a dozen proteins that have never before been described. These novel proteins may prove to be entirely new pharmaceutical targets for the dynamic field of PPAR-alpha stimulation. Furthermore, proteomic analyses allowed entire metabolic pathways to be monitored for treatment-induced changes. As shown in Figure 3, enzymes along the entire citric acid cycle pathway, as well as entry and exit points into a variety

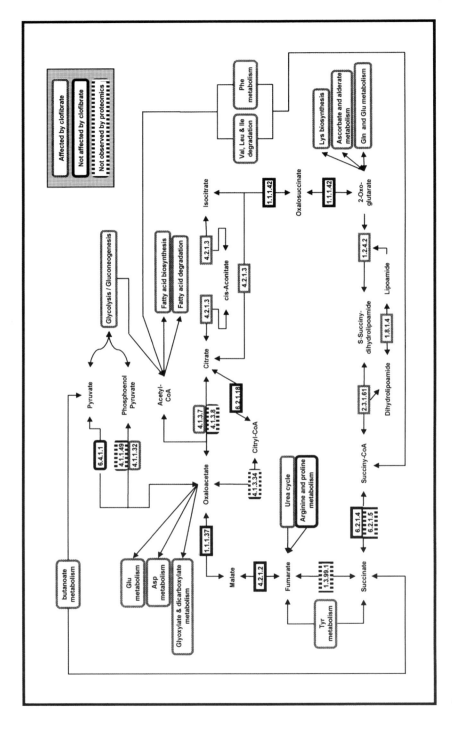

Figure 3. Proteomic analysis of the effects of clofibrate on the citric acid cycle in the rat liver.

of other metabolic pathways, were simultaneously monitored, leading to a considerably broader insight into this drug's effects on cellular metabolism. Such a system-wide view is essential for rational decision-making to advance lead compounds into development.

2.5.6 Preclinical toxicology

Toxicological screening is an important early stage in the development of a drug when the mode of action has been validated. Many initially promising drugs fail at this stage, but traditional toxicological screens (see Chapter 13) have been both laborious and difficult to interpret. Proteomics has great potential in early toxicological screening, since it is anticipated that drug toxicity will produce a limited array of proteomic signatures that can be documented and used to screen the toxicological properties of novel candidates. A number of companies are exploring the potential of this approach. LSP are compiling a Molecular Effects of Drugs database, while OGS has launched a collaborative program with Quintiles (Ledbury, UK) to validate this approach by focusing on a range of drugs for which toxicity has already been characterized in detail using standard measures. The OGS studies, which focus on the P450 enzyme series responsible for metabolizing most drugs (see Chapter 13), show that proteomics can detect and predict the toxic effects of drugs at significantly lower doses than would be required to demonstrate toxicity using standard techniques.

2.5.7 Clinical development

The applications of proteomics in the clinical development of drugs are outside the scope of this chapter, but have been discussed in detail elsewhere [19]. Proteomics has the capacity to reduce the duration of clinical trials, through the development of biomarkers of response as discussed above in section 4.2. It also has the ability to reduce the size of patient populations by enabling patients to be subtyped and pre-selected according to stringent molecular criteria. As a result of these developments, clinical development has the potential to become more efficient and less expensive, while pharmacological therapy in general is likely to become increasingly tailored to the genetic and molecular profiles of specific patient subgroups.

2.6 Conclusions

Proteomics has assumed a major role in modern drug discovery within just a few years of its inception. The ability of proteomics to screen proteins on a mass scale and to analyze their structure, location, relationships and functional organization using parallel, high-throughput techniques has greatly enhanced the ability of biotechnology to understand disease processes and identify and validate specific drug targets. Together with genomics, proteomics provides drug discovery programs with unprecedented insight into the molecular organization of cells in health and disease, an insight that is now driving the development of more rapid, robust and cost-effective drug discovery strategies.

2.7 References

1 Page MJ, Amess B, Rohlff C et al (1999) Proteomics: a major new technology for the drug discovery process. *Drug Discovery Today Development and Therapeutics* 4: 55–62
2 Page MJ, Amess B, Townsend RR et al (1999) Proteomic definition of normal human luminal and myoepithelial breast cells purified from reduction mammoplasties. *Proc Natl Acad Sci U SA* 96: 12589–12594
3 Celis JE, Gromov P (1999) 2D protein electrophoresis: can it be perfected? *Curr Op Biotech* 10: 16–21
4 Chevallet M, Santoni V, Poinas A et al (1998) New zwitterionic detergents improve the analysis of membrane proteins by two-dimensional electrophoresis. *Electrophoresis* 19: 1901–1909
5 Frederickson RM (1998) Macromolecular matchmaking: advances in two-hybrid and related technologies. *Curr Opin Biotechnol* 9: 90–96
6 Ostergaard M, Wolf H, Orntoft TF et al (1999) Psoriacin (S100A7): a putative urinary marker for the follow-up of patients with bladder squamous cell carcinomas. *Electrophoresis* 20: 349–354
7 Byrjalsen I, Mose Larsen P, Fey SJ et al (1999) Two-dimensional gel analysis of human endometrial proteins: characterization of proteins with increased expression in hyperplasia and adenocarcinoma. *Mol Hum Reprod* 5: 748–756
8 Alaiya AA, Franzen B, Moberger B et al (1999) Two-dimensional gel analysis of protein expression in ovarian tumors shows a low degree of intratumoral heterogeneity. *Electrophoresis* 20: 1039–1046
9 Faulk WP, Rose M, Meroni PL et al (1999) Antibodies to endothelial cells identify myocardial damage and predict development of coronary artery disease in patients with transplanted hearts. *Hum Immunol* 60: 826–832
10 Shetty J, Nabby-Hansen S, Shibahara H et al (1999) Human sperm proteome: immunodominant sperm surface antigens identified with sera from infertile men and women. *Biol Reprod* 61: 61–69
11 Arnott D, O'Connell KL, King KL et al (1998) An integrated approach to proteome analysis: identification of proteins associated with cardiac hypertrophy. *Anal Biochem* 258: 1–18
12 Edgar PF, Schonberger SJ, Dean B et al (1999) A comparative proteome analysis of hippocampal tissue from schizophrenic and Alzheimer's disease individuals. *Mol Psychiatry* 4: 173–178
13 Wimmer K, Thorval D, Asakawa J et al (1996) Two-dimensional separations of the genome and proteome of neuroblastoma cells. *Electrophoresis* 17: 1741–1751
14 Williams K, Chubb C, Huberman E et al (1998) Analysis of differential protein expression in normal and neoplastic human breast epithelial cell lines. *Electrophoresis* 19: 333–343
15 Hirano T, Franzen B, Uryu K et al (1995) Detection of polypeptides associated with the histopathological differentiation of primary lung carcinoma. *Br J Cancer* 72: 840–848
16 Ji H, Reid GE, Moritz RL et al (1997)A two-dimensional gel database of human colon carcinoma proteins. *Electrophoresis* 18: 605–613
17 Sarto C, Marocchi A, Sanchez J-C et al (1997) Renal cell carcinoma and normal kidney protein expression. *Electrophoresis* 18: 599–604
18 Knopp RH (1999) Drug treatment of lipid disorders. *N Engl J Med* 341: 498–511
19 Moyses C (1999) Pharmacogenetics, genomics, proteomics: the new frontiers in drug development. *Int J Pharm Med* 13: 197–202

Modern Methods of Drug Discovery
ed. by A. Hillisch and R. Hilgenfeld
© 2003 Birkhäuser Verlag/Switzerland

3 Bioinformatics

David B. Jackson[1], Eric Minch[2] and Robin E. Munro[2]

[1] Cellzome, Meyerhofstr. 1, D-69117 Heidelberg, Germany
[2] LION bioscience AG, Waldhofer Str. 98, Wieblingen, D-69123 Heidelberg, Germany

3.1   Introduction

The writer T.S. Eliot once mused, "Where is the knowledge we have lost in information?" [1]. From a biological perspective, the answer to this profound question
is today having far-reaching consequences for the future of biomedical research
and, in particular, the drug discovery process.

Like the greatest authors' works, a genome may be viewed as a superior work
of complex linguistics, sculpted by evolution, to tell the story of life.
Unfortunately, when it comes to genome linguistics we are much like a four-year-
old child attempting to read and comprehend a Shakespearean masterpiece. Thus,
we are not wholly illiterate, but we still have much to learn. Despite the fact that
the fundamental elements of evolution's alphabet are limited to a set of four characters, to say that the manner in which it has "written" the story of life is complex,
constitutes a gross understatement. With language comprehension being a pivotal
element in the purveyance of knowledge, it is not surprising that our efforts to
decipher the secrets of evolution's prose have witnessed a recent and rapid growth.

But why this sudden imperative? It was predicted in the early 1980s that the key
scientific pursuits of the 21st century would be those related to the disparate sciences of molecular biology and computing. Now, as we begin a new millennium,
we find that this prophecy is not without substance. To be more accurate, however, what we have recently witnessed is an amalgamation of these disciplines into
a single science; the science of bioinformatics. This fusion was in retrospect
inevitable, since biology is very much an information science. However, it was the
technological advances in both fields that provided the deciding cohesive force.
Improved cloning techniques, the advent of high-throughput chip technologies
and the profusion in genomic sequencing efforts led to an "information explosion"
that could only be handled by employing new advances in information technology and computing power. Bioinformatics was thus born.

In simple terms, bioinformatics may be thought of as a theoretical, computer-
based science dedicated to the extrapolation of biological knowledge from biological information. It operates via the acquisition, analysis and administration of
information, typically obtained by DNA sequencing, laboratory experiment, database searches, scientific publications, instrumentation or modeling.

It differs from its sibling, biological computation, in which strategies and structures derived from biology are applied to the design of computational algorithms
and architectures (including neural networks, genetic programming, DNA computation, etc.).

The aforementioned challenge to understand genome linguistics is now the *forte*
of bioinformatics. This is an important consideration since the challenge not only
holds the key to rapid drug discovery techniques, but also promises to change the
face of biology forever. In fact, despite its recent birth, bioinformatics is already
having profound implications for pharmaceutical and diagnostic concerns. Costs
incurred while pursuing a specific drug target and, indeed, the number of such
putative targets have increased enormously over the past decade. While the application of genomics and high-throughput methodologies early in the chain of drug

discovery have provided a proliferation in the number of putative drug-targets, we now rely on bioinformatic analyses to filter and assess the most relevant candidates.

In this chapter we attempt to provide the uninitiated with a synopsis of the key components of this new and vibrant science. In so doing, we hope to highlight the strength of the bioinformatic approach in modern drug discovery efforts. Unfortunately, an exhaustive review of all available algorithms is, given the present climate of fervent bioinformatic tool/database development, beyond the scope of any single book chapter. However, we will highlight some of the best-known algorithms and databases with particular emphasis placed on those that epitomize the most commonly used strategies and techniques.

3.2   Sequence comparison

The eagerly awaited transition from fluorescent slab gel based sequencers (e.g., Applied Biosystems 373/377, Licor-4000/4200 IR2) to capillary based methods has ushered in a new era in rapid genomic scale DNA sequencing. By mid 2002 the entire genomes of 695 viruses, 22 viroids, 13 archaeae, 57 bacteriae and 4 eucaryotae including one fungus, one plant, two animals and the human have been sequenced. Among the 3 billion base pairs of the human genome lies an estimated 65,000–75,000 genes, of which many could be the pharmaceutical "golden nuggets" of the future (see Chapter 1).

"Mining" such extensive genomic data to extract relevant biological information is the pivotal function of bioinformatics. The basic foundation upon which this data mining process lies involves the mutual arrangement and comparison of two or more sequences (DNA or protein), with the aim of defining regions of putative similarity and/or identity. This strategy, referred to as sequence alignment, allows us to extrapolate quantitative expressions of inter-relatedness and to infer possible structural, functional or evolutionary relationships. While alignments provide a powerful way to compare related sequences, they can be used in an effort to describe different facts. The alignment of two residues might be viewed as reflecting a common evolutionary origin, or could be viewed as representing a common structural motif, which, according to the forces of convergent evolution, may be completely independent of origin. Interestingly, there is no documented example to date where two sequences have clear sequence similarity, but dissimilar three-dimensional (3D) structures. Thus it is generally assumed that a common ancestor always implies a common 3D structure, a notion which forms the basis of homology based modeling, discussed later.

Because of their fundamental importance, numerous classes of comparison algorithms and scoring matrices have been developed, the majority of which are applicable to both DNA and protein. Sequence-comparison methods fall into two broad categories 1) pairwise methods (e.g., BLAST [2], FASTA [3]) in which two sequences are compared and 2) methods in which models or profiles are built from multiple sequence alignments and compared either with other sequences or profiles (e.g., HMMER).

   Pairwise comparison methods can be further classified as either global or local alignment methods, depending on the intended purpose (see Fig. 1). Global methods force a complete alignment of the input sequences, from start-to-finish. The stringent Needleman-Wunsch [4] and Sellers [5] algorithms are the best known exponents of this method. While of great benefit when comparing sequences where a relationship has already been established (e.g., phylogenetic analysis), global alignment strategies are of little use when performing database homology searches. For this purpose, "local" alignment algorithms such as the stringent Smith-Waterman [6] or the heuristic FASTA or BLAST methods are best suited. In contrast to the global strategies, homologous domains embedded in non-homologous sequences are easily identified, as are intron-exon definitions when a cDNA is compared against a genomic database.

   While both methods are similar in terms of the underlying algorithms, the statistics needed to assess their output differ significantly. For global alignments, the

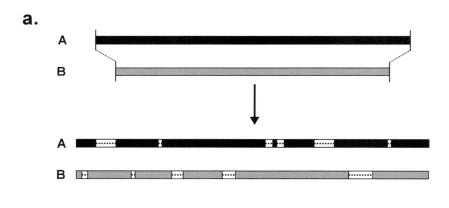

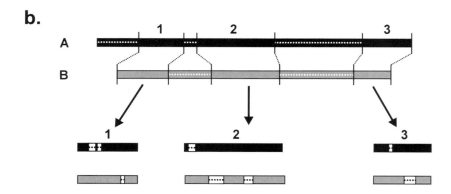

Figure 1. The basic difference between global (a) and local (b) alignment strategies. Global alignments incorporate the entire lengths of the two sequences, whereas local methods only align their most homologous parts. Arrows indicate the alignment process.

situation is ill-defined, since very little is known about the random distribution of their scores. One of the very few methods available generates an empirical score distribution from the alignment of random sequences of the same lengths as the two sequences being compared. A so-called Z-score for the alignment is then calculated from this distribution. An accurate significance estimate for the Z-score cannot currently be derived, due to the ill-defined nature of the distribution score.

For local alignments, the situation is thankfully simpler. In this case, the random score distribution for an optimal ungapped alignment has been proven to follow an extreme value distribution. While unproven for gapped local alignments, assessments suggest that such a distribution also holds true. Both the BLAST and FASTA sequence comparison algorithms report so-called raw scores for each comparison, including an assessment of their statistical significance, based upon the extreme value distribution, called E-values. For a given alignment, the E-value is dependent on 1) the length of the query sequence, 2) the size if the database searched and 3) the raw score.

3.2.1   Scoring matrices

Quantifying the degree of similarity between two sequences requires three components; 1) a scoring scheme, 2) a gap model and 3) an alignment optimizing algorithm. Scoring schemes are represented as matrices that define the value for aligning each possible pair of residues. For amino acids, the scoring scheme is a 20×20 matrix of numbers. Such matrices are usually symmetrical, since Asp aligned with Glu has the same score as Glu aligned with Asp. The most elemental scoring scheme is identity scoring. Here, the aligned units are scored as either identical (+1) or non-identical (0). Using this strategy, the normalized sum of the alignment scores is popularly quoted as "percentage identity." Although simple in nature, identity scoring does not reflect the three-dimensional scenario. Another strategy, this time specific to proteins, known as genetic code scoring considers the minimum number of nucleotide changes [1–3] that would be required to interconvert the codons for two amino acids. While this scheme is rarely employed for today's alignment tasks, it is often useful for constructing phylogenetic trees. Chemical similarity scoring schemes give greater relevance to the alignment of amino acids with similar physico-chemical properties. Here, amino acids of similar character (e.g., aspartate and glutamate) are given higher scores than those of different character (e.g., phenylalanine and glutamate). Despite the large number of alternative matrices available, it is those based on evolutionary aspects, or observed substitutions, that have become the most widely used and reliable to date.

One of the most popular observed substitution schemes, until recent times, was developed by the late Margaret Dayhoff and co-workers [7]. They devised a model of protein evolution from which they extrapolated a mutation matrix scoring scheme. The underlying principle of this approach assumes that the estimation of mutation rates for closely related proteins can be extrapolated to more distant relationships. Mutation rates were determined by examining several hundred global alignments between highly related proteins, and then calculating the rate at which each residue changed into another at a very short evolutionary distance, i.e., where one percent of residues had changed. This task was of course a bit of a "chicken and egg" problem, since in order to derive the scoring scheme one needs alignments, and *vice versa*. To overcome this problem they used very closely related sequences and hand aligned them. The assessment of the alignments was then simple. If at a particular position in a multiple alignment of, say, ten sequences a serine occurred eight times and a threonine twice, they inferred that serine was ancestral at that position, and for the evolutionary distance used in the alignment, serine to threonine mutations occurred 20% of the time. A general view of the mutation process was then obtained by calculating the average ratio of the number of changes a specific amino acid underwent to the total number of amino acids of that type in the database. Combining this information with the point mutation data led to a frequency table representing the observed substitution rates for a specific evolutionary distance. The table was then normalized such that for every 100 residues, an average of only one mutation occurred. This so-called 1 percent accepted mutation (1PAM) was then extrapolated to other distances. At 250PAMs (the general standard), for example, approximately one in five amino acids remain unchanged. This average, of course hides the fact that the amino acids differ in their mutability, with almost half of all tryptophans and cysteines remaining unchanged.

One factor that might question the credibility of this scheme is the small size of the dataset used. In an attempt to enhance the accuracy of the PAM matrices, Jones et al. [8] derived an updated scoring matrix termed PET92, using 2,621 protein families. Despite the larger dataset, few differences in the character of the matrices are discernible, with both PAM and PET92 reflecting replacements that maintain residue size and hydrophobicity. Similar results, in the form of the Gonnet matrix [9], were also obtained in a study that attempted to use more distant relationships.

One false premise of the PAM model is that relationships can be modeled accurately from alignments of closely related sequences. In reality these alignments strongly favor the least conserved positions, whereas the most conserved positions have little influence at all. In fact, it became apparent over time that the PAM matrix was of limited value in detecting motifs, patterns and more distant relationships.

An alternative mutation data matrix was thus developed by Henikoff and Henikoff employing local multiple alignments of more distantly related sequences, from the BLOCKS database [10]. Substitution frequencies for every pair of amino acids were then used to calculate a log odds BLOSUM (blocks substitution matrix) matrix. In addition to the inconsistencies for the rarer amino acids, important consistent differences were also uncovered such as the fact that hydrophilic matches and aromatic mismatches tend to be favored by the BLOSUM matrix in comparison with PAM. In general terms, BLOSUM matrices tend to perform better than PAM matrices for local similarity searches, with BLOSUM

62 being the default value for BLAST, and the less stringent BLOSUM 50 being the FASTA and Smith Waterman (see below) defaults.

Comparison of 3D structures also allows alignment of more distantly related proteins. Many groups have used this knowledge to extrapolate structure-based substitution matrices. While in theory this approach should provide the most accurate comparison matrices, the size of the available dataset limits its present applicability.

A word of caution before we proceed. The use of BLOSUM matrices per default is only a general rule. Time and experience have shown that no single matrix is the solution to all questions. One study [10], for example, compared the BLOSUM, PAM, PET, 3-D based, Gonnet and multiple PAM matrices, using ungapped BLAST. While concluding that BLOSUM62 is generally most effective, there were certain protein families for which all the other algorithms performed better. The moral of the story: failure to detect a significant alignment in a search does not always imply that there are no detectable homologues. In such instances, adjust the matrix and try again.

3.2.2   Affine gap penalties

As sequences evolve and diverge, they often accumulate insertions and/or deletions. Two parameters, a) the gap creation penalty and b) the gap extension penalty, have been devised to penalize the cost of these events.

GAP creation (length-independent) penalty: The application of a negative scoring penalty to the substitution matrix for the first residue in a sequence alignment gap. FASTA, for example, penalizes the opening of a gap by scoring a default of -12 for proteins and -16 for DNA.

GAP extension (length dependent) penalty: The application of a negative scoring penalty for every additional residue in a gap. FASTA applies a default score of -2 for proteins and -4 for DNA.

Thus, combining the two components implies that, for a gap of length x, the gap penalty P is expressed as:

$$P(x) = a + bx$$

where a is the gap opening penalty and b is the gap extension penalty.

This form of penalty function is referred to as affine and has many efficiency advantages over more elaborate penalty types.

In most cases the penalty for a gap is equal at all positions in an alignment. However, in cases where one would like to align say, a domain, to a longer sequence, then penalties at the ends of the domain should be reduced. This facilitates accurate alignment of the domain. Similar considerations also hold true when the secondary structure of a protein is known. For example, increasing the gap penalty within core secondary structure elements will reduce the likelihood of placing gaps there.

3.3   Database searches

A database search is the most powerful bioinformatics approach to highlighting the biologically significant features of a sequence. In its simplest form, a database search involves the submission and pairwise comparison of a query sequence against the individual entries of a sequence archive. How the search is implemented is dependent on the search algorithm employed, which in turn influences the sensitivity of the comparison.

3.3.1   BLAST

The BLAST (Basic Local Alignment Search Tool) suite of sequence alignment algorithms [2] was developed against a backdrop of advances in the statistical treatment of ungapped sequence alignments. Today, the BLAST2 family generally performs gapped alignments and is defined by seven algorithms, BLASTP (protein against protein database), BLASTN (DNA against DNA database), BLASTX (translated DNA against protein database), TBLASTN (protein against translated DNA database), TBLASTX (translated DNA against translated DNA database), position specific iterative (PSI-) BLAST [2], and pattern hit initiated (PHI-) BLAST [11].

3.3.2   BLASTP

BLASTP is now the most widely used program for sequence comparison, primarily because of its accurate statistics, rapid speed and high sensitivity. The algorithm operates in six intuitive steps:

1) A list of every possible three-residue sequence is extracted from the query. Each element of this list is commonly referred to as a "word."
2) By now comparing every element of this word list to the original query sequence using BLOSUM62, compile a list of words that score equal or above a preset threshold score (T) of 11. Such words are seen on average 150 times per sequence.
3) A so-called discrete finite automaton (DFA) is then built to recognize the list of identical and substituted words.
4) The database is searched using the DFA. All matching database words are thus identified.
5) Step five involves the so-called 'two-hit' approach. Applying BLOSUM 62, every word pair on the same diagonal separated by a distance of less than A is extended toward the N- and C- termini to produce a score that must always be above the threshold score (S). These locally optimal alignments are called 'high-scoring segment pairs' or HSPs. For all such sequences, a gapped alignment may be triggered.

6) The resulting alignment is reported if the expectation value (E-value; i.e., the possibility of such a homology occurring by chance in a search of any given database) is low enough to deem it of interest.

3.3.3   PSI-BLAST

Profile-based database searches are more sensitive to distant evolutionary relationships than standard pairwise similarity search strategies such as BLAST. Position Specific Iterated (PSI) BLAST [2] was developed with this fact in mind, combining the power of BLAST with a profile-based search method. Its logic is beautifully simple. First a gapped BLAST search is performed using the standard BLOSUM62 matrix. All hits scoring above a user-defined threshold are then multiply aligned. A position-specific scoring matrix (PSSM) is then constructed and used in place of BLOSUM62 for another round of searching. New hits can be added to the alignment and the process of profile-building repeated multiple times until no significant hits are found.

Quantitative studies have shown that PSI-BLAST can reveal three times as many homologies as conventional pairwise comparison methods, when proteins are less than 30% identical. This increased sensitivity, however, costs computing time, with the automated profile building procedure taking anywhere from 1 to 30 min, depending on the sequence length and the number of homologues detected.

While PSI-BLAST is not the first algorithm to employ profile-based database searches, it is unique in both its automation of the profile construction process, and in its ability to apply fast heuristic similarity search techniques to profile comparison. These features make the algorithm appealing to both expert and novice alike. Indeed the program has found many "fans" in the structural genomics field. Using every structure in the PDB (3D-structure) database, the program has successfully assigned putative structures to 19–39% of sequences from 11 prokaryotic genomes and 21–24% of sequences from *Saccharomyces cerevisiae* and *Caenorhabditis elegans* [12]. Moreover, many agree that recent advances in protein structure prediction owe more to the advent of PSI-BLAST than to improvements in existent structure prediction algorithms.

3.3.4   PHI-BLAST

Pattern Hit Initiated (PHI) BLAST [11] is the newest addition to the BLAST family of sequence comparison algorithms that implements a "hypothesis-directed" search method. It does this by restricting a BLAST search to those protein sequences within a database that contain a specified pattern. Thus it requires as input both a protein query sequence and a query seed pattern. The latter is specified according to the PROSITE [13] pattern search convention and must occur at least once in the query sequence and less than once every 5,000 database residues, i.e., the pattern must have four fully specified residues.

Every occurrence of the seed pattern in a database sequence is first matched with the parent pattern in the query sequence. An optimal local alignment is then constructed, the quality of which is evaluated and expressed as an expectation value. By coupling a pattern search to significant local alignments, PHI-BLAST allows the significance of a pattern found in a target sequence to be assessed within the context of the peripheral sequence.

3.3.5   FASTA

Like BLAST, the FASTA family of programs [3] is composed of seven algorithms; FASTA3 (protein against protein or DNA against DNA), ssearch3 (the Smith-Waterman algorithm), FASTX3/FASTY3 (translated DNA against protein, but only in three frames), TFASTX3/TFASTY3 (protein against translated DNA database), TFASTA3 (Protein against translated DNA database in six frames), FASTF3 (compares a mixed peptide sequence against a protein database), TFASTF3 (compares a mixed peptide sequence against a translated DNA database).

In comparing a protein sequence against a protein database, the FASTA3 algorithm employs a six-step strategy:

1) Locate regions shared by two sequences with the highest density of single (ktup = 1) or double (ktup = 2) identities (Fig. 2, panel a).
2) Take the ten highest regions of density and score them using BLOSUM50. The value of highest scoring region is assigned to the variable init1 and reported later (Fig. 2, panel b).
3) If several regions exist with a score greater than the CUTOFF value, establish if they can be joined to form an approximate alignment with gaps (Fig. 2, panel c).
4) The similarity score (initn) is calculated by adding the scores of the separate joined regions and subtracting the gap penalties (Fig. 2, panel d).
5) For sequences with scores greater than a threshold, an optimal local alignment is constructed and scored; the so-called opt-score.
6) Calculate Z-scores, rank the alignments accordingly and display them using the Smith-Waterman algorithm.

3.4   Multiple sequence alignment

Having compared your query sequence against a database, it is often desirable to convert the output of numerous alignments into a single multiple alignment, where three or more sequences are compared in unison. While it is theoretically possible to extend the use of dynamic programming methods to three or more sequences, constraints on CPU time and memory make this approach impractical. Instead, the most successful multiple alignment strategies rely on hierarchical methods that are fast, accurate and function in three steps: 1) For a given number

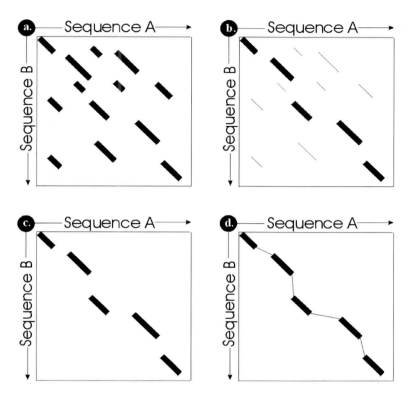

Figure 2. Graphical representation of the alignment procedure employed by FASTA. See accompanying steps for details.

of sequences, all unique pairwise comparisons are made and the similarity scores for the comparisons recorded in a table. These pairwise alignments may be scored according to percentage identity, normalized alignment score (where the raw score for alignment of two sequences is divided by the length of the alignment) or a statistical score from Monte-Carlo shuffling of the sequences. 2) Hierarchical cluster analysis is then performed on the table of pairwise scores generating a dendrogram of similarity groups. 3) By then following the dendrogram from leaves to root, the multiple sequence alignment is generated. In general, the denser the tree, the better the alignment.

One of the most sophisticated hierarchical multiple alignment programs is ClustalW [14]. ClustalW applies different pair-score matrices when aligning sequences of differing similarity. The program also modifies gap-penalties depending upon the composition and degree of conservation at each position. Instead of using a single substitution matrix, ClustalW incorporates a progressive series, such that the matrix chosen is appropriate for the level of dissimilarity encountered. The AMPS program [15], in contrast, implements pairwise global alignment with assessment of statistical significance by Monte-Carlo methods. It also implements position-specific weights and a variety of profile-based searching and alignment methods. However, the main function of AMPS is to produce multiple sequence alignments by a hierarchical method.

The advantages of programs such as ClustalW and AMPS are many. While multiple alignment is of little help when alternate domain orders exist in otherwise closely related proteins, the approach has been instrumental in identifying putative domain structures in proteins which were previously thought to have little in common. Correlated mutation analysis has also proved useful in providing information about both intra and intermolecular interactions. Such analyses examine the varying mutability of particular residues in a multiple alignment thus correlating less mutable ones with a putative function.

One of the most recent applications of multiple sequence alignments is in the generation of "position-specific scoring matrices." The synonymous terms "position-specific scoring matrix", "position-dependent weight matrix" and PSSM (pronounced "possum"), refer to a "profile" based scoring method first described by Gribskov et al. [16]. A PSSM consists of columns of scores for each amino acid derived from corresponding columns of a multiple sequence alignment. The terms "profile" and "hidden Markov model" (HMM) are often confusingly used in place of PSSM. While both terms allude to the fact that each is a PSSM, a subtle difference exists. A profile is a PSSM constructed using the average score method. An HMM is a PSSM constructed using an iterative alignment procedure that has the additional feature of determining position-specific gap penalties. Unlike profiles, HMMs can be trained from aligned or unaligned sequences. Several available software packages implement HMMs and are generally classified according to whether they generate "profile" models or "motif" models. SAM [17] and HMMER implement the former, while PROBE [18], META-MEME [19] and BLOCKS [20] are examples of the latter.

3.5 Biological databases

Biology is an information science. Harvesting the benefits of this information demands efficient methods for data administration and retrieval, the so-called biological database. Until recently, biological databases were designed and built by biologists, primarily in flat file format. This non-unified effort has led to a great heterogeneity in data structures, vocabulary, annotation detail and querying methods. Moreover, the enormous volumes of data presently being submitted to raw sequence databases (i.e., primary databases, such as EMBL or Genbank) and secondary databases (i.e., databases derived from primary sequence information, e.g., PROSITE, or PDB) has made management difficult. A number of efforts are currently underway to address these shortcomings, principally by means of database redesign and database integration systems.

The move away from flat-file repositories to standard relational or object oriented architectures is an important step, enabling the use of flexible querying languages such as SQL (structured query language). The relational model, however, still suffers from the use of different database management systems, a fact that hinders efficient cross database querying. However, object oriented models of data in relational databases permit inter-database communication. When the objects in two or more databases are similar, standards such as the "Common Object Request Broker Architecture" (CORBA) can be used to communicate between them in a flexible and dynamic manner.

One of the most important developments in biological database querying was the introduction of integrative systems that allow queries over numerous hetero-geneous databases. SRS (Sequence Retrieval System; Fig. 3) [21] and ENTREZ from NCBI were developed with this goal in mind, allowing easy hypertext based navigation between entries in different member databases. SRS version 6 provides

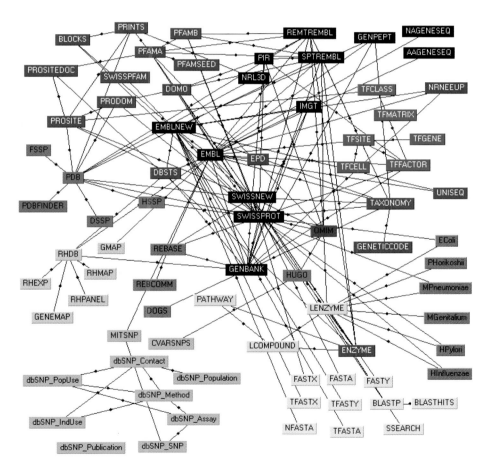

Figure 3. A selection of the most commonly used databases with particular emphasis on the integrative powers of SRS (Sequence Retrieval System).

three different querying interfaces, suiting novice and expert users alike. Because the system provides intricate database linking, one can easily progress from sequence information, to metabolic information (e.g., KEGG [97]), to possible disease associations (e.g., OMIM the "Online Mendelian Inheritance in Man" database [22]) to proprietary data on lead-compounds, and examine the results using either pre-defined or user-defined views.

With the fervent progress in database development, it has become increasingly difficult to obtain a current overview of available databases. While centrally curated catalogues such as DBCAT [23] and the former LiMB [24] help users locate databases, escalating efforts are required to keep-up with the current flux of development. SRS deals with this problem by generating a database of "DATA-BANKS" by traversing public SRS installations around the world, providing a unified system for accessing ~400 different biological databases. Currently, the public network of SRS servers provides access to ~1300 databank copies, distributed over 40 sites in 26 countries.

3.5.1   Sequence databases

Sequence databases serve as repositories for archiving experimentally determined DNA/Protein sequence information. The International Nucleic Acid Sequence Data Library, for example, is the primary repository for nucleic acid sequence data, formed in collaboration between the EMBL, NCBI and DDBJ (DNA data bank of Japan). Protein sequence data is also often supplied in the form of conceptual translations. In terms of sequence quality, the database is extremely heterogenous with high quality cDNA sequence data present alongside short segments of DNA sequence derived from the high throughput sequencing of specifically constructed cDNA libraries, better known as expressed sequence tags (ESTs). Redundancy is also a problem, a factor that has led to the establishment of so-called sequence cluster databases. Unigene, STACK (Sequence Tag Alignment and Consensus Knowledgebase) and EGAD (the Expressed Gene Anatomy database) are three such examples, grouping sequences that are sufficiently similar to each other and providing functional annotation and original tissue source where appropriate.

The well known SWISSPROT [25] database is a curated protein sequence database that strives to provide a high level of sequence annotation (e.g., protein function, domain structure, post translational modifications, disease implications, etc.), a minimal level of redundancy and high level of integration with other databases. Maintaining such a high quality database is a labor-intensive process that excludes the possibility of including all new protein sequence data. For this reason, a supplementary database called TrEMBL [25] (translation of EMBL) database was established. TrEMBL consists of entries in SWISSPROT- like format derived from the translation of all coding sequences (CDS) in the EMBL nucleotide sequence database. PIR (the Protein Information Resource) [26] is similar in many respects to SWISSPROT, but is more redundant and less well annotated.

Patent databases such as DERWENT also provide invaluable sequence information. Despite the obvious lag time between patent application and appearance
in the database, averaging at around 18 months, the information contained within
these databases is extremely well annotated. Moreover, annotations tend to focus
on the applications and intellectual property rights of a sequence together with
other features such as novelty and variants. By incorporating patented sequences
into a search strategy, one can rest assured that almost all relevant sequence material has been covered.

Genome databases are another important class of sequence database. At the end
of 2002, the TIGR (The Institute for Genomic Research) microbial genome database reported a total of 88 completely sequenced genomes, with many more in
progress. For eukaryotes, the situation is less profuse with data for three completely sequenced organisms available. ACeDB [27], for example, was developed
as a database of genome mapping information for the nematode worm
Caenorhabditis elegans and is an acronym for "A *Caenorhabditis elegans*
DataBase." ACeDB allows the retrieval of data at various levels, from whole chromosomes down to individual genes. Many groups are presently adopting ACeDB
to organize genomic data from various species.

3.5.2   Domain databases

In contrast to the explosion in protein sequence data, the number of newly
described protein families has steadily declined. This, of course, offers the hope
that someday nearly all protein-coding genes can be assigned to a protein family
on the basis of database search results. Unfortunately, searches of sequence databases often yield numerous seemingly disparate hits, making the challenge of
deducing family membership increasingly difficult. The multi-module nature of
many proteins further complicates the situation. A simple solution to this problem
is to query a sequence against a database of protein families, domains and modules.
The impetus to develop such databases has been so great that many groups have
been motivated to construct several disparate (yet overlapping) databases, each differing in the breath, scope, method of construction and degree of annotation.

PROSITE [28], PRINTS [29], Pfam-A [30] and SMART [31] are among the
best established curated databases, with expert judgment discerning protein family membership. PROSITE entries correspond to a set of protein sequences derived
from unpublished multiple alignments of the sequences in the family, grouped by
an expert using biological information and provided together with documentation.
Each family is represented by one or more simple patterns or weight matrices corresponding to shared modules or domains. A PRINTS family, in contrast, comes
in the form of a fingerprint consisting of a set of ungapped multiple alignments
corresponding to the shared modules or domains. PRINTS alignments can be used
to derive patterns or weight matrices for searching with a variety of algorithms.

A Pfam-A seed entry is initially represented by a gapped multiple alignment
constructed semi-manually. This seed is then automatically adjusted by searching

against sequence databases to add more sequences to the family. In contrast to the procedures used to construct PROSITE and PRINTS, which concentrate on the conserved regions of a family's sequences, Pfam's gapped alignments may include long regions of uncertain alignment between conserved regions.

All of the above represent domain collections that cover a wide spectrum of cellular functions. This broad scope comes at the cost of optimal sensitivity, specificity and annotation quality. For this reason, the developers of the SMART (Simple Modular Architecture Research Tool) database have collected gapped alignments of signaling and extracellular domains that are imperfectly covered in the aforementioned databases. Three iterative methods (HMMER, MoST and WiseTools) were used to detect distant homologues that were then scored for statistical significance by searching with BLAST. Hidden Markov models of each domain were then constructed. SMART domains are annotated by providing links to i) recent literature via Entrez, ii) homologues with known three-dimensional structure via PDBsum [32], and iii) the domain or motif collections of PROSITE, Pfam and BLOCKS.

In addition to curated compilations, protein family databases based solely on sequence similarity have been constructed by automated clustering of protein sequence databanks. Despite the fact that these databases contain many more entries than curated compilations, groupings are made solely on the basis of sequence similarity with many entries including only a few sequences. Moreover, family designations change after clustering each new release of a database.

The most advanced uncurated effort is the current version of ProDom [33], which uses PSI-BLAST to cluster Swiss-Prot. Each ProDom entry corresponds to a single module or domain which is represented by a gapped multiple alignment, together with a consensus sequence derived from it. The database is re-constructed with each release of Swiss-Prot. DOMO is a similarly constructed collection of protein families which tries to group sequences that share multiple modules into a single entry [34]. Like ProDom, a DOMO entry is represented by a gapped multiple alignment.

While the large number of databases ensures maximal coverage of all possible domains, there is a considerable degree of redundancy between them. Moreover, using these databases to annotate new sequences is difficult because different family representations and search algorithms are employed. To address this problem, a recent venture, in the form of the InterPro consortium [35], plans to integrate PROSITE, Prints and Pfam with the possibility of others at a later stage. In so doing they will not only improve the functional annotation of TrEMBL entries, and thus their speed of promotion to SWISSPROT status, but every other functional annotation effort as well. The BLOCKS+ [20] database was a similar initiative that used the automated block making PROTOMAT system to delineate sets of blocks representing conserved regions and the LAMA program to detect overlap of family groupings. As a result, the BLOCKS+ database has about twice the coverage of the original BLOCKS database.

3.5.3   Structural classification databases

There are presently a few hundred classified protein folds, with a maximum estimate of 1,400 from sequence analysis. Despite this, we generally categorize a structure using the simple classification of all-alpha, all-beta, or alpha-beta. Databases such as SCOP [36] and CATH [37] classify protein folds at many levels, but the primary classification comes with the aforementioned assignment.

SCOP is a highly comprehensive description of the structural and evolutionary relationships between all known protein structures. The hierarchical arrangement is constructed by mainly visual inspection in conjunction with a variety of automated methods. The principal levels are "family," "superfamily" and "fold." The family category contains proteins with clear evolutionary relationships. Members of the superfamily have probable common origin; there may be low sequence identity, but structure and functional details suggest a common origin. The fold level groups together proteins with the same major secondary structure elements in the same arrangement with identical topological connections.

CATH is a hierarchical domain classification of structures with a resolution better than 3.0. The database is constructed by automatic methods wherever possible. The four levels in the hierarchy are: class (C), architecture (A), topology (T) and homologous superfamily (H); a further level called sequence families (S) is sometimes included. "C" classifies proteins into mainly alpha, mainly beta and alpha-beta; "A" describes the shape of the structure, or fold; "T" describes the connectivity and shape; "H" indicates groups thought to have a common ancestor, i.e., homologous; "S" structures are clustered on sequence identity.

3.5.4   Comparative databases

It is now clear that the majority of microbial proteins are highly conserved, with some 70% containing ancient conserved regions. Proteome *versus* proteome comparisons are now utilizing this fact to establish protein orthology, i.e., direct protein counterparts in different genomes typically retaining the same physiological function. The Clusters of Orthologous Genes (COGS) [38] approach identifies the closest homologs in each of the sequenced genomes for each protein, even at levels of seemingly insignificant similarity. The system is built upon the simple notion that any group of at least three genes from distant genomes, which are more similar to each other than they are to any other genes from the same genomes, is most likely to belong to an orthologous family. Each COG consists of individual proteins or groups of paralogs from at least three different lineages (Fig. 4, left).

The original COGS database was derived by comparing seven complete proteomes. As of late 2002, the database has an additional 43 proteomes and now represents 30 major phylogenetical lineages. Another effort called MGDB (microbial genome database for comparative analysis) is dedicated to orthologue identification, paralogue clustering, motif analysis and gene order comparison. Given the present pace of microbial genome acquisition, the number of representatives will

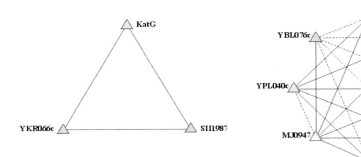

Figure 4. A minimal COG consists of three genes from three different lineages and can be illustrated by a triangle (left). COGs were expanded by combining triangles that shared sides (right). Each of the 21 genomes on the COGs web-site is represented by a different color. Reciprocal best matches between genomes are represented by a solid line. Paralogous relationships, where one genome sequence has a best match to a sequence in another genome but the reverse is not true are indicated by dashed line. (reproduction permission from NCBI news and COGs service).

no doubt increase enormously over the coming years making the COGs and MGDB databases invaluable resources for the annotation of microbial sequences.

3.5.5   Mutation databases

Numerous classifications of mutation databases are available, including general, locus specific, national, ethnic and artificial mutation databases. A comprehensive list is presently maintained by the Mutation Research Centre, Melbourne, Australia. Perhaps the best known and most comprehensive human mutation databases is OMIM (Online Mendelian Inheritance in Man) [22] from the NCBI. The OMIM database is a catalog of human genes and genetic disorders authored and edited by Dr. Victor A. McKusick and his colleagues at Johns Hopkins, and developed for the world wide web by NCBI. The database contains textual information, pictures, and reference information pertaining to human disease genes and variants. The Human Gene Mutation Database (HGMD) is another example, representing data on published germline mutations in nuclear genes underlying human inherited disease [39].

Another important class of mutation database is dedicated to single nucleotide polymorphisms (SNPs). SNPs are the most common genetic variations and occur once every 100 to 300 bases. The first such database on human SNPs, is called HGBASE [40]. Like all SNP databases, HGBASE (Human Genic Bi-Allelic SEquences) facilitates genotype-phenotype association studies using the rapidly growing number of known SNPs. DbSNP is another prime example and is publicly available via the NCBI [41].

3.5.6   Bibliographic databases

Bibliographic databases that catalog abstracts from scientific literature are an invaluable tool in modern research efforts. While many such as EMBASE (medical) and BIOSIS (zoological) are commercial in nature, others such as PUBMED are free and can be accessed via NCBI's ENTREZ system. None of these abstracting services promise definitive coverage of all publications. However, as automated text-mining efforts improve, so too will the importance of these databases.

3.6   The bioinformatics tool-box

Bioinformatics includes within its tasks the challenge of understanding genome linguistics, from sequence to structure to function. In this lies the key to intelligent approaches to drug discovery, gene therapy, and genetic and metabolic engineering. The stakes are high, and so too are the intellectual challenges, which certainly accounts for the bewildering array of algorithms that are presently available. Here, we examine some of these strategies and their implications.

3.7   Gene characterization

One of the most pressing and fundamental problems in this era of profuse DNA sequencing is the accurate prediction of gene structure, and in particular, open reading frames. While numerous *in silico* methodologies have appeared over the past decade, extracting this knowledge has proved less than easy. Currently, the correlation between predicted and actual genes is around 70% with just 40–50% exons predicted correctly. The most successful methods rely on the recognition of three key factors; 1) structural components such as initiation codons, splice signals and polyadenylation sites, 2) the compositional tendencies of coding regions, and 3) homologous sequences.

Detecting functional components in genomic DNA is very much a difficult exercise in pattern recognition. The process is problematic because the structures that delineate a gene are ill-defined, highly unspecific and diverse. However, a number of methods have been applied to the task, the most simple of which use exact word, regular expression or plain consensus sequence (e.g., a poly-adenine stretch) searches. Weight matrices provide an extra level of complexity allowing each position to match any residue but with different scores, while the most sophisticated methods employ neural networks. Alone, these methods are insufficient for gene identification since they result in a computationally untreatable array of potential products. Instead, developers have opted to combine this approach with those dedicated to detection of coding potential.

While identifying prokaryotic coding sequence is a relatively easy task, with typically single contiguous stretches of open reading frame coding for a single protein, the situation for eukaryotes is complex. Applied methods utilize statistical models of nucleotide frequencies and dependencies present in the endogenous codon information. Markov models and neural networks are two well recognized approaches, being used by the GeneMark [42] and Grail [43] programs, respectively. The program BestORF is also based on the Markov chain model but is dedicated to ORF (open reading frame) prediction from EST sequences, an often difficult task due to sequencing errors and partial clones. The use of hidden Markov models (HMMs) has also proved powerful, being employed by many algorithms such as Xpound (for human gene prediction) [44], Ecoparse (used in the annotation of the *Escherichia coli* and *Mycobacterium tuberculosis* genomes) [45], VEIL (the Viterbi Exon-Intron Locator) [46] and GeneMark.hmm [47].

The Procrustes system [48] applies a novel approach, relying both on splice site recognition and homology. The program accepts as input, one genomic DNA sequence and one or several protein sequences. The system finds the chain of exons with the best fit to the target proteins and outputs the amino acid sequence of the predicted protein together with the alignment between the predicted and target proteins.

Identifying functional elements within the non-coding sequences (95–97% of the genome) will be a major challenge for the interpretation of human chromosome sequences. Current knowledge, however, is limited to transcription factor binding sites, origins of replication and untranslated RNA genes. Even for well-studied regulatory elements such as promoters, current computational methods remain unconvincing. The solution to this discrepancy will come from 1) identification of evolutionary conserved regions by comparative sequence analysis of homologous genes in different species, and 2) Chip-based transcription analysis (see Chapter 1). In fact, the advent of chip-based expression analysis assays brings a growing need to link this information to promoter characterization. MatInspector [49] is one such tool that utilizes a large library (>250 entries) of predefined of matrix descriptions for transcription factor binding sites to locate putative promoter elements in sequences of unlimited length. Information about the transcription factors connected to these matrices can then be retrieved from the TRANSFAC database [50].

It is still difficult to assess the accuracy with which a genome can be annotated using these different computational tools. The main problem is the lack of large genomic sequences in which all regions have been unquestionably mapped. Moreover, the accuracy of available tools is usually assessed using sets of simple gene sequences and because they represent a very biased sample, may provide a poor estimation of the true accuracy when applied to newly sequenced genomic regions.

Thus, despite the many available tools, reliable prediction of complex exon assemblies is still a distant goal. Significant improvements in the performance of algorithms relying on statistical information will remain unlikely, unless of course a major breakthrough comes in our understanding of splicing mechanisms.

3.8   EST clustering

While a working draft of the human genome provides an important foundation for gene discovery, it will take a considerable period of time to reconstruct the exact amino acid sequence of all human proteins from these data. Crucial to this effort will be a combination of both genomic and expressed sequence tag (EST) data analysis. As the name suggests, ESTs are short segments of DNA sequence derived from the high throughput sequencing of specifically constructed cDNA libraries. It is by no means surprising that many of today's more successful biotechnology concerns have reaped the benefits of ESTs in the drug discovery process. ESTs provide a rapid route to gene discovery, allow *"in silico* northern blot" analysis and reveal alternative splice variants. This all comes at a price, however. Due to the high-throughput nature of the process, EST sequences tend to be poor in quality. In order to improve the situation, pre-processing (e.g., masking), clustering and post-processing of the results is required.

Clustering information is a prerequisite of all EST clustering tasks. While shared annotation later provides joining information, the most accurate criteria for cluster membership is sequence identity. Small scale clustering projects involving at most a few thousand sequences can easily apply standard tools of contiguous assembly (i.e., the fusion of multiple overlapping DNA sequence clones into one contiguous sequence) or multiple alignment. However, the task of clustering the millions of publicly available ESTs is a venture of greater magnitude. A number of purpose-built clustering methods have risen to the challenge. ICA tools [51], for example, represents an alignment based approach that uses a BLASTN [2] type of algorithm to examine if one sequence is a sub-set of another. A component thereof, N2tool, is a dedicated clustering tool that uses an indexed file format and local alignment to compare all sequences to each other and identify those which share regions of similarity. D2-cluster is a non-alignment based clustering method that uses the approach of comparing word composition within two sequence windows, thus identifying sequences that are greater than 96% identical over a window of 150 bases. The algorithm is used to produce the initial loose clusters in the STACK clustering system. The well known TIGR-ASSEMBLER has also been used for EST clustering and assembly. The system uses BLAST and FASTA to identify all sequence overlaps and stores them in relational database. So-called "transitive closure groups" are then formed and subjected to assembly using the TIGR-ASSEMBLER. Systems such as Genexpress Index [52], Unigene (at the NCBI) and the Merck Gene Index [53] also group sequences into clusters based on sequence overlap above a given threshold.

3.9   Protein structure prediction

The prerequisite information for carving a protein's structure is inherent both in its amino acid sequence and its native solution environment. Not only does the amino acid sequence contain endogenous folding information, it presumably also facilitates the many accessory cellular factors that participate in ensuring its desired structure. However, the solution to the accurate automated *ab initio* prediction of 3D structure using solely the endogenous aspect, remains an unsolved conundrum. The results of the first four CASP (Critical Assessment of protein Structure Prediction; a comparative assessment within the structural biology community of the current state of the art in protein structure) experiments have also embraced this notion. While theoretical approaches such as molecular dynamic simulations have been examined, they remain practically hindered by inadequate computing power and knowledge base limitations. These facts have led to an ever-increasing void between sequence information and the structural knowledge it harbors (see also Chapter 8). Offsetting this imbalance can to some extent be achieved using secondary structure prediction algorithms, of which there are many.

3.9.1   Secondary structure prediction

Secondary structure prediction has a long and varied history, roughly defined by three generations of strategy. First generation algorithms examined the three-state structural propensities (defined here as H for helices, E for extended strands, and L for non-regular loops) of single residues, using a very restricted information base (e.g., Chou-Fasman [54]). The three-state method is the most commonly used scoring scheme when assessing the relative performance of secondary structure prediction algorithms (an alternative eight-state model is also available). This simple scheme reflects the percentage of correctly predicted residues, denoted as Q3:

$$Q3 = 100\infty \frac{\sum_{i=1}^{\cdot} C_i}{N}$$

where C_i is the number of correctly predicted residues in state i (H, E, L), and N is the number of residues in the protein. For first generation algorithms, scores were typically below the 50% level.

Growing bodies of data, information about local physicochemical influences, the use of statistical inference, and greater understanding of tertiary structure all contributed to the enhancements of second generation algorithms (e.g., Garnier-Osguthorpe-Robson [55]). At this stage, secondary structure prediction made an important step away from simply looking at the structural propensities of single residues to include information about the influence of adjacent residues. Predictions thus became a process akin to pattern recognition, relying on the fact that segments of contiguous sequence, typically within a window of 13–21 residues contain an endogenous bias for certain secondary structure states. However, despite these conceptual advances, second generation algorithms, in common with their predecessors, produce results that are often erroneous. Undersized α-helices, near random prediction of β-sheets and a Q3 value <70%

are very much the norm. Note that despite this fact, many bioinformatics providers continue to offer these algorithms as standards.

The advent of third-generation algorithms, heralded by the development of PHD [56], has seen the inclusion of evolutionary information using sequence alignments, coupled with pattern recognition strategies such as neural networks or nearest neighborhood methods. The evolutionary aspect is important because of the implicit information contained about long-range intra-protein interactions, derived from correlated mutation/conservation at key residues.

PHDsec is the most accurate and best known exponent of this multi-level approach. Here, the query sequence is first screened against a non-redundant database of protein sequences. Homologues are then extracted and compared using the

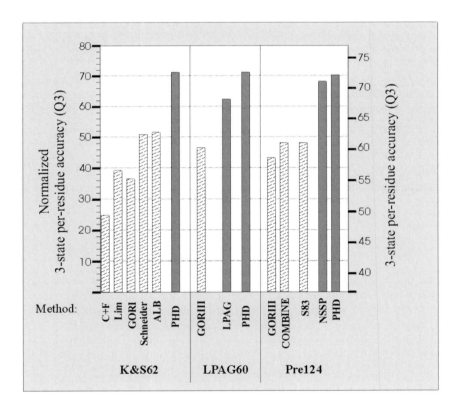

Figure 5. Three-state per-residue accuracy of various prediction methods. Shaded bars: methods of 1st and 2nd generation; filled bars: methods of 3rd generation. The left axis showed the normalized three-state per-residue accuracy, for which a random prediction would rate 0%, and an optimal prediction by homology modeling would rate as 100%. Only methods were included for which the accuracy had been compiled based on comparable data sets, the sets in particular are: K&S62, 62 proteins taken from [102]; LPAG60, 60 proteins taken from [103]; Pre124, 124 unique proteins taken from [104]. The methods were: C + F Chou & Fasman (1st generation) [54]; Lim (1st) [105]; GORI (1st) [55]; Schneider (2nd) [106]; ALB (2nd) [107]; GORIII (2nd) [108]; LPAG (3rd) [103]; COMBINE (2nd) [109]; S83 (2nd) [110]; NSSP (3rd) [58]; PHD (3rd) [56].

MaxHom algorithm [57]. MaxHom generates a pairwise profile-based multiple alignment by compiling a length-dependent cut-off for significant pairwise identities. The choice of alignment algorithm is important, since the accuracy of the multiple alignment is a limiting factor. A profile of position-specific residue substitutions is then compiled and fed into an optimized multi-tier neural network. For typical globular proteins, this results in a prediction accuracy of $72 \pm 11\%$ (see Fig. 5 for comparative analysis). Also included in the PHD package are algorithms for the prediction of surface accessible residues (PHDacc; prediction accuracy = $\sim 75 \pm 7\%$), and transmembrane domains (PHDhtm; prediction accuracy = $\sim 98\%$).

The NNSSP program [58], in contrast, combines a neural-network and nearest-neighbor method to obtain a reputed three-state prediction accuracy of $\sim 67.6\%$ when tested on a non-homologous set of 126 proteins. Improved accuracy is attainable using multiple sequence alignment information. Others such as DSC [59], employ physicochemical information from pre-aligned sequences coupled with GOR secondary structure decision constants. The PREDATOR algorithm [60] is unique in that it relies on both pairwise sequence alignment information and on the recognition of potentially hydrogen-bonded residues in the amino acid sequence. For single sequences, the algorithm demonstrates a Q3 value of 68%, with potential improvements made if a multiple sequence alignment is also employed.

Each of the aforementioned strategies were recently combined with two other algorithms (MULPRED and Zpred) to generate a consensus-based prediction approach. The result, a prediction server called JPRED [61], provides predictions reputed to be 1% better than the best single performer, PHD.

3.9.2   Tertiary structure prediction

The ultimate aim of protein structure prediction is to take a protein with unknown structure and, from its sequence alone, predict the tertiary, or 3D, structure. Despite the simplicity with which the basic problem can be stated, over the 40 years that people have been considering it, no method has ever proved to be generally (some would say, even partly) successful. The intellectual challenge of the problem, despite its apparent intractability, has ensured that many have been (and still are) willing to look at it. Although no general theory has resulted, all this effort has not been in vain as there are now many methods that, although they cannot predict a full tertiary structure, can provide insight into the sort of structure that a sequence might adopt.

Generally, when a protein has high sequence identity to a protein with known 3D-structure, models can be generated using homology modeling (see Chapter 8). Where no sequence homology can be determined, then threading can be employed to find appropriate folds. Only when there is no structure with any sequence homology and no confident threading prediction must purely *ab initio* methods be used.

3.9.3   Homology modeling

Homology or comparative modeling uses information derived from the homology between a target sequence of unknown structure and a sequence whose structure has been solved, to provide accurate predictions of the targets structure. Fragment based homology modeling is a further refinement of this technique where models of proteins can be constructed from separate fragments of other proteins. Areas where there are inserted residues and no structure in the homologue can be built using fragment matching.

The COMPOSER algorithm [62] is a comparative modeling program that derives an average framework from a series of homologous structures and then uses that as a base for constructing a structure from homologous fragments. Another method, MODELLER [63], optimally satisfies structural restraints derived from an alignment with one or more structures. These restraints are expressed as probability density functions (PDF's) for each feature, where a feature may be solvent accessibility, hydrogen bonding, secondary structure, etc. at residue positions and between residues.

3.9.4   Fold recognition/threading

Fold recognition, or threading, is a process whereby a sequence with unknown structure is compared to a database of structures with different folds. In making the comparison, the structure that the query sequence finds a best fit to is taken to be the most likely fold that the sequence will adopt. Fold recognition falls broadly into two categories, one using pairwise energy/interaction potentials and the other performing a 1D to 3D comparison.

Pairwise potentials are any measure which can be used to classify a residue-residue interaction, or atom-atom interaction. The THREADER algorithm [64] takes an empirical potential map of a protein and fits (or threads) the query sequence onto the structure of the known protein. The targets are compared to a database of non-homologous proteins, performed in 3-D space. The THREADER output for the query sequence can be ranked according to several scores. Structures that score significantly well may be correctly associated with the target sequence. A gene threading method (genTHREADER [64]) that can be applied to whole proteomes (translated genomes) can be accessed on a web server, but only for individual sequences. This method takes into account evolutionary relationships to filter out false positive predictions, one of the most serious problems with threading algorithms. A Multiple Sequence Threading (MST) algorithm [65] has also been developed which compares multiple sequence information with a database of structures to determine the correct fold. Models can also be produced which are based on threading homology rather than sequence homology.

Multiple sequence information is also used in the simpler 1D/3D fold recognition methods which perform a secondary structure prediction on the sequence of interest and then compares that secondary structure with all the secondary structures in sequences with known structures to find a possible match. Methods that fall into this category include: TOPITS [66], MAP [67] and H3P2 [68].

3.9.5   Ab initio *modeling*

Ab initio modeling, or *de novo* folding, is not based on any template structures, but rather on a secondary structure assignment and various sets of constraints. One advantage of the *ab initio* method is that models are not restricted to a known fold and so can be used to model proteins with no known fold in the databank. One of the first steps in any *ab initio* prediction is to try and identify the secondary structure elements. Combinatorial methods try to explore all the combinations of arrangements of the secondary structure elements. These methods generally try to pack hydrophobic residues in the core of the protein (if it is a globular protein). Some use a framework or lattice on which to base the secondary structure elements. Methods such as these are often preferred as they impose a reduced number of conformations and reduce complexity. Distance geometry is a method which is more commonly applied to homology modeling and NMR structure solution. One such distance geometry method is incorporated into a program called DRAGON [69] for *ab initio* prediction. A simplified model chain is folded by projecting it into gradually decreasing dimensional spaces whilst subjecting it to a set of pre-defined restraints, primarily secondary structure. In this way the geometry space is successfully explored to produce a protein backbone. The method generates many folds in a short time using an embedding algorithm incorporated into the program. Simple models such as these can then be refined with more complex modeling and side chain minimization to generate structures. Programs such as PROCHECK [70] exist to validate protein models and structures.

Computing power and the complexity of the problem still limit the use and success of *ab initio* methods in protein structure prediction. Moreover, when the time comes that all protein folds have been characterized then there will be little need for *ab initio* modeling, as both fold recognition and homology modeling are far more likely to produce higher quality models. We would, however, still have to ask ourselves if we had definitely identified all possible protein folds.

3.9.6   *Prediction of structural elements*

Coiled coils and leucine zippers
The coiled-coil is a common structural motif, often employed in nature to stabilize α-helices in proteins. As a consequence, many structural proteins involved in the maintenance of cytoskeletal integrity possess this motif, as do all transcription factors of the bZIP and bHLH-LZ families, where the coiled coil acts as a specific dimerization interface.

Two well-known algorithms, COILS2 [71] and PAIRCOIL [72], lend themselves to the prediction of this important domain. COILS2 elicits a profile search against a database of known parallel two-stranded coiled-coils. By comparing the resultant score to the distribution of scores in globular and coiled-coil proteins, the program returns the probability that the sequence will adopt a coiled-coil conformation. COILS is specific for solvent-exposed, left-handed coiled coils. As such, it cannot detect other types of coiled-coil structure, such as buried or right-handed coiled coils and, according to the original designer, it often over-predicts [73]. PAIRCOIL is a newer method based on the correlated occurrence or absence of residues throughout the heptads. In general, PAIRCOIL does not perform as well as COILS2, though in practice it is particularly adept in the prediction of longer coiled-coils.

Leucine zipper (LZ) domains represent a particular coiled-coil sub-type, typically comprising between four and six heptads. Prior to the existence of suitable predictive programs, the detection of leucine zipper domains relied on the identification of the consensus sequence L-x(6)-L-x(6)-L-x(6)-L (where L is leucine and x is any amino acid). Advances were initially made with the development of the TRESPASSER (Two RESidue Pattern Analysis for Sequence StructurE Relationships) program [74]. The presence of a leucine repeat (three heptads or more) as a prerequisite for positive prediction limits the large scale application of this approach, since many LZ proteins possess residues other than leucine at a given position in the structure. To address this shortcoming, the developers of the 2ZIP algorithm [75] combined the predictive powers of COILS2 with an approximate search for the characteristic leucine repeat. In so doing, they developed the most accurate tool available for leucine zipper detection.

Helix-turn-helix

The best known exponent of helix-turn-helix (HTH) prediction is HTHScan. The program uses a profile derived from Pfam Release 2.0 [30] and constructed using MEME 2.1 [19], to detect the presence of H-T-H motifs in protein sequences.

Transmembrane domains

Intrinsic membrane proteins comprise some 20–30% of the eukaryotic proteome. Transiting the membrane by means of α-helices, each containing approximately 20 amino acids of hydrophobic character, their functions are diverse. Some act as growth factor receptors, transducing extracellular signaling events and controlling processes such as cell proliferation, differentiation and apoptosis. Others, such as the G-protein coupled receptors, are of immense pharmaceutical interest being responsible for functions as diverse as olfactory/visual transduction, hormone reception, and the purveyance of neuromodulatory signals. Not surprisingly then is the search for putative membrane proteins an integral part of any genome characterization project. Luckily, predicting α-helical transmembrane domains is a far easier task than predicting the same domain in a globular context. This derives from constraints placed by the lipid bilayer, which cause a hydrophobic bias. This bias is so strong that even a simple strategy, such as calculating a propensity scale using a sliding window and cutoff, performs admirably well. However, an additional level of complexity is introduced if we wish to assess the topology of a membrane protein, i.e., whether the N-terminal segment persists in the intra or extra cellular environment. One general rule of thumb, known as the "positive-inside rule," suggests that the basic amino acids arginine and lysine are found on the cytoplasmic side.

Numerous algorithms of varying strategy have been developed in search of transmembrane proteins. TopPred [76], for example, applies a dual empirical hydrophobicity cutoff to the output of a sliding trapezoid window, thus producing a list of putative transmembrane domains. A potential problem with this technique comes from applying a fixed hydrophobicity threshold, since many α-helical bundle type transmembrane proteins use the non-hydrophobic residues for inter-helical contacts. The PHDhtm program [77] uses evolutionary information, in the form of a profile-based multiple sequence alignment coupled with neural networks and post-processing, to produce what has repeatedly proven to be the most accurate transmembrane prediction strategy. Dynamic programming and a set of statistical tables derived from well-characterized membrane proteins form the basis of the Memsat program [78]. This program uses separate propensity curves for amino acids found in the head and tail regions of the lipid bilayer. The TMAP algorithm [79] also utilizes this strategy, this time in combination with the frequency biases of 12 residue types. Recently, an entirely new approach has been reported, that summons the help of HMMs ("hidden Markov models," see Multiple alignments). The algorithm, TMHMM [80], encompasses many of the conceptual and methodological aspects of the aforementioned methods. Its originality, however, derives from the probabilistic framework of HMMs, which closely resemble the biological system. This avoids the need for specialized post-processing or dynamic programming methodologies. In terms of accuracy TMHMM produces a comparable single TM prediction accuracy to PHDhtm, however, the overall accuracy of TMHMM is not as high as for PHDhtm when topology is also considered.

3.10 Cellular localization prediction

Cellular localization is an integral aspect of protein function. This is best exemplified by the fact that evolution has often used localization as a function control mechanism. Segregating putative protein sequences into protein classes related to their location can thus provide a suitable framework for functional hypotheses. Studies suggest that two components are important in targeting proteins to their required location; 1) total amino acid composition and 2) specific targeting signals. Those algorithms developed to date consider these points to varying degrees and also differ in the number of location classes considered.

The ProtLock algorithm [81] uses a statistical analysis of the total amino acid composition to discriminate among five classes of location: integral membrane proteins, anchored membrane proteins, extracellular proteins, intracellular proteins and nuclear proteins. Others such as PSORTII [82] employ a combination of these factors. PSORTII is a general-purpose localization prediction program, being applicable to both eukaryotic and prokaryotic sequences. For eukaryotes it predicts the probable location of a query sequence from a set of eleven possible target sites.

Proteins destined for the extracellular environment, mitochondria or chloroplasts normally possess N-terminal sorting signals, referred to here as signal peptides, targeting peptides and transit peptides, respectively. These sequences tend to start with a small region rich in charged amino acids followed by a longer hydrophobic part (in fact, this domain is often predicted as a transmembrane region by certain algorithms). The protease sensitive cleavage site normally occurs within 20 amino acids of this domain. Working on the premise that these sorting signals are found at the start of a protein and that they are contiguous in nature, pattern recognition strategies employing neural networks or HMMs have been developed for their identification. SignalP [83] was the first method to apply neural networks to the problem of signal peptide recognition. This was an important advance, particularly for pharmaceutical concerns, as secreted proteins are prime therapeutic candidates. Three versions of the program are available based on three different training sets (eukaryotes, Gram-negative and Gram-positive), each reflecting the significant differences in the characteristics of signal peptides from these organisms. The program actually employs two different neural networks, one trained to recognize the cleavage site (C-score), the other classifying each residue as a signal peptide residue or not. The prediction of cleavage site location is formalized using the Y-score, which takes into consideration where the C-score is high and the S-score changes from a high to a low value, an important consideration where multiple putative cleavage sites are predicted. More recently a HMM based version of SignalP has appeared possessing advantages over its predecessor in its ability to discriminate between signal peptides and signal anchors. Both the HMM and neural network based components have now been combined in SignalP version 2. Analogous methods have also been implemented for the prediction of chloroplast transit peptides and mitochondrial target peptides, called ChloroP [83] and MitoP [83] respectively.

With the recent advances in protein localization prediction the possibilities for an integrated cellular localisation prediction program grow ever stronger. While the PSORTII algorithm is certainly a step in the right direction, what is required is a system that integrates all the best available methods and provides an output of likelihoods based on the results of all strategies. Such a system should not only rely on amino acid composition and short signal peptides, but also use the knowledge implicit in domain databases where clear hints to cellular localization are already available.

3.11 Phylogenetic analysis

One application of tree-construction algorithms in molecular biology is to determine phylogenetic relationships among organisms on the basis of nucleotide or peptide sequences; another area of application, not discussed here, is in hierarchical clustering, e.g., of genes in expression array studies. Since the 1950s it has been assumed that phylogenetic relationships can be inferred from sequence similarity. Species with very similar sequences for one or more genes or proteins can be assumed to have diverged from a relatively recent common ancestor, while the common ancestor for those with more different sequences is relatively more ancient. The basis for this assumption is simply that the mutational events underlying speciation can be regarded as random and cumulative. Two facts complicate this picture: first, mutations can be recurrent, and second, they can be convergent. Since transitions (purine-purine or pyrimidine-pyrimidine) are much more frequent than transversions (purine-pyrimidine), almost all multiple mutations are recurrent, in that they revert to the original non-mutant type. Thus polymorphic loci in mutational hot spots (such as the hypervariable mtDNA D-loop) are not very informative for longer time scales. Similarly, microsatellite markers, whose mutation rate is an order of magnitude higher than other polymorphisms, are about as likely to gain as to lose a repeat (except at the extremes of their allelic range).

Convergent mutation (or homoplasy) occurs often enough at the phenotype level to be a problem in phenetic phylogenetic reconstruction: the anatomy of the eye, for example, shows unexpected similarities among widely separated species. At the molecular or sequence level, however, this becomes less of a problem (although the fundamental distinction between genotype and phenotype becomes less clear when we can directly measure DNA sequence!)

The two main types of tree-building algorithms are those based on discrete characters (e.g., maximum likelihood and maximum parsimony) and those based on distance matrices (e.g., neighbor joining and average linkage). It is always possible, of course, to transform character data into distances, but not *vice versa*, so character data retain more information than do distances. On the other hand, the distance matrix methods are computationally much less demanding than the character-based methods. The practical and theoretical advantages and disadvantages of these are still under debate. Felsenstein's Phylogeny Page [84] has pointers to 161 different software programs and packages related to phylogenetic reconstruction. The review by Wills [85] covers many of the pros and cons, and includes recommendations for the appropriate use of different methods. Setubal and Meidanis [86] provide a somewhat more general and rigorous introduction. Since even the distance matrix methods are overwhelmed when given thousands of cases and/or characters, much of the recent work has been directed towards the development of approximation or sampling methods. This includes, among others, the disk covering method [87], routed insertion [88], and perfect matchings [89].

3.12 Metabolic simulation

Metabolic modeling is entering an exciting era. Today, we have detailed and fairly accurate representations of various metabolic pathways in different organisms. We have holistic approaches which encompass entire organisms, albeit in less detail [90]. Metabolic reconstruction promises eventually to deliver models which are both detailed and integrative [91]. Metabolic engineering promises to deliver methods for designing modifications to organisms tailored to the production of inexpensive, efficient products. As the scope and power of these methods have expanded, so have the horizons of our expectations.

Yet many problems and difficulties still remain. Although genomics has produced a flood of information, the problem of identifying relevant facts requires sophisticated methods of data mining and knowledge base construction. In this section we focus on techniques for metabolic modeling based on appropriate datasets.

The overall goal of metabolic modeling is to reflect as realistically as possible the underlying physico-chemical events of metabolism. Events relevant to metabolism can range from—at the low end—conformational changes taking a fraction of a nanosecond to—at the high end—intercellular signaling mechanisms or cell cycle regulatory processes occurring over a period of many hours, encompassing a dynamic range of at least fourteen orders of magnitude. Including yet finer detail of reaction mechanisms or broadening the scope to developmental (or even evolutionary) processes increases this range.

At one extreme, then, a molecular dynamics based approach is conceivable. At the other extreme we can imagine a knowledge-based textbook level representation. These two extreme examples correspond respectively to the quantitative and qualitative approaches in simulation methodology.

3.12.1 Quantitative approach

Continuous kinetic models were the first to be applied to limited systems of chemical reactions, as there was already an extensive history of dynamical systems simulation using differential equations. Use of this paradigm has continued, especially with increasing computational power over the last decades, and some of the most recent efforts are based on pure quantitative models.

Several computer packages have been specifically developed for quantitative metabolic modeling (i.e., for kinetic simulation and optimization), including MIST [92] and Gepasi [93]. The Gepasi system is well-maintained, quite flexible, easy to use, and exemplifies the state of the art for quantitative modeling. It allows specification of a variety of reaction mechanisms, can search for solutions and estimate parameters based on empirical data, and characterize steady states using metabolic control analysis (MCA) and linear kinetic stability analysis. An even more recent example applying this paradigm is E-CELL [94]. E-CELL is, like Gepasi, a general-purpose biochemical simulation environment, which has been used to model the primary metabolism of a whole cell (an *in silico* chimera of *Mycobacterium genitalium* and *Escherichia coli*), resulting in a counterintuitive prediction of the glucase starvation response which still awaits empirical verification.

3.12.2   Qualitative approach

Using the data of EcoCyc, Heidtke and Schulze-Kremer have developed a qualitative simulation framework (BioSim) [95] and a model description language
(MDL), to derive qualitative models from the EcoCyc knowledge base. The Lisp-
based frame definitions within EcoCyc were translated to Prolog, then the classes
were defined within MDL as being either objects (compounds, elements,
enzymes, genes, proteins) or processes (reactions, pathways). Thus a model could
be constructed from a query to the EcoCyc knowledge base. An interpreter was
then applied to the model to generate qualitative behavior (i.e., indications of the
increase or decrease in quantity of a substance.

It should also be mentioned in this context that a good many efforts are under
way in reconstructing and representing genetic regulatory networks [96]. Although
these do not model metabolism *per se*, they do constitute qualitative models whose
function will eventually be necessary for accurate metabolic models.

The drawback of the qualitative approach is that the quantitative aspects of the
system have been abstracted away, so that accurate modeling of gradual change is
impossible; even to approximate it is difficult without considerable complication
of the knowledge- and rule-base. Variations in concentration of metabolites, in
rate of reactions, or in delay between events, cannot be both easily and accurately modeled using purely qualitative methods.

3.12.3   Semi-quantitative approach

The predominant model used in the semi-quantitative approach is the Petri net formalism. A Petri net is a bipartite graph i.e., a graph in which the vertices are partitioned into two subsets, such that vertices from each set are connected only to
vertices in the other set. For purposes of metabolic modeling, one set of vertices
(called places) is used to represent metabolites; the other set of vertices (called
transitions) is used to represent reactions. The characteristic of Petri nets which is
useful for modeling is that each of the places contain "tokens," and when every
place incident to a transition is occupied by a token, the transition fires, which
changes the state of each of the places connected to the transition.

Petri nets have been extended in many different ways, e.g., (black/white versus
colored, delayed or timed, capacity, self-modifying, stochastic, etc.). By incorporating one or more of these extensions, it is possible to model at any level or combination of levels from qualitative to quantitative, i.e., one can include as much
quantitative information as available on concentrations, rates, etc.

While flexible, the formal frameworks of the semi-quantitative approach are
still unfamiliar, and not yet very well standardized. In addition, the more accurately quantitative the model is to be, the more extensions must be made to the
supporting formalism, making it somewhat complex and unwieldy. Nonetheless,
the representational power and generality of this approach means that it will probably be used more often in the future.

3.12.4 Hybrid approaches

A hybrid model contains two or more classes of objects, such that the interactions among the members of each class are defined according to a set of rules peculiar to that class (constituting an integral model), while the interactions between classes (the "glue" linking the various integral representations) are specified by *ad hoc* rules peculiar to the model. Hybrid models are thus non-generic, special-purpose models which must be hand-crafted for each target system.

McAdams and Shapiro [97] present an outstanding example of this approach, in which they simulate the kinetics and signal logic of the bacteriophage lambda lysis/lysogeny decision circuit. Their model involves twelve genes organized in five operons with seven promoters, and both mRNA, direct gene products, and posttranscriptional modifications. The kinetics of transcription and signaling are modeled using numerical integration of differential equations, while the switching logic is superimposed on the signal levels using both sigmoid-thresholded and edge-triggered boolean gates. Signal delays are produced by distance between genes on the DNA, transcription rates, concentration thresholds for signaling proteins, and rates of protein degradation. The model is remarkably detailed and accurate, and allows verification of the logic as well as identification of the control function of different components.

One advantage of hybrid models such as this is that (like semi-quantitative models) they can represent aspects of continuous kinetics at short time scales and discrete switching mechanisms at longer time scales. A further advantage, distinguishing them from the semi-quantitative approach, is that they utilize algorithms for these different time scales, which have been highly optimized over decades of use, which are both efficient and familiar. The practical disadvantage is that they must be recreated by hand for each particular target system.

Currently, the main obstacles to metabolic modeling include the practical problems of obtaining relevant data, and the theoretical problem of allowing both quantitative and qualitative interpretations of the same events at different levels. The practical problems will inevitably yield, bit by bit, to sustained effort. The theoretical problem can be circumvented, for a time, by hybrid models or by relying on increases in computing power, but its solution requires a new paradigm involving a change in perspective whose nature is yet unknown.

3.13 Automated sequence annotation

While manual annotation of sequences undoubtedly produces the highest quality of functional assignment, the process is becoming less feasible, with the rates of sequence acquisition far outpacing manual annotation efforts. A number of automated methods have been developed to address this shortcoming. Systems such as Genequiz [99] (marketed as bioSCOUT), PEDANT [100] and MAGPIE [101] dedicate a selection of the aforementioned algorithms and strategies to the analysis of sequence data. Features such as secondary and tertiary structure, sequence statistics, motifs, expression patterns and function are all extractable at the push of a button.

While predicted characteristics such as isoelectric point (pI) and molecular weights are static entities, at least in the absence of post-translational modifications, functional assignments based on tentative sequence homology must always be treated with caution. Couple this with the fact that database annotations are often erroneous and one can appreciate that such automated methods should be treated as facilitators of the manual approach, at least for the foreseeable future. Indeed, the bioSCOUT system supports this contention by assigning a confidence level to its annotation, thus allowing the manual annotator to focus only on difficult cases for which the confidence is low. Moreover, an automated update procedure allows the results of subsequent database searches to be immediately incorporated into the function prediction process, with so-called "feature reports" updated accordingly. Another important feature is the unique "alerting" capability of the system. Not only can a sequence be automatically searched against database updates every evening, but strategies can be combined using boolean operators to include (or exclude) specified PROSITE patterns, keywords and even HMMs. The result is a system that attempts to overcome the stated limitations of the automated process, by staying constantly up-to-date with new sequence data and associated annotations.

3.14 Conclusions and perspectives

We began with a literary analogy, simplistically linking a genome to a work of great linguistic merit. From this perspective, one of the most obvious future developments points to the emergence of genome libraries, housing tomes of complete genomic and expression data from diverse species. Comparative analyses of these data, as exemplified by the COG's initiative, will eventually provide a global view of protein function both at the molecular and organismal level. Entire proteomes will help to unravel missing links in functional pathways, to explore alternative pathways, and to expand our understanding of principle mechanisms and of evolutionary cross-links. Similar comparative analyses, initially using ESTs, will also provide data on single nucleotide polymorphisms (SNPs) and their associated pharmacogenetic implications. The information gained through such analyses will also improve the speed and accuracy of the annotation process, particularly for automated methods. Interestingly, while estimates for sequencing error rates are available, the accuracy of functional annotation has been less well scrutinized, though assumed to be considerable. In an attempt to offset a possible domino effect in database errors, the move towards cross-community annotation efforts may grow. The PRESAGE database [111] represents a prototypic example that encourages the biological community to become increasingly involved in the curation of sequence annotation databases. Many large-scale annotation efforts may in the future adopt this strategy.

Methods for the automated extraction of biological information from scientific literature will also improve, raising exciting possibilities. This approach will allow text-based *in silico* "two-hybrid" analyses (e.g., "protein X phosphorylates protein Y") that may be extended to automated pathway reconstruction. This will be instrumental in providing specific synopses of the huge body of knowledge on molecular interactions and pathways that exists in the literature, thus further aiding the drug discovery process.

The recent advent of chip-based expression analysis beckons the need for systematic construction of transcriptome expression maps. These maps will provide an essential tool in the study of gene function and disease. Furthermore, issues of integrated database access, data normalization, visualization and analysis are continually being improved, with many bioinformatics concerns rising to the commercial promise (e.g., LION's "arraySCOUT"). Coupling this to the advances made with metabolic and signaling pathway reconstruction will bring us ever closer to an accurate "*in silico*" simulation of the cell.

Time will tell if it is possible to accurately predict 3D-structure using solely the endogenous aspect. While homology modeling has proved successful, methods for fold recognition may in time become more reliable with fewer false positives and negatives. Important use can and should be made of experimental data in the modeling process. Some groups who have access to large computing resources are now creating many homology models and threadings for whole databases. These can then be used as a tentative structural database for drug design. The accuracy of such models and predictions may be difficult to judge, but the pharmaceutical industry may gain important clues for drug design (see Chapter 8). Emphasis in the future will be placed on better methods for modeling function from structure, thus enabling the rational design or modification of protein and ligand alike. Significant numbers of specifically designed therapeutic proteins may derive from this process.

3.15 References

1 T.S. Eliot choruses from *The Rock*
2 Altschul SF, Madden TL, Schaffer AA et al (1997) Gapped BLAST and PSI-BLAST: a new generation of protein database search programs. *Nucleic Acids Res* 25: 3389–3402
3 Pearson WR, Lipman DJ (1988) Improved tools for biological sequence comparison. *Proc Natl Acad Sci USA* 85: 2444–2448
4 Needleman SB, Wunsch CD (1970) A general method applicable to the search for similarities in the amino acid sequence of two proteins. *J Mol Biol* 48: 443–453
5 Sellers PH (1974) On the theory and computation of evolutionary distances. *SIAM J Appl Math* 26: 787–793
6 Smith TF, Waterman MS (1981) Comparison of bio-sequences. *Adv Appl Math* 2: 482–489
7 Dayhoff MO, Schwartz RM, Orcutt BC (1978) Atlas of protein sequence and structure. *Nat Biomed Res Foundation, Washington D.C., USA* 5, Suppl 3: 345–352
8 Jones DT, Taylor WR, Thornton JM (1992) A new approach to protein fold recognition. *Nature* 358: 86–89
9 Gonnet GH, Cohen MA, Benner SA (1992) Exhaustive matching of the entire protein-sequence database. *Science* 256: 1443–1445

10 Henikoff S, Henikoff JG (1993) Performance evaluation of amino acid substitution matrices. *Proteins* 17: 49–61

11 Zhang Z, Schaffer AA, Miller W et al (1998) Protein sequence similarity searches using patterns as seeds. *Nucleic Acids Res* 26: 3986–3990

12 Teichmann SA, Chothia C, Gerstein M (1999) Advances in structural genomics. *Curr Opin Struct Biol* 9: 390–399

13 Bairoch A (1991) PROSITE: a dictionary of sites and patterns in proteins. *Nucleic Acids Res.* 19 Suppl: 2241–2245

14 Thompson JD, Higgins DG, Gibson TJ (1994) CLUSTAL W: improving the sensitivity of progressive multiple sequence alignment through sequence weighting, position specific gap penalties and weight matrix choice. *Nucleic Acids Research* 22: 4673–4680

15 Barton GJ (1994) The AMPS package for multiple protein sequence alignment. *Methods Mol Biol* 25: 327–347

16 Gribskov M, McLachlan AD, Eisenberg D (1987) Profile analysis: Detection of distantly related proteins. *Proc Natl Acad. Sci USA* 84: 4355–4358

17 Hughey R, Krogh A (1996) Hidden Markov models for sequence analysis: extension and analysis of the basic method. *Comput Appl Biosci* 12: 95–107

18 Neuwald AF, Liu JS, Lipman DJ et al (1997) Extracting protein alignment models from the sequence database. *Nucleic Acids Res* 25: 1665–1677

19 Grundy WN, BaileyTL, Elkan CP et al (1997) Meta-MEME: Motif-based hidden Markov models of protein families. *Comput Applic Biosci* 13: 397–406

20 Henikoff JG, Henikoff S, Pietrokovski S (1999) New features of the Blocks Database servers. *Nucleic Acids Res* 27: 226–228

21 Etzold T, Argos P (1993) SRS – an indexing and retrieval tool for flat file data libraries. *Comput Appl Biosci* 9: 49–57

22 Online Mendelian Inheritance in Man, OMIM (TM). McKusick-Nathans Institute for Genetic Medicine, Johns Hopkins University (Baltimore, MD) and National Center for Biotechnology Information, National Library of Medicine (Bethesda, MD), 2000. World Wide Web URL: http://www.ncbi.nlm.nih.gov/omim/

23 Discala C, Benigni X, Barillot E (2000) DBcat: a catalog of 500 biological databases. *Nucleic Acids Res.* 28: 8–9

24 Lawton JR, Martinez FA, Burks C (1989) Overview of the LiMB database. *Nucleic Acids Res* 17: 5885–5899

25 Bairoch A, Apweiler R (1999) The SWISS-PROT protein sequence data bank and its supplement TrEMBL. *Nucleic Acids Res* 27: 49–54

26 Barker WC, Garavelli JS, Huang H et al (2000) The protein information resource (PIR). *Nucleic Acids Res* 28: 41–44

27 Walsh S, Anderson M, Cartinhour SW (1998) ACEDB: a database for genome information. *Methods Biochem Anal* 39: 299–318

28 Hofmann K, Bucher P, Falquet L et al (1999) The PROSITE database, its status in 1999. *Nucleic Acids Res* 27: 215–219

29 Attwood TK, Flower DR, Lewis AP et al (1999) PRINTS prepares for the new millennium. *Nucleic Acids Res* 27: 220–225

30 Sonnhammer EL, Eddy SR, Birney E et al (1998) Pfam: multiple sequence alignments and HMM-profiles of protein domains. *Nucleic Acids Res* 26: 320–322

31 Ponting CP, Schultz J, Milpetz F et al (1999) SMART: identification and annotation of domains from signalling and extracellular protein sequences. *Nucleic Acids Res* 27: 229–232

32 Laskowski RA (2001) PDBsum: summaries and analyses of PDB structures. *Nucleic Acids Res* 29: 221–222

33 Corpet F, Gouzy J, Kahn D (1998) The ProDom database of protein domain families. *Nucleic Acids Res* 26: 323–326

34 Gracy J, Argos P (1998) DOMO: a new database of aligned protein domains. *Trends Biochem Sci* 23: 495–497

35 Apweiler R, Attwood TK, Bairoch A et al (2000) InterPro-an integrated documentation resource for protein families, domains and functional sites. *Bioinformatics* 16: 1145–1150

36 Hubbard TJ, Ailey B, Brenner SE et al (1999) SCOP: a Structural Classification of Proteins database. *Nucleic Acids Res* 27: 254–256

37 Orengo CA, Pearl FM, Bray JE et al (1999) The CATH Database provides insights into protein struc-

ture/function relationships. *Nucleic Acids Res* 27: 275–279

38 Tatusov RL, Koonin EV, Lipman DJ (1997) A genomic perspective on protein families. *Science* 24: 631–637

39 Cooper DN, Ball EV, Krawczak M (1998) The human gene mutation database. *Nucleic Acids Res* 26: 285–287

40 Brookes AJ, Lehvaslaiho H, Siegfried M et al (2000) HGBASE: a database of SNPs and other variations in and around human genes. *Nucleic Acids Res* 28: 356–360

41 Sherry ST, Ward MH, Kholodov M et al (2001) dbSNP: the NCBI database of genetic variation. *Nucleic Acids Res* 29: 308–311

42 Borodovsky M, McIninch J (1993) GeneMark: Parallel Gene Recognition for both DNA Strands. *Computers & Chemistry* 17: 123–133

43 Xu Y, Einstein JR, Mural RJ et al (1994) An improved system for exon recognition and gene modeling in human DNA sequences. *Proc Int Conf Intell Syst Mol Biol* 2: 376–384

44 Thomas A, Skolnick M (1994) A probabilistic model for detecting coding regions in DNA sequences. *IMA J Math Appl Med Biol* 11: 149–160

45 Cole ST, Brosch R, Parkhill J et al (1998) Deciphering the biology of Mycobacterium tuberculosis from the complete genome sequence. *Nature* 393: 537–544

46 Henderson J, Salzberg S, Fasman K (1997) Finding genes in DNA with a Hidden Markov Model. *J Comput Biol* 2: 127–141

47 Lukashin AV, Borodovsky M (1998) GeneMark.hmm: new solutions for gene finding. *Nucleic Acids Res* 26: 1107–1115

48 Gelfand MS, Mironov AA, Pevzner PA (1996) Gene recognition via spliced sequence alignment. *Proc Natl Acad Sci USA* 93: 9061–9066

49 Quandt K, Frech K, Karas H et al (1995) MatInd and MatInspector: new fast and versatile tools for detection of consensus matches in nucleotide sequence data. *Nucleic Acids Res* 23: 4878–4884

50 Heinemeyer T, Chen X, Karas H et al (1999) Expanding the TRANSFAC database towards an expert system of regulatory molecular mechanisms. *Nucleic Acids Res* 27: 318–322

51 Parsons JD (1995) Improved tools for DNA comparison and clustering. *Comput Appl Biosci* 11: 603–613

52 Pietu G, Eveno E, Soury-Segurens B (1999) The genexpress IMAGE knowledge base of the human muscle transcriptome: a resource of structural, functional, and positional candidate genes for muscle physiology and pathologies. *Genome Res* 9: 1313–1320

53 Williamson AR (1999) The Merck Gene Index project. *Drug Discov Today* 4: 115–122

54 Chou PY, Fasman GD (1974) Prediction of protein conformation. *Biochemistry* 13: 222–245

55 Garnier J, Osguthorpe DJ, Robson BJ (1978) Analysis of the accuracy and implications of simple methods for predicting the secondary structure of globular proteins. *J Mol Biol* 120: 97–120

56 Rost B (1996) PHD: predicting one-dimensional protein structure by profile-based neural networks. *Methods Enzymol* 266: 525–539

57 Schneider R, Sander C (1996) The HSSP database of protein structure-sequence alignments. *Nucleic Acids Res* 24: 201–205

58 Salamov AA, Solovyev VV (1995) Prediction of protein secondary sturcture by combining nearest-neighbor algorithms and multiply sequence alignments. *J Mol Biol* 247: 11–15

59 King RD, Sternberg MJ (1996) Identification and application of the concepts important for accurate and reliable protein secondary structure prediction. *Protein Sci* 5: 2298–2310

60 Frishman D, Argos P (1995) Knowledge-based secondary structure assignment. *Proteins* 23: 566–579

61 Cuff JA, Clamp ME, Siddiqui AS et al (1998) JPred: a consensus secondary structure prediction server. *Bioinformatics* 14: 892–893

62 Sutcliffe MJ, Hayes FR, Blundell TL (1987) Knowledge based modelling of homologous proteins, Part II: Rules for the conformations of substituted sidechains. *Protein Eng* 1: 385–892

63 Sali A, Overington JP (1994) Derivation of rules for comparative protein modeling from a database of protein structure alignments. *Protein Sci* 3: 1582–1596

64 Jones DT, Tress M, Bryson K et al (1999) Successful recognition of protein folds using threading methods biased by sequence similarity and predicted secondary structure. *Proteins* 3: 104–111

65 Taylor WR (1997) Multiple sequence threading: an analysis of alignment quality and stability. *J Mol Biol* 269: 902–943

66 Rost B (1995) TOPITS: threading one-dimensional predictions into three-dimensional structures. *Proc Int Conf Intell Syst Mol Biol* 3: 314–321

67 Russell RB, Copley RR, Barton GJ (1996) Protein fold recognition by mapping predicted secondary

structures. *J Mol Biol* 259: 349–365

68 Rice DW, Eisenberg D (1997) A 3D-1D substitution matrix for protein fold recognition that includes predicted secondary structure of the sequence. *J Mol Biol* 267: 1026–1038

69 Aszodi A, Munro RE, Taylor WR (1997) Protein modeling by multiple sequence threading and distance geometry. *Proteins* Suppl 1: 38–42

70 Laskowski RA, MacArthur MW, Moss DS et al (1993)PROCHECK: a program to check the stereochemical quality of protein structures. *J Appl Cryst* 26: 283–291

71 Lupas A (1996) Prediction and Analysis of Coiled-Coil Structures. *Methods Enzymol* 266: 513–525

72 Berger B, Wilson DB, Wolf E et al (1995) "Predicting Coiled Coils by Use of Pairwise Residue Correlations". *Proc Natl Acad Sci USA* 92: 8259–8263

73 Lupas A (1997) Predicting coiled-coil regions in proteins. *Curr Opin Struct Biol* 7: 388–393

74 Hirst J, Vieth M, Skolnick J et al (1996) Predicting leucine zipper structures from sequence. *Protein Eng* 9: 657–662

75 Bornberg-Bauer E, Rivals E, Vingron M (1998) Computational approaches to identify leucine zippers. *Nucleic Acids Res* 26: 2740–2746

76 Claros MG, von Heijne G (1994) TopPred II: an improved software for membrane protein structure predictions. *Comput Appl Biosci* 10: 685–686

77 Rost B, Fariselli P, Casadio R (1994) Refining neural network predictions for helical transmembrane proteins by dynamic programming. *Comput Appl Biosci* 10: 685–686

78 Persson B, Argos PJ (1997) Prediction of membrane protein topology utilizing multiple sequence alignments. *Protein Chem* 16: 453–457

79 Jones DT, Taylor WR, Thornton JM (1994) A model recognition approach to the prediction of all-helical membrane protein structure and topology. *Biochemistry* 33: 3038–3049

80 Sonnhammer EL, von Heijne G, Krogh A (1998) A hidden Markov model for predicting transmembrane helices in protein sequences. *Proc Int Conf Intell Syst Mol Biol* 6: 175–182

81 Cedano J, Aloy P, Perez-Pons JA et al (1997) Relation between amino acid composition and cellular location of proteins. *J Mol Biol* 266: 594–600

82 Nakai K, Horton P (1999) PSORT: a program for detecting sorting signals in proteins and predicting their subcellular localization. *Trends Biochem Sci* 24: 34–36

83 Nielsen H, Brunak S, von Heijne G (1999) Machine learning approaches for the prediction of signal peptides and other protein sorting signals. *Protein Eng* 12: 3–9

84 Felsenstein J (1989) PHYLIP—Phylogeny Inference Package (Version 3.2). *Cladistics* 5: 164–166; also see http://evolution.genetics.washington.edu/phylip.htmll

85 Wills C (1994) Phylogenetic analysis and molecular evolution. In: DW Smith (ed): *Biocomputing: Informatics and Genome Projects*. Academic Press, San Diego, 175–201

86 Setubal J, Meidanis J (eds) (1996) *Introduction to Computational Molecular Biology*. PWS Publishing Co., Boston

87 Huson DH, Vawter L, Warnow TJ (1999) Solving large scale phylogenetic problems using DCM2. In: Lengauer T, Schneider R (eds): *Proceedings of the Seventh International Conference on Intelligent Systems for Molecular Biology*. AAAI Press, Menlo Park (CA), 118–129

88 Strimmer K, von Haeseler A (1997) Likelihood-mapping: a simple method to visualize phylogenetic content of a sequence alignment. *Proc Natl Acad Sci USA* 94: 6815–6819

89 Diaconis PW, Holmes SP (1998) Matchings and phylogenetic trees. *Proc Natl Acad Sci USA* 95: 14600–14602

90 Karp PD, Riley M, Paley SM et al (1996) EcoCyc: an encyclopedia of Escherichia coli genes and metabolism. *Nucleic Acids Res* 24: 32–39; see also http://ecocyc.pangeasystems.com/ecocyc/

91 Bork P, Dandekar T, Diaz-Lazcoz Y et al (1998) Predicting function: from genes to genomes and back. *J Mol Biol* 283: 707–725

92 Ehlde M, Zacchi G (1995) MIST: a user-friendly metabolic simulator. *Comput Appl Biosci* 11: 201–207

93 Mendes P. (1993) GEPASI: a software package for modelling the dynamics, steady states and control of biochemical and other systems. *Comput Appl Biosci* 9: 563–571

94 Tomita M, Hashimoto K, Takahashi K et al (1999) E-CELL: Software environment for whole cell simulation. *Bioinformatics* 15: 72–84; also see E-Cell Project http://www.e-cell.org/

95 Heidtke KR, Schulze-Kremer S (1998) BioSim – a new qualitative simulation environment for molecular biology. In: J Glasgow, T Littlejohn, F Major, R Lathrop, D Sankoff, C Sensen (eds) (: *Proceedings of Sixth International Conference on Intelligent Systems for Molecular Biology*. AAAI Press, Menlo Park (CA), 85–94

96 D'haeseleer P, Liang S, Somogyi R (1999) Gene expression data analysis and modeling. *Tutorial session at Pacific Symposium on Biocomputing,* Hawaii, January: 4–9; also see http://www.cgl.ucsf.edu/psb/psb99/genetutorial.pdf

97 McAdams HH, Shapiro L (1995) Circuit simulation of genetic networks. *Science* 269: 650–656

98 Kanehisa M, Goto S (2000) KEGG: kyoto encyclopedia of genes and genomes. *Nucleic Acids Res* 28: 27–30

99 Scharf M, Schneider R, Casari G et al (1994) GeneQuiz: a workbench for sequence analysis. *Proc Int Conf Intell Syst Mol Biol* 2: 348–353

100 Frishman D, Mewes H-W (1997) PEDANTic genome analysis. *Trends in Genetics* 13: 415–416

101 Gaasterland T, Sensen CW (1996) MAGPIE: automated genome interpretation. *Trends Genet* 12: 76–78

102 Kabsch W, Sander C (1983) How good are predictions of protein secondary structure? *FEBS Lett* 155: 179–182

103 Levin JM, Pascarella S, Argos P et al (1993) Quantification of secondary structure prediction improvement using multiple alignment. *Protein Eng* 6: 849–854

104 Rost B, Sander C (1994) Combining evolutionary information and neural networks to predict protein secondary structure. *Proteins* 19: 55–72

105 Lim VI (1974) Structural Principles of the Globular Organization of Protein Chains. A Stereochemical Theory of Globular Protein Secondary Structure. *J Mol Biol* 88: 857–872

106 Schneider R (1989) Sekundärstrukturvorhersage von Proteinen unter Berücksichtigung von Tertiärstrukturaspekten. Diploma thesis, Universität Heidelberg, Germany

107 Ptitsyn OB, Finkelstein AV (1983) Theory of protein secondary structure and algorithm of its prediction. *Biopolymers* 22: 15–25

108 Gibrat J-F, Garnier JRobson B (1987) Further developments of protein secondary structure prediction using information theory. New parameters and consideration of residue pairs. *J Mol Biol* 198: 425–443

109 Garnier J, Gibrat J-F, Robson B (1996) GOR method for predicting protein secondary structure from amino acid sequence. *Meth Enzymol* 266: 540–553

110 Kabsch WSander C (1983) Segment83. Unpublished

111 Brenner SE, Barken D, Levitt M (1999) The PRESAGE database for structural genomics. Nucleic Acids Res 27: 251–253

Modern Methods of Drug Discovery
ed. by A. Hillisch and R. Hilgenfeld
© 2003 Birkhäuser Verlag/Switzerland

4 High-throughput screening technologies

Ralf Thiericke

CyBio Screening GmbH, Winzerlaer Straße 2a, D-07745 Jena, Germany

4.1 Introduction

In the past 5 years the world of high-throughput screening (HTS), which targets lead discovery for both pharmaceutical and agrochemical applications, is in a highly rapid developmental stage. Today, restructuring the drug discovery programs to develop highest performance is a key issue of pharmaceutical companies. In an increasingly competitive environment prioritizing innovation has become of major strategic importance. Driving forces are the pressure to discover more innovative and even better drugs, as well as time to market. Driven by technological innovation the hybrid "science" HTS [1, 2] depends on and merges latest developments in molecular biology, chemistry, pharmacology, laboratory automation, bioinformatics, and computing to meet "economics," "big numbers," "high speed," and "high information content" necessary for the discovery of new and innovative drugs.

For an effective HTS infrastructure the fundamental working fields of target discovery and validation, screen design, assay technology, detection method, sample generation and handling, laboratory automation and robotics, as well as of data management have to be integrated, coordinated and optimized. At present, latest technology and laboratory automation in all parts of the early stages of drug discovery play crucial roles in the success of HTS strategies [3]. Enabling technologies that characterize HTS are listed (Tab. 1) [4]. However, drug discovery is

Table 1. Enabling technologies that characterize HTS [4].

- Increase in compound library diversity and size
- Exponential increase in targets from genomics/proteomics
- Miniaturization
- Automation
- Integrated systems
- More sensitive and efficient assay and detection systems
- Cellular assay system improvements
- Sensitive alternatives to radioactive assays
- Computational methods for assay simulation
- Data management improvements and innovations
- Outsourcing and customization
- Lead optimization tools

---

extremely complex and the goals, company resources, products portfolio, as well as future expectations of the different players vary widely.

This chapter is intended to provide an overview about the processes, technology, and instrumentation that are in use for HTS from a personal, more technological point of view. Latest technologies as well as future trends like miniaturization, higher degree of parallel processes, uHTS, and chip-technologies are given priority. However, space does not allow to present a complete summary in the rapidly growing field of HTS technologies, especially in-house developments and the high number of latest technologies developed and marketed by start-up biotech companies on a proprietary basis. Furthermore, for novel technologies in, for example, target discovery (Chapters 2 and 3), compound library design (Chapters 5 and 7), or combinatorial chemistry (Chapter 6) see the corresponding chapters in this book.

4.2   Present status

Laboratory automation has grown from a novel technology with its various difficulties and problems of about 10 years ago to powerful, more reliable, and easy to handle tools today. Applications range from sample generation and preparation, genomics [5] and proteomics [6] for target finding and validation (see Chapters 2 and 3), automated multiple parallel synthesis stations, robotic storage devices, parallel liquid handling systems, workstations and fully integrated robots for primary biological screening, to automation of analytical problems (sample purity, mass detection etc.). Today, about 40% of all HTS operations are automated. With the increasing number of hits, however, technology development as well as automation strategies are expanded to the steps after hit-discovery, like hit-profiling, pharmacological testing, re-synthesis in gram quantities and *in vivo* profiling.

Today, HTS for drug discovery (up to 10,000 wells analyzed per day) has been established in most of the pharmaceutical companies, and overall, HTS is already conducted in more than 500 laboratories worldwide [7]. Often, there exists the infrastructure to perform more than 100,000 test samples in the 96- or 384-well format in reasonable time like a week. A number of companies have reached this throughput per day (ultra high-throughput screening, uHTS; 100,000 wells analyzed per day) [4, 8]. Screening capacity usually is 20 to 50 different assays per year with a strong upward trend. Data management like acquisition, analysis, report generation, and mining is established.

About 50 to 60% of all assays in HTS are already running in the 384-well format. The 1536-well format is close to moving from an experimental stage to becoming more routine. Companies report on more than 1,000,000 stock compounds ready to be screened. The sample collections are stored in specially designed compound depositories mostly in the 96-well format. However, pharmaceutical companies already began to run depositories in the 384-format and to establish respective logistics for handling this format in both HTS primary screening and in secondary profiling. A wide diversity of equipment necessary for reliable and routine use are commercially available. In addition, there exists sufficient knowledge to establish and run new and efficient HTS laboratories.

4.3   Logistics

Although the different technology areas of high-throughput screening have their own developmental rates, an effective lead discovery unit in the pharmaceutical and agrochemical industry has to be carefully balanced. Optimized coordination of technology, processes and people are of fundamental importance and is one of the main obstacles in HTS at present [9]. Criteria are the capacities, efficiencies and quality in both the various infrastructural elements of HTS, and the efficient feeding of the drug development pipeline with most promising new compounds. Furthermore, an important part of these strategies has to deal with the processes surrounding primary screening (HTS), e.g., target search and evaluation, secondary screening, in vivo testing, pharmacological profiling, data mining, quality control, re-synthesis, structure-activity relationships, or support of laboratory automation.

Depending on the different organization forms of companies HTS exists as part of biological or chemical groups, or has been developed in a more centralized hybrid structure. Some companies reach factory-like HTS facilities with dedicated staffs [10]. The reader is directed to an article by Tomlinson et al. [11] describing the planning and establishment of a new HTS site at GlaxoSmithKline (GSK) in more detail.

Focusing on HTS technology management, a capacity balanced flow of compounds, assays, data, and hit follow-up has to be established. Once organized and practiced for a given number of samples to be screened with a significant turnover of different assays, which defined the capacities of the data management, the workflow already has to meet the next generation in knowledge and/or technology. Thus, there exists a permanent need to supply and enhance the HTS laboratories

with the latest technologies. Technology watch and integration into already existing and well running platforms need a significant amount of personnel capacity.

Another important aspect is, which of the necessary activities of the HTS field
should be conducted in-house in pharmaceutical companies, and which should
made be derived from cooperation and alliances [12]. At present, there exists a
remarkable trend in outsourcing of defined areas of drug discovery, and big pharmaceutical companies have turned to biotechnology companies that specialize in
equipment and services for HTS. Examples are the acquisition of samples, either
from synthesis (see, e.g., Chapter 6) or from nature (see Chapter 5, access to recent
results from biochemistry or genomics, new assay technologies, and pharmacological profiling, etc. In addition, there is a high demand in alliances with niche companies with access to a key technology. Dove [13] summarizes a number of these
major HTS products, its inventors, as well as its major partners/customers. However,
all these outsourced activities have efficiently to be integrated and coordinated.

4.4   Sample sourcing

New compounds, or the broadest structural diversity for screening purposes, is in
great demand [14]. In some pharmaceutical companies sample collections from
synthesis generated over past decades now total hundreds of thousands of compounds. Although being cost- and time-consuming processes the compounds are
transferred into microplate formats. For a certain percentage special automated
dosage units for dry powders allows to perform, for example, over 270 dosages
per day (REMP, Switzerland); (Fig. 1).

For the various operations in organic synthesis (e.g., pipetting, liquid dispensing, mixing, filtration, heating, cooling, inert-gas atmosphere) various automated
equipment is now available from various suppliers [15, 16]. The user can choose

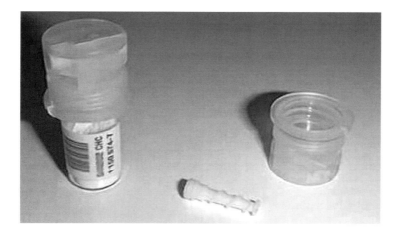

Figure 1. Specially designed automated dosage units (REMP, Switzerland)

from modular or integrated, from open or closed architecture, as well as from standardized or customized robotic systems. Furthermore, small space workstations with liquid handling devices and a specially designed reaction block fitting in chemical hoods are popular. Major efforts are spent on solid-phase synthesis and on an integration of liquid synthesis approaches in synthesis cascades towards new libraries (see Chapter 6). Although split-mix techniques are a bit out of fashion, further developments, for example, towards radio-tagged synthesis approaches (microchips integrated into a single bead) and appropriate automation of the processes have realized new capabilities.

Because of quality reasons, i.e., compound purity and broadest structural diversity, an evolution in combinatorial chemistry from ideas based on numbers, through chemical diversity (Chapter 9), to focused libraries (Chapter 10) and drug-like properties (Chapter 12) can be recognized. In win-win approaches companies share compound libraries. From both combinatorial synthesis and multiple parallel organic synthesis approaches [17] (see Chapter 6) the logistics are adapted mostly to the 96-well format. Quality control (physico-chemical analysis via HPLC-DAD, HPLC-MS, and other techniques) plays a crucial role. Therefore, some strategies involve purification of the samples deriving from these synthesis approaches, which is rather time-consuming and cost-intensive. Although lacking in parallel processes, automation in chromatography is state of the art.

Secondary metabolites from plants, animals and microorganisms have been proven to be an outstanding source for new and innovative drugs and show a striking structural diversity which supplements chemically synthesized compounds or libraries in drug discovery programs [18] (see also Chapter 6). In addition, genetic engineering with the aim of combinatorial biosynthesis can add to structural diversity from natural sources [19]. Unfortunately, extracts from natural sources are usually complex mixtures of compounds often generated in time-consuming and for the most part manual processes. A novel tool for automated sample preparation, called CyBi™-XTract from CyBio AG, faces the challenges of sample preparation from nature, especially quality and quantity purposes. The main feature of the CyBi™-XTract is a high efficiency parallel device for solid-phase extraction (SPE) based on positive pressure technology. Due to controlled flow rates it allows 96-fold parallel chromatographic fractionation of complex compound mixtures from natural sources yielding concentrated samples with a tremendously reduced complexity (higher sample quality). The integration of the CyBi™-XTract into a fully automated storage and retrieval system for source plates, resin blocks and deep-well plates for fraction collection, combined with barcode readers and data management sets a new standard in drug discovery from nature (higher sample quantity) [20, 21].

4.5 Compound depository

Compound collection (synthetics, pure natural products, and extracts from natural origin) is a key asset in the drug discovery attempts of pharmaceutical companies (see Chapter 7). These collections have to be stored under defined conditions and logistics. Each compound should be ready and available in time for the various needs of primary and secondary screening, as well as first pharmacological studies, either in defined plate formats (e.g., 96-, 384-, and 1536-wells), or as a single probe. Therefore, the companies are going to install highly automated storage and retrieval systems for compound management [22] in which the samples and/or plates are uniquely identified by plate barcode and their position in the storage plates/racks. For example, REMP (Switzerland) (Fig. 2) developed and established storage and retrieval systems for the individual needs of the pharmaceutical and chemical industry. The systems are based on standardized storage racks

Figure 2. Compound depository (REMP, Switzerland)

which holds up to 384 sealed microtubes with samples at a pre-defined dilution [23]. In addition, all types of microtubes, deep well plates, as well as other containers can be stored, organized, and handled.

Retrieval of samples in microtubes can be managed by robotic systems as well [23]. For example, at Hoffmann La Roche two robots are capable of picking over 6,000 tubes a day from any storage plate and transferring them to standardized distribution plates with the same 384-well format (Fig. 3). Furthermore, it is possible to arrange special compound series or sub-libraries. The entire storage and retrieval system is kept at –20°C and at a constant, low humidity.

Figure 3. Specially designed storage and retrieval (cherry picking) plates in the 384-well format (REMP, Switzerland; F. Hoffmann La Roche).

4.6 Bioassay automation and robotics

For a throughput of about 10,000 wells to be analyzed per day (HTS) well established automation concepts are on the market that allow reliable routine applications [15]. Excellent instrumentation for routine low-volume liquid handling, sensitive detection, robotic plate handling, and more specialized devices, like washing stations, plate sealer, piercer, and incubators in the 96-, and 384-well format is now readily available and have been integrated into HTS campaigns in various setups worldwide. A list of suppliers of HTS equipment and services is presented in

[24]. In addition, technologies are continually being developed and suppliers are teaming up to make available more integrated equipment and consumable sets.

Robotic systems used in biological screening can be divided into fully integrated systems and workstations [25]. Robotic systems are defined as those systems that utilize articulated robotic arms to move items (e.g., plates) and usually have access to all points in the three dimensions of the operation area of the robotic arm. A "workstation" is defined as an independent system that is highly specialized to perform limited functions in most cases without highly-sophisticated robotic arms. The question of whether a fully automated robotic system or workstation best fit the needs of HTS depends on the individual priorities, although a major trend is to use multiple workstation-type devices. However, the design of an automated system should be as simple as possible as long as it performs the desired task.

For robotic arms, positional accuracy is limited by the accuracy of the servo motors used. Several tens of microns can be achieved for positional accuracy with precision in the same order. However, millimeter accuracy is probably sufficient for 96-, and 384-well plates, and a tenth of that for 1536-well plates. As the well size decreases the positional accuracy becomes more important. However, the liquid handling parameters have to be estimated carefully because of problems with capillary forces, mixing, bubble forming, etc. in reaction volumes down to 1 to 5 µl [26]. In terms of overall speed a conveyor belt system will move a plate faster than a robotic arm.

The major differences in the two systems, workstations and fully integrated robots, arise in terms of hours of use, ease of use, repair, and cost. In a number of cases the price for a workstation is significantly lower than for fully automated robotic systems, allowing to purchase multiple workstations. Therefore, failure of a fully automated robotic system that causes work stoppage for several hours/days can be circumvented with a back up workstation. The redundancy afforded by a multiple workstation concept also allows parallel processing of different assays resulting in reliable results day in and day out. The workstation concept provides greater flexibility and is easier to be programmed and handled. Even though robotic systems are touted as being fully automated "walk away" devices, the reality is that they need "human" attention. Comparison of a robotic system *versus* the use of workstations, especially concerning throughput, handling, personnel, and price is discussed by K. R. Oldenburg [25]. As an example, a workstation developed by Cybio AG (Jena, Germany) is depicted (Fig. 4).

In case of testing more than 100,000 compounds a day the Zymark Allegro™ robot system moves the bar upwards to uHTS [27]. The system is a series of fully independent modules, each of which performs a specific task. The modules are essentially workstations being interconnected by more simple robot arms and the system is fully enclosed which allows to control the plate environment, i.e., temperature, humidity, or gas composition in the atmosphere. Because of its modularity, it can be quickly reconfigured to fulfill needs of different assay types. A concept with a conveyer belt system (assembly line style) has recently been presented by Thermo CRS with its Dimension4™. Furthermore, CyBio has devel-

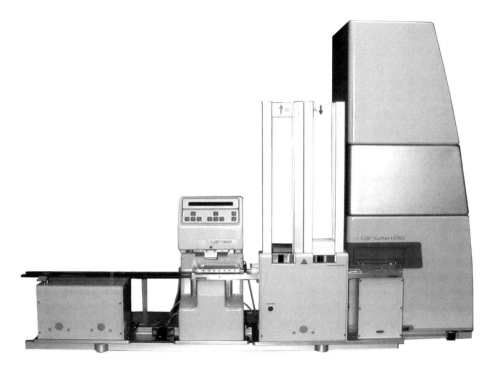

Figure 4. Screening workstation from CyBio AG (Jena, Germany)

oped the CyBi™-Screen machine for ultrahigh throughput screens with a compact design, modular approach, and format flexibility (96-, 384-, and 1536-well plates). At the core of the CyBi™-Screen machine is a newly developed transport mechanism that passes the plates between hardware components along a straight line while keeping to one level—i.e., following the shortest possible route [28].

4.7   Microplate technologies

Global standardization of the microplates, especially of higher density formats pertaining to dimensions, well-to-well spacing, external footprint, plate height, lid, etc. allows to benefit from the choice of equipment from different manufacturers [29]. With the aim to shrink the test volumes from the 96-well plate (e.g., 100 µl test volume) to the 384-well format (20 µl) down to 1–5 µl (1536-well plate), a first aim for decreasing the cost of targets, compounds, and reagents was reached. The majority of HTS laboratories have already established the 384-format obviously being the workhorse in near future. A number of different groups report on their experience with routine assays in the 1536-well format [30, 31]. However, a number of assays are still performed (and in some cases depend on) the 96-well format. As reported in a recent study, the use of the 384-well format will increase from ca. 55.6% in 2001 to ca. 60% in the 2003. Assays performed in the 1536-well format will already be about 12.2% in 2003 [4]. Aurora BioSciences tantalizes with a comprehensive uHTS system based on 3456-well plates. In addition, higher density formats like 9,600 wells, or open but addressable formats are in an experimental stage. The scope and impact of miniaturized methods as applied to HTS are summarized in detail by J. Burbaum [32].

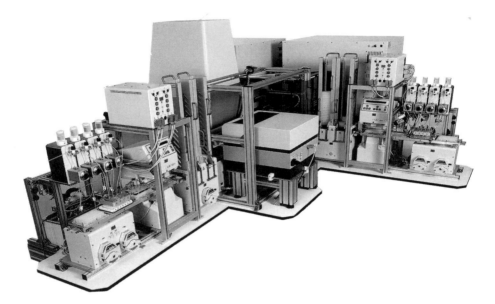

Figure 5. CyBi™-Screen machine, an uHTS system from CyBio AG (Jena, Germany)

4.8   Detection

New technologies make reading of high-density microplates more convenient and sensitive. In order to follow a biochemical principle in microplates several techniques are in practise, e.g., fluorescence, homogeneous time-resolved fluorescence (HTRF), fluorescence polarization (FP), fluorescence resonance energy transfer (FRET), fluorescence correlation spectroscopy (FCS), chemoluminescence, and scintillation counting [33]. Most instruments on the market are able to perform more than one type of measurement [34]. For standard fluorescence detection limits are down to the pg/well range with, for example, fluorescent routine screens allow reading of 96-, 384-, and 1536-well plates either from top or bottom. However, non-parallel approaches for reading is a time-consuming task for the higher density plate formats.

Another option is to produce images of whole microplates at once. For fluorometric imaging Molecular Devices developed a plate reader called FLIPR which allows both adding compounds simultaneously to the wells (96-, and 384-format), and scanning the whole microplate for kinetic measurements [35]. This device is well suitable to analyze functional fluorescent-based assays in living cells. Main applications are the determination of intracellular Ca^{2+}-content and pH-value, as well as membrane-potentials. Depending on the type of assay, a throughput up to 100 microplates is possible in principle [36]. With the NightOWL system from EG&G Wallac based on a CCD camera luminescence and fluorescence applications are possible, while a fluorometric microvolume assay technology (FMAT) is available from PE Biosystems that analyzes 1 mm^2 areas in each well up to the 1536-format.

Based on a single photon counting CCD camera with a special light intensifier CyBio AG has developed the high resolution luminescence imaging system CyBi™-Lumax 1536. This family of readers is not limited to a microplate format. Integration of a parallel liquid handling device (CyBi™-Lumax 1536 D/SD) enables measurement of flash luminescence which can be used to determine intra- and extracellular calcium ion concentration, e.g., *via* the photoprotein aequorin.

A novel screening system has jointly been developed by Carl Zeiss Jena GmbH and F. Hoffmann-La Roche which resulted in ultra high throughput (uHTS) application with simultaneous 96-well parallel reading of fluorescence, luminescence, and adsorption on the basis of 96 miniature lenses. A 1536-well plate can be read in 16 steps [23].

With the move toward miniaturization causing both decreases in sample signal, and increase in background signal will require the development of new technologies. Fluorescence correlation spectroscopy (FCS) allows to determine binding properties at the level of single molecules in volumes of a few femtoliters and therefore lead discovery by miniaturized HTS (Evotec OAI, Hamburg, Germany) [37]. For single molecule detection both optical (single molecule microscopy/ spectroscopy) and scanning probe technology (microscopy, manipulation and force spectroscopy) might make their way to HTS. Apart from fluorescence, surface plasmon resonance (binding properties), infrared spectroscopy (non-invasive measurements in living cells) NMR (see chapter 9), as well as colorimetric and amperometric methods are in the stage to be adapted to HTS [13].

The need for understanding how targets and potential hit compounds affect cell function in early stages of drug discovery requires novel technologies for cell-based assays [38]. So-called high-content screening (HCS) has been introduced which allows to gain temporal and spatial cellular information about how a drug interacts with cells based on multi-wavelength fluorescence measurements (e.g., microscope imagers) [39].

4.9 Chip technologies in drug discovery

At present there exists an impact of microsystem technologies on biomedical and life sciences instrumentation [40]. Thus, the 1536-well plate, which has been available for a few years, already has competition from lab-on-chip technologies which use capillary and electrokinetic forces to move and mix liquids with reagents, targets and/or test compounds [41]. Orchid Biocomputer (Princeton, NJ), Caliper (Mountain View, CA) and others have developed special chips on the basis of microfluidics systems. On the other hand, however, miniaturization toward chip technologies pose a number of its own problems. Evaporation results in significant variation of concentration in the sub-microliter region. Liquid dispensing in nanoliter drops is possible in principle, but has problems in the pick-up mode [42]. Capillary action causes well-to-well contamination of small reaction caves, and surface adhesion/chemistry as well as surface tension has to be circumvented [43].

Processing fluidics rather than electrons, specialized microchips have been developed that allow chemical and biochemical analysis [44]. Microchannels created in the surfaces of chips allow non-mechanical pumping of liquids by both electrophoretic and electroosmotic principles. Miniaturizing and integrating liquid-handling and biochemical-processing devices such as pumps, valves, volume-measuring, reactors, extractors, and separation possibilities led to create novel drug discovery tools. For example, on-chip measurement of enzymatic reactions, organic synthesis, ion sensing, DNA amplification, and immunossays, as well as separation procedures including electrophoresis, chromatography, and solid-phase biochemistry have been demonstrated. Knowledge contributed from both genomics and proteomics could have far-reaching implications in such areas as biology and medicine, clinical diagnostics, high-throughput drug discovery, molecular toxicology, and pharmacodynamics modeling [44].

4.10 Data handling

By screening all compounds available against all possible targets will tremendously increase the growth of databases. A typical HTS program generated 200,000 data points a year in the early 1990s, while 5 years later 5 to 10 million, and in the year 2000, 50 million data points have to be handled by a pharmaceutical company [45]. This leads to one of the biggest obstacles in HTS: The collecting, deconvoluting, analyzing, sorting, and storage of the information generated [46, 47]. The stored information itself as well as the ability to mine the burgeoning number of databases by linking HTS data to chemistry, and hit follow-up data like HTS pharmacokinetics and toxicological experiments [1, 48] are key to success and of strong value as a corporate asset. Data-integration technologies are maturing rapidly [49] and the software in use has to provide enough flexibility [50]. Collecting the chemical, biological, preclinical, and clinical data in centralized databases is one of the strategies [51]. Another is to create integrated data systems for both, target discovery, and drug discovery [52, 53], and to combine data from relevant processes of drug discovery and development.

4.11 Economics

In order to sustain the historical revenue growth of about 10% of the pharmaceutical industry, the major global players have to launch five to six new chemical entities (NCE) with sales higher than US$350 billion per year in the near future [54] (see also Chapter 1). This is the pressure to fill the drug development pipeline with new and innovative leads from drug discovery. In this process, quality plays a crucial role because a significant component of the overall cost of drug development is the high failure rate of clinical candidates never reaching the market place.

Increasing the capacity and decreasing the costs are the major arguments for HTS miniaturization and a higher degree of parallel processing. In any drug company, the return on investment in and the use of HTS should properly be assessed in terms of the number of hits/leads identified. In general, less expensive assays and new technologies are expected to be the major players for reducing the per-well cost and the overall cost of lead identification.

4.12 Future aspects

In the near future, most pharmaceutical companies will have access to a portfolio of HTS and uHTS technologies carefully adapted to the various needs. Conceivably, there will be an excess in the screening capacity, particularly if uHTS is widely implemented. The challenge is to generate high-quality hits which can be developed into new drugs for the market and not what technologies and machines have been installed and used in HTS programs. The aspect of virtual screening may become more of a reality [55] (see also Chapters 8 and 10). New applications of technology developments emerge in areas such as toxicology, lead optimization, pharmacokinetics, secondary screening, and compound profiling. The future of drug discovery is to converge and co-develop informatics (both bioinformatics and chemoinformatics), medicinal chemistry, genomics, proteomics, and pharmacokinetics, as well as technology hardware [46, 56].

The question arises how to speed up and optimize steps following drug discovery. The next bottleneck are clear visible: Testing the lead compounds to determine profiles for adsorption, diffusion, metabolism, and toxicity and excretion (ADME, see Chapter 12), expensive and time-consuming animal tests [57] (see Chapter 13).

4.13 References

1 Elgen RM (1999) High throughput screening: Myths and future realistics. *J Biomol Screening* 4: 179–181
2 Devlin JP (ed) (1997) *High throughput screening.* Marcel Dekker, New York
3 Oldenburg KR (1998) Current and future trends in high throughput screening for drug discovery. *Ann Rep Med Chem* 33: 301–311

4 a) Fox S, Farr-Jones S, Yund MA (1999) High throughput screening for drug discovery: Continually transitioning into new technologies. *J Biomol Screening* 4: 183–186
   b) Fox S, Farr-Jones S, Sopchak L, Wang H (2002) Fine-tuning the technology strategies for lead finding. *Drug Discovery World*, Summer 2002, 23–30

5 Spence P (1998) Obtaining value from the human genome: A challenge for the pharmaceutical industry. *Drug Discovery Today* 3: 179–188

6 Celis JE (1999) Proteomics: Key technology in drug discovery. *Drug Discovery Today* 3: 193–195

7 Fox S, Farr-Jones S, Yund MA (1998) High throughput screening: Strategies and suppliers. *High Tech Business Decisions, Moraga, CA*

8 Kell D (1999) Screensavers: Trends in high-throughput analysis. *Trends Biotechnol.* 17: 89–91

9 Studt T (1999) Drug development bottlenecks not cured by technology alone. *Drug Discovery & Development*, Jan. 40–41

10 Archer R (1998) Towards the drug discovery factory. *J Assoc Lab Automat* 3: 4

11 Tomlinson JJ, Butler BT, Frezza J, Harris CO, Smith AA, Kwight WB (1997) The planning and establishment of a high throughput screening site. *Proceedings of the International Symposium on Laboratory Automation and Robotics (ISLAR)* 20–25

12 Grindley JN (1998) Success rates for strategic alliances – are they good enough? *Drug Discovery Today* 3: 145–146

13 Dove A (1999) Drug screening – beyond the bottleneck. *Nature Biotechnology* 17: 859–863

14 Borman S (1999) Reducing time to drug discovery. *Chem & Engineer News* March 8: 33–48

15 Thiericke R, Grabley S, Geschwill K (1999) Automation strategies in drug discovery. *Drug discovery from nature*: 56–71, Springer Verlag, Berlin

16 Brown RK (1999) Content and discontent: A combinatorial chemistry status report. *Modern Drug Discovery* July/Aug.: 63–71

17 Paululat T, Tang Y-Q, Grabley S, Thiericke R (1999) Combinatorial chemistry: The impact of natural products. *Chimica Oggi, Chemistry Today* 17: 52–56

18 Grabley S, Thiericke R (eds): (1999) *Drug discovery from nature*. Springer Verlag, Berlin

19 Borchardt JK (1999) Combinatorial biosynthesis. Planning for pharmaceutical gold. *Modern Drug Discovery* July/Aug.: 22–29

20 Thiericke R, Schmid I, Moore T, Ebert G (2000) Drug discovery from nature. A device for automated sample preparation. *Am Biotechnol Lab* 18: (10) 66

21 Thiericke R (2000) Drug discovery from nature. Automated high-quality sample preparation. *J Automated Meth & Management in Chemistry* 22: 149–157

22 Wedin R (1999) Taming the monster haystack. The challenge of compound management. *Modern Drug Discovery* Jan./Feb.: 47–53

23 Fattinger C, Gwinner E, Gluch M (1999) Looking for new drugs with speed, precision and economics. *BioWorld* 4: 2–4

24 Fox S, Yund MA, Farr-Jones S (1998) Seeking innovation in high-throughput screening. *Drug Discovery & Development* Nov.: 32–37

25 Oldenburg KR (1999) Automation basics: Robotics vs. workstations. *J Biomol Screening* 4: 53–56

26 Berg M, Undisz K, Thiericke R, Zimmermann P, Moore T, Posten C (2001) Evaluation of liquid-handling conditions in microplates. *J. Biomol Screening* 6: 47–56

27 Wildey MJ, Homon CA, Hutchins B (1999) Allegro™: Moving the bar upwards. *J Biomol Screening* 4: 57–60

28 Thiericke R, Schmid I, Gropp T, Ebert G (2000) Automation for ultrahigh throughput screens. *Genetic Engeneering News* 20: (14) 32

29 Ferragamo T (1999) The importance of microplate standardization. *J Biomol Screening* 4: 175

30 Marshall S (1999) SmallTalk ´99 focuses on automated assay systems. *Drug Discovery & Development* Sept.: 34–35

31 Berg M, Undisz K, Thiericke R et al (1999) Miniaturization of an enzyme assay (β-galactosidase) in the 384- and 1536-well plate format. *J Assoc Lab Autom* 4: 64–67

32 Burbaum JJ (1998) Miniaturization technologies in HTS: How fast, how small, how soon. *Drug Discovery Today* 3: 313–322

33 Wedin R (1999) Bright ideas for high-throughput screening. *Modern Drug Discovery* May/June: 61–71

34 Karet G (1999) Microplate readers keep pace with miniaturization. *Drug Discovery & Development* May: 44–48

35 Schroeder K, Naegele B (1996) FLIPRTM: A new instrument for accurate, high-throughput optical

screening. *J Biomol Screening* 1: 75–80

36 Hafner F (1999) FLIPR: High Throughput mit lebenden Zellen. *BioTec* 6: 32–34

37 Auer M, Moore KJ, Meyer-Almes FJ et al (1998) Fluorescence correlation spectroscopy: Lead discovery by miniaturized HTS. *Drug Discovery Today* 3: 457–465

38 Bridges AJ (1998) Cell signaling: Signal transduction and gene transcription. *Drug Discovery Today* 3: 443–445

39 Giuliano KA, DeBiasio RL, Dunlay RT et al (1997) High-content screening: A new approach to easing key bottlenecks in the drug discovery process. *J Biomol Screening* 2: 249–259

40 Gwynne P, Page G (1999) Microarray analysis: The next revolution in molecular biology. *Science* 285: 911–938

41 Mir KU (1998) Biochips: From chipped gels to microfluidic CDs. *Drug Discovery Today* 3: 485–486

42 Rose D, Lemmo T (1997) Challenges in implementing high-density formats for high throughput screening. *Lab Autom News* 2: 12–19

43 Berg M, Undisz K, Thiericke R et al (2000) Miniaturization of a fuctional transcription assay in yeast (human progesterone receptor) in the 384- and 1536-well plate format. *J Biomol Screening* 5: 71–76

44 Marshall S (1998) Lab-on-a-chip: Biotech´s next california gold rush. *Drug Discovery & Development* Nov.: 38–43

45 Drews J (1999) Informatics: Coming to grips with complexity. *Pharmainformatics:* 1–2

46 Divers M (1999), What is the future of high throughput screening? *J Biomol Screening* 4: 177–178

47 Kyranos JN, Hogan JC (1999) High-throughput characterization of combinatorial libraries. *Modern Drug Discovery* July/Aug.: 73–81

48 Cargill JF, MacCuish NE (1998) Object-relational databases: The next wave in pharmaceutical data management. *Drug Discovery Today* 3: 547–551

49 Recupero AJ (1999) Data integration accelerates discovery. *Drug Discovery & Development* Nov./Dec.: 59–62

50 Studt T (1999) Software flexibility speeds discovery process. *Drug Discovery & Development* Nov./Dec.: 68–69

51 Karet G (1999) One database for everyone. *Drug Discovery & Development* Nov./Dec.: 71–74

52 Brocklehurst SM, Hardman CH, Johnston SJ (1999) Creating integrated computer systems for target discovery and drug discovery. *Pharmainformatics*: 12–15

53 Grund P, Sigal NH (1999) Applying informatics systems to high-throughput screening and analysis. *Pharmainformatics*: 25–29

54 Drews J (1998) Innovation deficit revisited: Reflections on the productivity of pharmaceutical R & D. *Drug Discovery Today* 3: 491–494

55 Walters WP, Stahl MT, Murcko MA (1998) Virtual screening – an overview. *Drug Discovery Today* 3: 160–178

56 Marshall S (1999) Forecasting the future of the Biotech industry. *Drug Discovery & Development* Sept.: 39–41

57 Lipper RA (1999) E pluribus product. *Modern Drug Discovery* Jan./Feb.: 55–60

5 Natural products for lead identification: Nature is a valuable resource for providing tools

Susanne Grabley and Isabel Sattler

Hans-Knöll-Institut für Naturstoff-Forschung e.V., Beutenbergstrasse 11a D-07745 Jena, Germany

5.1   Introduction

About 30% of drugs on the worldwide market are natural products or are derived from natural products. A similar ratio accounts for clinical candidates currently under development. Though recombinant proteins and peptides account for an increasing market volume, the superiority of low-molecular mass compounds in human diseases therapy remains undisputed mainly due to more favorable compliance and bioavailability properties.

5.2   The impact of natural products on the drug market

In the past, new therapeutic approaches often evolved from research involving natural products [1]. Prominent examples from medicine that impressively demonstrate the innovative potential of natural compounds and their impact on progress

Table 1. History of commercialization of modern drugs derived from nature (for chemical structures see Fig. 1).

Year of Intro-duction	Drug	Natural Product [a]; Commercialized as	Indication	Company [b]
1826	manufacturing of morphine	natural compound (p)	analgesic	E. Merck
1899	acetylsalicylic acid (Aspirin®)	salicin (p) synthetic analogue	analgesic, antiphlogistic, etc.	Bayer
1941	penicillin	natural compound (m)	antibacterial	Merck
1964	first cephalosporin antibiotic (cephalothin)	semi-synthetic derivative based on 7-ACA (m)	antibacterial	Eli Lilly
1983	cyclosporin A	natural comound (m)	immunosuppressant	Sandoz
1987	artemisinin	natural compound (p)	antimalaria	Baiyunshan
1987	lovastatin	natural compound (m)	antihyperlipidemic	Merck
1988	simvastatin	lovastatin (m); semi-synthetic derivative	antihyperlipidemic	Merck
1989	pravastatin	mevastatin (m); semi-synthetic derivative	antihyperlipidemic	Sankyo/ BMS
1994	fluvastatin	lovastatin, mevastatin (m); synthetic analogue	antihyperlipidemic	Sandoz
1990	acarbose	natural compound (m)	antidiabetic (type II)	Bayer
1993	paclitaxel (Taxol®)	natural compound (p) as a semi-synthetic derivative of baccatin III (p)	anticancer	BMS
1993	FK 506 (tacrolimus)	natural compound (m)	immunosuppressant	Fujisawa
1995	docetaxel (Taxotère®)	10-deacetyl baccatin III (p); semi-synthetic derivative	anticancer	Rhône-PR
1996	topotecan, irinotecan [c]	camptothecin (p); semi-synthetic derivatives	anticancer	SKB, Pharmacia & Upjohn
1996	miglitol	1-deoxynojirimycin (m, p); synthetic analogue	antidiabetic (type II)	Bayer
1999	orlistat	lipstatin (m); synthetic analogue	obesity	Roche

[a] m = microbial metabolite, p = plant metabolite.
[b] BMS = Bristol-Myers Squibb, Rhône-PR = Rhône-Poulenc Rorer, SKB = SmithKline Beecham, Roche = Hoffmann-LaRoche.
[c] Irinotecan was launched in Japan by Yakult Honsha and Daiichi Pharmaceutical first in 1994.

in drug discovery and development are listed in Table 1 (for chemical structures see Fig. 1).

Today, approaches to improve and accelerate the joint drug discovery and development process are expected to arise from both innovation in drug target elucidation and lead compound finding. Structural and functional analysis of the human genome will provide access to a dramatically increased number of new

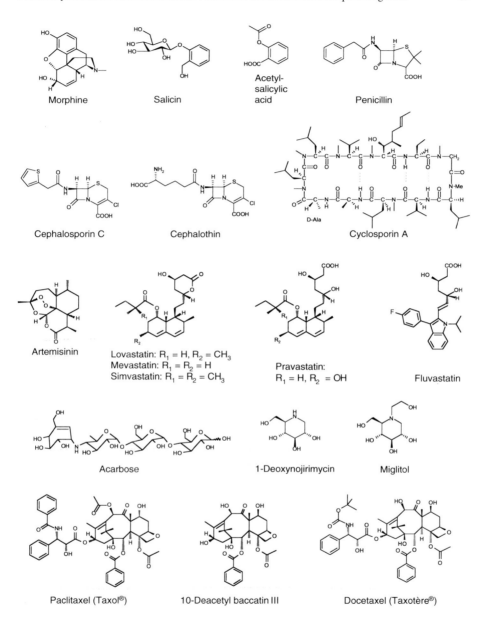

Figure 1a. Chemical structures of the compounds listed in Table 1.

potential drug targets that has to be evaluated (see Chapters 2, 3 and 8). Improvements in high-throughput screening enable the testing of increasing numbers of targets and samples with the consequence that already 100,000 assay

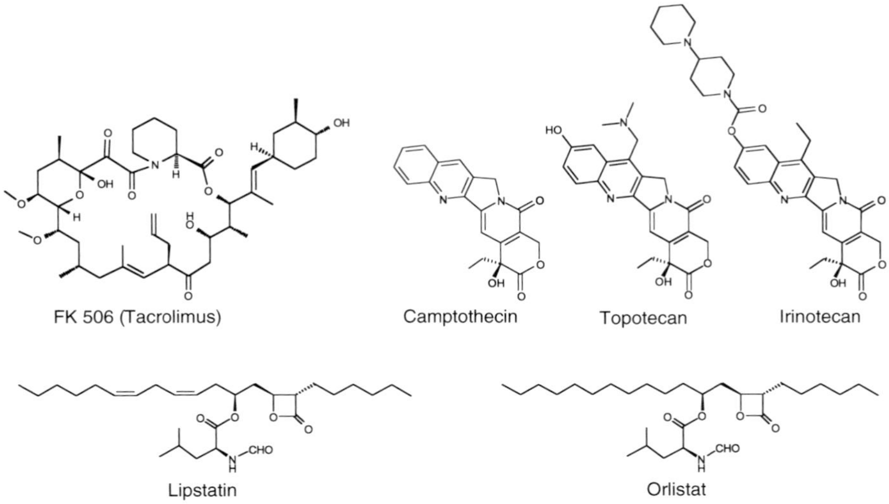

FK 506 (Tacrolimus) Camptothecin Topotecan Irinotecan

Lipstatin   Orlistat

Figure 1b. Chemical structures of the compounds listed in Table 1.

points per day are possible (see Chapter 4). However, accelerated identification of valuable lead compounds by random screening approaches can only be achieved by new concepts to generate large varieties of structurally diverse test samples (see also Chapters 6 and 7). Research on natural products shows that nature has evolved a diversity of chemical structures that is not accessible even by the most sophisticated synthetic concepts. Moreover, natural products have often opened up completely new therapeutic approaches. They substantially contributed to identifying and understanding novel biochemical pathways and, consequently, proved to make available not only valuable drugs but also essential tools in biochemistry and molecular cell biology [2, 3]. Table 2 summarizes selected natural products currently evaluated as drug candidates (for their chemical structures see Fig. 2). examples

In fact, natural products are currently undergoing a phase of reduced attention because of the enormous effort which is necessary to isolate the active principles from the complex matrices of the natural source and to elucidate their chemical structures. However, we presume that comprehensive compound collections comprising pure natural substances, their derivatives and analogues as well as libraries generated by standardized fractionation of extracts from natural origin represent most valuable tools for current drug discovery programs. Also improving access to large quantities of the compounds is a key prerequisite for detailed biological studies as well as for clinical development and commercialization.

Table 2. Selected natural products evaluated as drug candidates (without anti-infectives). (For references see [4–12]; for chemical structures see Fig 2).

Natural Product (Source)	Comments	Target	Indication	Status
CC 1065 (streptomyces)	high toxicity; bizelesin under evaluation	DNA	anticancer	clinical trials
epothilone (myxobacterium)	synthetic analogues and derivatives under evaluation	microtubuli	anticancer	
fumagillin (fungi)	TNP-470 and other derivatives under evaluation	angiogenesis	anticancer (solid tumors, Kaposi's sarcoma)	TNP-470 in clinical trials
staurosporine (streptomyces)	UCN 01 = 7-hydroxystaurosporine under evaluation	protein kinase C	anticancer	clinical trials
flavone (plant)	synthetic analogue flavopiridol under evaluation	kinases	anticancer	clinical trials
aplidine (dehydro-didemnin B) (marine; tunicate)	accessible by synthesis; didemnin B was discontinued	GTP binding receptor EF-1a and others	anticancer	Phase I
bryostatin 1 (marine; bryozoan)	manufactured by aquaculturing	protein kinase C (leukemia)	anticancer	Phase II
discodermolide (marine; sponge)		microtubuli	anticancer (immune suppressant)	adv. preclinical trials
dolastatin 10 (sea hare)		microtubuli	anticancer	preclinical trials
ecteinascidin 743 (marine; ascidian)		DNA-minor groove (G rich)	anticancer (ovarian, other solid tumors)	Phase I
eleutherobin (marine; soft coral)		microtubuli	anticancer	
halichondrin B (marine; sponge)		microtubuli	anticancer	adv. preclinical trials
squalamine (marine; shark)		sodium-hydrogen exchanger	anticancer	clinical trials
calanolide A, B (tree)		DNA polymerase action on reverse transcriptase	AIDS (HIV I)	clinical, preclinical trials
manoalide (marine; sponge)	biochem. tool; synthetic analogues towards reduced toxicity under evaluation	phospholipase A_2, Ca^{2+}-release	anti-inflammatory	clinical trials
epibatidine (frog)	high toxicity; synthetic analogues under evaluation	nicotinic acetylcholine receptors	analgesic	pharmacological studies
pseudopterosins (marine soft coral)	extracts in cosmetic formulations	phospholipase A_2, lipoxygenase	analgesic, anti-inflammatory	adv. preclinical trials
huperzine A (moss)	TCM derived plant extracts on the market	cholinesterase	Alzheimer's disease	clinical trials in China

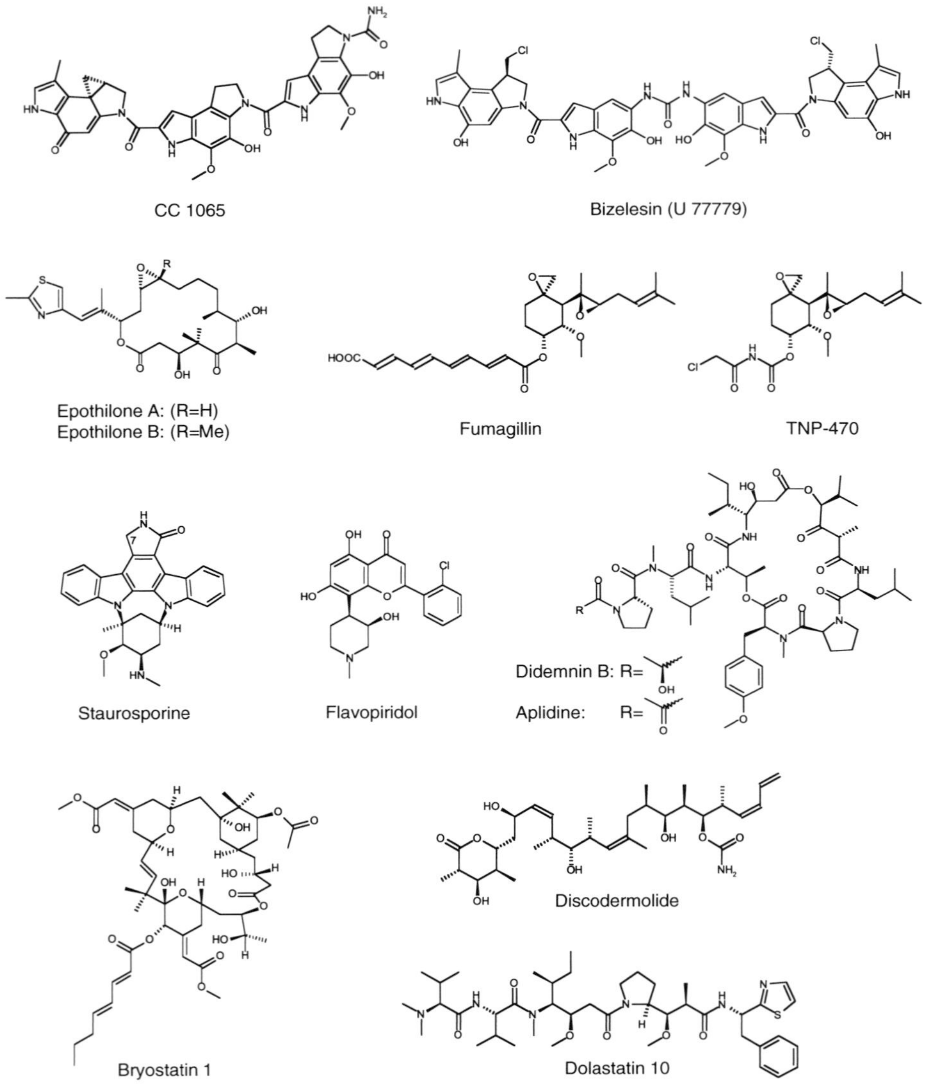

CC 1065

Bizelesin (U 77779)

Epothilone A: (R=H)
Epothilone B: (R=Me)

Fumagillin

TNP-470

Staurosporine

Flavopiridol

Didemnin B: R=

Aplidine: R=

Bryostatin 1

Discodermolide

Dolastatin 10

Figure 2a. Chemical structures of the natural products listed in Table 2.

5.3 Natural sources for drug discovery

5.3.1 General aspects

The drugs used by ancient civilizations were extracts of plants or animal products
with a few inorganic salts. Especially plants have played an important role in med-

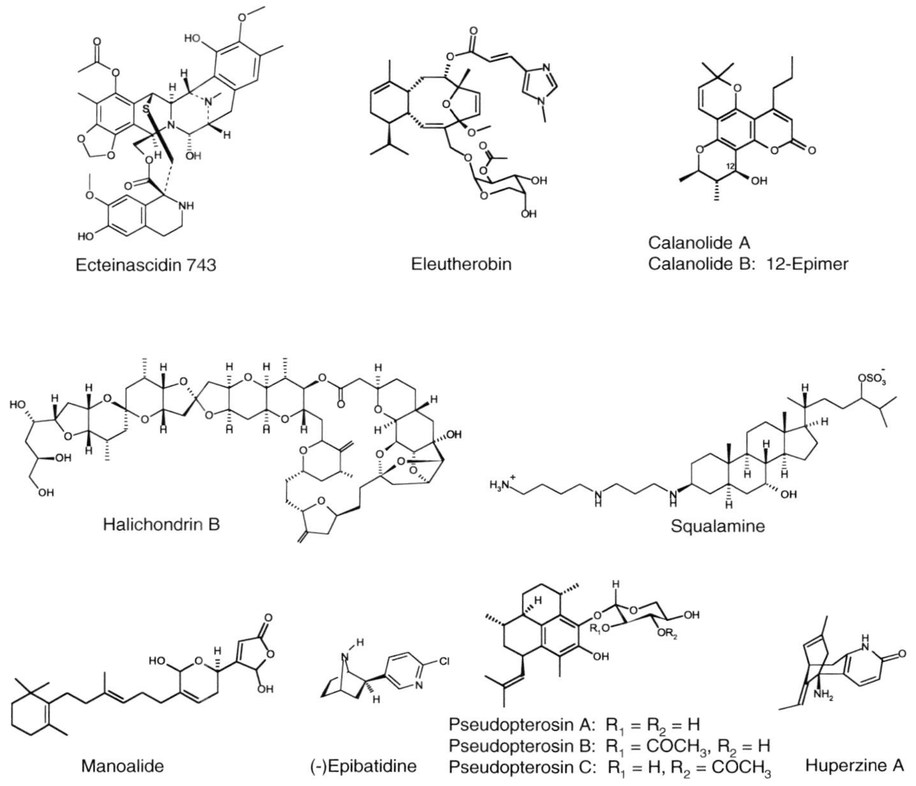

Ecteinascidin 743

Eleutherobin

Calanolide A
Calanolide B: 12-Epimer

Halichondrin B

Squalamine

Manoalide

(-)Epibatidine

Pseudopterosin A: R₁ = R₂ = H
Pseudopterosin B: R₁ = COCH₃, R₂ = H
Pseudopterosin C: R₁ = H, R₂ = COCH₃

Huperzine A

Figure 2b. Chemical structures of the natural products listed in Table 2.

ical use, e.g., in India where the Ayurveda gave access to a broad variety of medicines from plants reported since around 1000 BC. The earliest prescriptions in Chinese medicine based on natural products date back to about 500 BC, and some of the classical Chinese formulae handed down in the years between 25 and 220 are still in use. Moreover, with the focus of developing and commercializing improved TCM (TCM: Traditional Chinese Medicine) derived drugs that meet the requirements of western health care standards, identification, purification and structure elucidation of the bioactive principles are increasingly addressed.

In contrast to plants, microorganisms were not known to biosynthesize secondary metabolites (low-molecular mass compounds that derive from biosynthetic pathways which are not required for maintenance and growth) useful for medicinal application until the discovery of the penicillins. However, the accidental discovery of penicillin from the culture broth of *Penicillium notatum* in 1928 by Alexander Fleming and its introduction in 1941/42 as an efficient antibacterial drug revolutionized medicinal chemistry and pharmaceutical research by stimulating completely new strategies in industrial drug discovery. In the following decades, microorganisms attracted considerable attention as a new source for pharmaceuticals. From various screening programs encompassing huge numbers of microbial extracts, an unexpected diversity of natural compounds performing a broad variety of biological activities became evident. Despite the superior role of phytogenic drugs in the past and the tremendous number of plant metabolites described in literature, today, secondary metabolites obtainable from the culture broth of microorganisms dominate applied natural products research (for reviews see [13–16]). Secondary metabolites are small organic molecules that derive from biosynthetic pathways which are not required for maintenance and growth of the respective organism. They generally are formed by only a few steps branching off main biochemical pathways. Referring to plants, they often contribute to biological defense strategies. However, so far their function and benefit for the producing microorganisms mostly remain unknown.

According to numbers of natural products described in literature and commercial relevance, secondary metabolites deriving from organisms that occupy terrestrial habitats are far ahead. In addition to plants and microorganisms animals gave rise to a number of bioactive compounds. In contrast to plants and microorganisms, structural diversity of natural compounds from animals seems to be very limited. However, various compounds belonging to the structural class of peptides and proteins served as leads for the development of innovative drugs interacting with new therapeutic targets. Thus, the commercial success of the synthetic heterocyclic inhibitors of angiotensin-converting enzyme (ACE-inhibitors) impressively demonstrates the potential of transferring concepts evolved in nature into drugs with considerable therapeutic value [17]. A mixture of peptides isolated from the snake *Bothrops jararaca* together with protein structure information on carboxypeptidase A, a closely related homologue of ACE (see Chapter 8), gave access to a novel type of anti-hypertensive drugs with captopril as its first representative commercialized in 1980.

From the taxonomic point of view, the life forms of marine organisms are significantly more diverse than the life forms of terrestrial organisms. Most of the phyla of life described so far are found in the ocean, whereas only about 60% of them occur on land. Furthermore, in contrast to animals from terrestrial habitats, vertebrates and in particular invertebrates from the marine environment are a rich source for complex natural products deriving from numerous biosynthetic pathways. However, recent results indicate that microorganisms, so called microbionts, play a significant role in the biosynthesis of compounds isolated by extraction of marine macroorganisms such as fishes, shellfishes, sponges, tunicates or corals.

5.3.2   *Microorganisms*

Among the bacteria, streptomycetes and myxobacteria [5] play a dominant role with respect to secondary metabolism. About 70% of the natural compounds from microbial sources derive from actinomycetes, mainly from strains of the genus

Streptomyces that can easily be isolated from soil samples. The genus *Streptomyces* comprises more than 500 species that perform an outstanding diversity in secondary metabolism yielding a yet increasing variety of new chemical structures. About 110 genera of the order of actinomycetales are known. So far, a number of these genera have been neglected with respect to the investigation of their potential for the biosynthesis of bioactive secondary metabolites.

Besides the prokaryotic actinomycetes and myxobacteria, certainly, fungi are one of the most significant groups of organisms to be exploited for drug discovery purposes. In particular, fungi imperfecti have provided mankind with lots of different bioactive secondary metabolites, many of them having entered clinical applications, such as the β-lactam antibiotics, griseofulvin, cyclosporin A, or lovastatin. Currently, most new natural products described in literature are isolated from fungi [18]. Therefore, fungi are an outstanding source for the isolation of structurally diverse small molecules that are highly qualified to supplement compound libraries for drug discovery.

Many fungi are presumed to occupy unsuspected niches in nature including cohabitation with larger life forms, such as higher plants, or the marine environment. Given the huge number and varieties of higher plants, the number of their associated microfungi, mainly belonging to the Ascomycotina and Deuteromycotina, is expected to be enormous. It is estimated that today only a few percent of the world's fungi are known. With respect to drug discovery, mycelium cultures of fungi are of major interest. Their metabolites are easily accessible in large quantities by fermentation processes, thus providing sufficient amounts for extended screening programs as well as pre-clinical and clinical studies.

Most microorganisms investigated for their secondary metabolism are heterotrophic. However, microalgae, an assemblage of prokaryotic (cyanophyta) and eukaryotic microorganisms with oxygenic photosynthesis, have to be considered as well. Until now, microalgae have been studied mainly according to their toxigenic potential and their impact on poisoning the animal and human food chain [19]. Due to inadequate sterile *in vitro* cultivation conditions, microalgae have been neglected with respect to the systematic investigation of their potential to biosynthesize structurally diverse non-toxic bioactive secondary metabolites. Therefore, we developed an apparatus for the parallel cultivation of 50 microalgal isolates in the 100 mL-scale. Our set up guarantees defined growth conditions with respect to light, medium, temperature and carbondioxide supply [20]. In order to obtain milligram quantities of microalgal metabolites, scale-up to 20 L glass vessel bioreactors with an inside illumination device was realized.

5.3.3   Plants

Although today, microorganisms dominate applied natural products research, higher plants remain a major source for new bioactive compounds due to the complexity and variability of their secondary metabolism. Therefore, secondary metabolites isolated from plant extracts are essential with respect to the generation of structural diversity for compound libraries used in drug discovery. Plant metabolites such as alkaloids or terpenoids are structurally unique and modifications of the biosynthetic pathways yield a tremendous diversity of derivatives. Studies addressing the variability of secondary metabolism in dependence on the place of origin have demonstrated the impact of the habitat. Therefore, current efforts focus on the investigation of plants from yet unexplored locations. However, in case of hit finding a number of problems arise referring to back-tracing and accessing sufficient quantities for more detailed biological or pharmacological studies. In cases of complex structures, multi-step chemical synthesis cannot solve the problem, as was shown by well known examples of phytogenic drugs such as morphine, codeine, reserpine, vincristine, the cardiac glycosides or, more recently, paclitaxel and camptothecin.

However, in various cases valuable precursors have been be made accessible from plants at moderate cost, thus, contributing to improved manufacturing processes, supplemented by chemical synthesis, and biocatalysis or biotransformation, respectively. Precursors sometimes are of considerable advantage because they also give access to unnatural analogues or derivatives valuable for both, the generation of compound libraries for screening purposes, and the optimization of biological properties.

5.3.4   The marine environment

Recent trends in drug discovery from natural sources emphasize investigation of the marine environment yielding numerous, often highly complex chemical structures. In most cases, *in vitro* cultivation techniques for the supply of sufficient quantities for biological activity profiling and clinical testing are missing. Focus on marine biotechnology is currently strengthened by findings that marine microorganisms are substantially involved in the biosynthesis of marine natural products initially isolated from macroorganisms such as invertebrates [7, 21–24].

Taking the number and the chemical diversity of metabolites provided by their terrestrial counterparts as an indicator, cultures of marine fungi promise to be a superior source for drug discovery. However, the true potential of marine fungi has not surfaced yet, as no unique secondary metabolites have been reported so far. This is possibly caused by the predominant isolation and cultivation of ubiquitous fungi even from samples collected in the marine environment. Therefore, investigation of marine fungi that have undergone their evolution in the ocean ("obligatory marine fungi") should be strengthened. These fungi, presumably, differ in their biosynthetic pathways from ubiquitous fungi. However, defining appropriate conditions for strain isolation and large scale cultivation will be essential for benefiting from marine fungi within industrial HTS programs.

It remains open whether marine natural products will play a major role in drug discovery in the future. Today, toxic principles dominate the spectrum of biological activities isolated from marine sources. This may partly be due to the major application of cytotoxicity directed screening assays. However, it has to be considered that defense strategies are necessary to survive in the highly competitive marine environment, thus resulting in a tremendous diversity of extremely toxic compounds affecting targets that are involved in eukaryotic cell signaling processes. The strong toxic properties of marine metabolites often prevent their application in medicine. On the other hand, a number of metabolites proved to be valuable tools in biochemistry, cell and molecular biology. For example, the water-soluble polyether type neurotoxin maitotoxin (Fig. 3)—the most toxic non-peptide compound—serves as a unique pharmacological tool for studying calcium transport [25]. Currently, various other marine natural products that exhibit considerable toxic potency, are hopeful candidates for clinical use mainly in anti-cancer therapy (see Tab. 2). However, today testing of marine metabolites within compound libraries, as well as in clinical studies is hampered by insufficient supply of material.

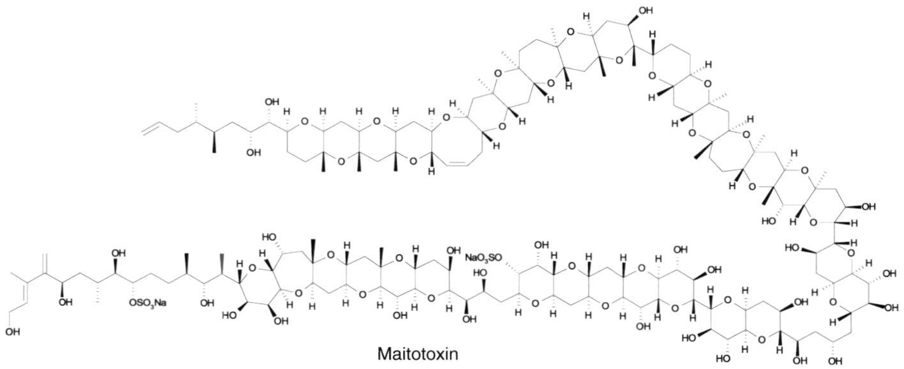

Figure 3. Maitotoxin: A water-soluble polyether-type neurotoxin of 3,422 Dalton produced by a marine dinoflagellate.

5.4   Approaches to exploit natures structural diversity

Low-molecular mass natural products from bacteria, fungi, plants, and invertebrates, either from terrestrial or marine environments, represent unique structural diversity. In order to get access to this outstanding molecular diversity, various strategies like the (target-directed) biological, physico-chemical, or chemical screening have been developed.

In contrast to a biological screening, the physico-chemical, and chemical screening approaches *a priori* provide no correlation to a defined biological effect. Here, the selection of promising, new secondary metabolites out of natural sources is based on physico-chemical properties (see also Chapter 12), or on chemical reactivity, respectively. In both strategies, the first step is a chromatographic separation of compounds from the complex mixtures obtained from plants, bacteria, fungi, or animals. In a second step, physico-chemical properties or chemical reactivities of the separated secondary metabolites are analyzed. Both strategies have proven to be efficient supplemental and alternative methods, especially with the aim to discover predominantly new secondary metabolites that can contribute to the development of valuable natural compound libraries [26, 27] (see Chapter 7).

5.4.1   The physico-chemical screening approach

This approach is characterized by the following steps: Mycelium extracts, culture filtrates, or crude extracts of microbial broths as well as samples obtained from plant and animal extraction are subjected to standardized reversed-phase HPLC (high-performance liquid chromatography) by making use of various coupled detection techniques. Most commonly, HPLC is coupled to a multi-wavelength UV/VIS-monitor (diode array detection, DAD). Comparison of the data (retention time and UV/VIS-spectra) to those of reference substances acts as selection criteria. However, success of this strategy depends upon the amount and quality of pure references in the database. Based on the UV/VIS monitoring, the HPLC-DAD screening is well suited for screening towards metabolites which bear significant chromophores. In combination with the efficient separation *via* HPLC (high-performance liquid chromatography) this screening procedure can advantageously be applied to plant material which contains numerous colored compounds.

   The data obtained from HPLC-DAD analysis are often helpful in de-replication, e.g., early identification and exclusion of known or otherwise unsuitable compounds during high-throughput biological screening programs. However, the necessity of the presence of a UV/VIS detectable chromophor in the metabolite to be analyzed limits its possible application. Therefore, alternative or supplemental detection methods like mass spectrometry (LC-MS), or nuclear magnetic resonance (LC-NMR) have gotten considerable attention.

5.4.2   The chemical screening approach

The TLC (thin-layer chromatography) based chemical screening approach has been developed for the investigation of metabolites from microbial cultures [26, 27]. In order to apply it in a reproducible way, standardized procedures for sample preparation including a 50-fold concentration are required. The obtained concentrates from both the mycelium and the culture filtrate, are analyzed by applying a defined amount to high-performance thin-layer chromatography (HPTLC) silica-gel plates which are chromatographed using different solvent systems. The metabolite pattern of each strain is then analyzed by visual detection (colored substances), UV-extinction/fluorescence, and colorization reactions obtained by staining with different reagents. The advantage of this combination of reagents lies in the broad structural spectrum of metabolites accessible through staining. The procedure mainly focuses on the chemical behavior and reactivity of the components and renders a good visualization of the secondary metabolite pattern (metabolic fingerprint) produced by each strain. In addition, recent technical developments have strongly enhanced the ease of procedures and quality of results from thin layer chromatography. Thus keeping the method compatible with the requirements of modern laboratory practice in the industrial as well as in the academic setting. For example, apparatuses for sample application by spraying the test solution onto the plate allow automatic processing and yield nicely resolved chromatograms due to initial concentration of the sample in thin lines rather than in spreading circular spots as in "traditional" manual procedures. Important improvements have also been achieved in documentation by scanning or video readout of chromatograms.

In comparison to a TLC-based screening, the chromatographic resolution, and sensitivity of HPLC based physico-chemical screening is of superior quality. On the other hand, TLC (thin-layer chromatography) allows a parallel, quick, and inexpensive handling of samples, and is superior in the mode of detection (UV/VIS and staining). As well as eluted compounds from HPLC separation, spots from TLC can easily be subjected to subsequent physico-chemical analysis (MS, IR, NMR etc.) *via* scraping off and elution from the silica gel materials.

Even with ongoing developments in HPLC-coupled analytical techniques, especially HPLC--pseudo high resolution ESI-mass spectrometry, and their widespread application in drug discovery from nature, we consider TLC (thin-layer chromatography) a very useful tool for natural products identification. The method is characterized by a high degree of parallelization as well as cost-saving and simple experimental procedures. These features make it an excellent technique especially in the initial steps of a natural products screening program where one has to deal with large sets of samples. Also in de-replication procedures after biological screening of extracts it is an efficient tool for the initial assessment of samples in order to prioritize them in a hit list.

5.4.3   The biological screening approach

The future potential of physico-chemical and chemical screening approaches lies in the possibility to tap the outstanding structural resources from nature and to build collections of pure natural products, new and known, which can advantageously be used for broad biological screening. Natural compound collections of substantial structural diversity contribute to improved lead discovery, and efficiently supplement synthetic libraries (e.g., from classical or combinatorial synthesis) (see Chapter 6 and 7). Often, it is more practical to run a biological screening with pure compounds from natural sources rather than with crude natural extracts. In order to extend the opportunities arising from testing pure compounds, a collection of natural products, their derivatives and analogues was established under the leadership of our institute.

5.5   Methods and technologies to build high-quality test sample libraries

5.5.1   General aspects

In drug discovery by biological screening, the selection usually is based on a wanted biological effect aiming at a defined pharmaceutical application (target-directed biological screening) [13]. Biological screening has been developed to a powerful concept culminating in HTS which integrates and makes use of recent findings in molecular and cell biology. Today, success of a biological screening program primarily depends on the functionality and therapeutic value of the bioassays running in the first screening, and on the time and effort required for the identification of first promising lead compounds, backed by secondary testing results, in order to start with lead optimization procedures.

   At present, libraries from classical and combinatorial chemistry are the major compound source for HTS programs in drug discovery (see Chapter 4). On the other hand, nature has been proven to be an outstanding source for new and innovative lead candidates. Due to the complexity of cellular metabolism, extracts from natural sources usually contain numerous different components covering a wide range of concentrations. Therefore, integration into drug screening approaches adds additional efforts concerning practical handling (enhanced viscosity, suspended particles) or de-replication for identifying the active component. Though more cost-intensive, dealing with crude or enriched extracts from natural sources is of remarkable interest. Consequently, there exists a need for high-quality samples from natural sources with less complexity in their composition achieved by distribution into several fractions. Up to the present, extraction procedures for sample preparation are usually performed with a low grade of automation. Therefore, a novel approach should account for automation protocols of routinely performed standardized procedures that involve all additional steps of fractionation and concentration of highly diluted crude extracts. For preparing fractionated natural extracts the following criteria have to be fulfilled: i) wide scope in chemical adsorption and elution characteristics, ii) sufficient recovery rates of interesting metabolites, iii) satisfying resolution of chromatographic separation, and iv) feasibility and reproducibility of the practical procedure.

   The first commercially available apparatus for the automated extraction and separation of plant material in a preparative scale is the HPLC-based workstation SEPBOX™ by AnalytiCon AG (Potsdam, Germany) in cooperation with Merck KGaA (Darmstadt, Germany). The SEPBOX™ concept which is also applicable to the separation of secondary metabolites from microbial culture broths allows efficient fractionation of crude material under standardized conditions [28]. Within less than 24 h 1 to 5 g of crude extract is fractionated into up to 300 fairly pure components that can be collected in a microplate compatible format and subsequently be directly used in screening programs.

5.5.2   *Automated chromatographic solid-phase extraction*

Facing the needs for high-quality test sample preparation as the basis for building libraries that fulfill the requirements for HTS programs, we developed a novel and efficient automated sample preparation method based on a multistep fractionation method by chromatographic solid-phase extraction (SPE). Our approach, which advantageously does not need HPLC-techniques, evolved from a procedure for sample preparation from microbial broths with XAD-16 resins that had been developed for chemical screening. The advanced protocol with novel polystyrol-based resins allows chromatographic mixture fractionation through variation of the organic solvent content in the eluent. Our protocol allows to generate single-step fraction samples as well as multiple-step elution fractionations from a single source. The automated multi-step procedure which is performed with modified RapidTrace® modules from Zymark GmbH (Idstein, Germany) shows highly reliable performance and requires only minor manual intervention [29]. In analogy to the SEPBOX™ concept, our multi-step SPE process can be adopted to the separation and purification of mixtures deriving from combinatorial chemistry or any chemical synthesis. In cooperation with CyBio Instruments GmbH (formerly OPAL Jena GmbH; Jena, Germany), an apparatus is currently under development that will provide samples in the 96-well microplate format.

The synergy of the analytical power of chemical screening and the enhanced quality of fractionated extracts allows to add significant physico-chemical information to "hit-lists" out of target-directed screening approaches. Therefore, we consider the integration of TLC-analysis into secondary biological screening and hit-verification as a remarkable tool in lead structure finding strategies, e.g., for assigning the active principle through fast and efficient de-replication, and for speed-up isolation and purification procedures. More recently, coupling techniques with mass spectrometry, or even TLC-FID coupling have been described which obviously can be integrated in our concepts.

Both, generation of natural compound libraries, and access to a wanted bioactive principle from fractionated high-quality samples rely on the efficiency of isolation procedures resulting in pure compounds for further evaluation. Therefore, in cooperation with the HKI Pilot Plant for Natural Products we addressed the transfer of our SPE fractionation process into a pilot scale chromatography in a preparative MPLC system on Amberchrom 161c. Fractionation characteristics by stepwise elution with water/methanol-mixtures can be "translated" into water/methanol gradients with nearly identical separation characteristics.

5.6   A comprehensive library for lead discovery: The natural products pool

As traditional natural products screening is done by testing crude extracts, followed by the crucial work of back-tracing the active compounds from the hit-extracts, a lot of experience is required to exclude both, false positive results and known effectors. Considerable efforts arise for gaining access to sufficient quan-

tities of raw material towards reproduction, compound isolation, structure elucidation and subsequent verification of biological activity. The complete process
proved to be highly time and capacity consuming. As a consequence, screening
with pure compounds, rather than with crude extracts has to be considered. For
small research and development (R & D) units, however, the problem arises to get
access to sufficient numbers of natural compounds covering substantial structural
diversity.

Therefore, our concept to build up a comprehensive Natural Products Pool for
industrial drug discovery purposes has gained considerable interest [30–32].
Academic research groups supply their compounds to this pool, thus having them
tested in target-directed bioassays. However, proprietary rights of the suppliers on
their compounds are not affected. In addition, the suppliers receive financial
incentives for each compound provided to the Natural Products Pool. In case of
hit identification, information is committed to the provider and possibly bilateral
arrangements are made in order to enable further studies. A contract regulates supply, delivery, and use of the Natural Products Pool. So far, a number of hits have
been identified and subsequent bilateral arrangements to provide additional material for hit verification have been realized.

Within an initial 3-year period, the project was supported by BMBF (German
Federal Ministry of Education, Science, Research and Technology), and German
enterprises with key activities in lead discovery. Now, beyond the BMBF supported period, the Natural Products Pool is exclusively funded by our industrial
partners. In order to maintain our concept of providing about 800 compounds per
year, acquisition of natural products from abroad is strengthened.

The Natural Products Pool aims at gaining importance in current industrial lead
discovery programs, thus strengthening the role of natural products in drug discovery. However, the HKI will also benefit from the compound collection in cooperation with compound suppliers for own lead finding purposes, and by providing
it to academic research groups in biochemistry as well as cell and molecular biology for their investigations, e.g., bioassay systems targeting cell signaling
processes.

Within the starting phase 3,500 natural compounds, derivatives, and analogues
have been obtained in amounts of 10 to 20 mg from about 40 German academic
groups, and AnalytiCon AG (Potsdam, Germany), as well. In order to be compatible with standards of modern screening programs, the Natural Products Pool is
organized in the 96-well microplate format (see Chapter 4). The Pool is delivered
to each industrial partner in quantities of 1 milligram per compound, and is
accompanied by a database adhering to industrial standards. The database comprises information about chemical/physical data, known biological activities, references and suppliers.

The range of producing organisms covers microorganisms, such as streptomycetes, rare actinomycetes, myxobacteria, fungi imperfecti, and basidiomycetes,
mosses, a broad variety of higher plant species, some marine organisms and few
animals. Referring to structural diversity, the Natural Products Pool contains representatives of most biosynthetic pathways. Furthermore, nature derived structur-

al diversity is supplemented by synthetic analogues, and derivatives of secondary metabolites. At present, the spectrum of the molecular masses of the compounds centers around 300 to 400 Dalton.

5.7   Combinatorial libraries based on natural products

Today, chemical libraries consisting of more than 100,000 compounds are evaluated in biological assays, typically searching for inhibiting effects towards a particular target (see Chapter 4). In order to generate these huge numbers of compounds, the "classical" strategies of organic and medicinal chemistry have been overflowed by technologies commonly called "combinatorial" chemistry (solid phase, solution phase, as well as split- and split-pool strategies, see Chapter 6). Recent concepts address the synthesis of medium-sized compound libraries consisting of single components.

Since its implementation several years ago, some distinct problems of combinatorial synthesis have surfaced, e.g.: i) purity of the samples often is insufficient, ii) parallel processing often yielded unexpected reaction products, iii) physicochemical analysis of the products generated is necessary, iv) reproduction and purification, especially in larger scales for more detailed pharmacological studies, is often difficult and a logistical problem, and v) over all, structural diversity is not as broad as expected.

Regarding the enhancement of structural diversity, there exists a need in development of combinatorial chemistry in the direction of more sophisticated synthesis concepts (e.g., multi-step synthesis, larger molecules, stereochemical approaches, synthesis of more reactive compounds, making use of complex templates and building blocks). Future success in combinatorial chemistry will substantially depend on both, the quality and the structural diversity of compound libraries submitted to HTS. With respect to the latter, exploiting natural sources for combinatorial synthesis strategies comes into focus [33, 34].

In order to take advantage of molecular diversity from nature for combinatorial chemistry, two major strategies are addressed. On the one hand, combinatorial synthesis allows efficient and systematic structural variation of a given natural product which is used as a kind of template for synthetic "decoration" [33]. The second strategy of integrating natural products into combinatorial chemistry involves total synthesis approaches of natural products. This approach does not depend on the availability of natural products and allows to generate broader structure variation of the basic skeleton of a natural product *via* divergent synthesis approaches and the use of various reagents and building blocks [33].

It should be highlighted that synthetic combinatorial libraries typically focus on low-molecular mass compounds ranging from 200 to 500 Dalton. The same molecular range, in which the majority of known secondary metabolites from natural sources are found. A promising approach is to generate compound libraries on the basis of natural products ranging from 500 to 900 Dalton. These molecules are expected to be more suitable for targets involving protein-protein, protein-DNA, and protein-RNA interaction which play important roles e.g., in cell regulation and differentiation processes.

So far, a restricted number of natural products, e.g., steroids have been used as templates for combinatorial synthesis [33]. A new approach for obtaining increased diversity in the search for new lead compounds, kombiNATURik™, has been co-developed by the German enterprises AnalytiCon AG (Potsdam, Germany) and Jerini Bio Tools GmbH (Berlin, Germany). The kombiNATURik™ program starts from natural compounds which are further diversified by solid-support chemistry introducing, for example, peptide or carbohydrate moieties. Those libraries generally comprise several hundreds to thousands of single molecules derived from multi-parallel synthesis [33]. Our own concept towards single compound libraries of natural products derivatives addresses a similar approach. However, the complexity of our libraries is restricted to around 300 members obtained by automated solid phase synthesis in parallel. Scale-up to about 200 mg quantities for hit validation and more detailed biological studies can easily be realized.

Besides organic chemistry, biological methods for structure modification and subsequent compound library generation can be a supplementary tool for derivatization of structurally complex molecules from both, natural and synthetic sources. The various methods can be categorized into those employing the native biosynthetic machinery of a producing organism, those involving a manipulation of the biosynthetic pathways on the enzymatic or genetic level, and finally, the application of individual biosynthetic enzymes. The experimental demands on the application of the various methods range from "simple" feeding of biosynthetic precursors into standard cultivations to more sophisticated approaches involving genetic engineering of biosynthetic enzymes. Genetics are applied in the cell-based combination of biosynthetic genes from different strains or the *in vitro* reconstitution of biosynthetic pathways with over-expressed enzymes [35]. All of the various methods of biological derivatization become possible due to a relaxed substrate specificity of some of the biosynthetic enzymes, especially those of microbial secondary metabolism.

However, nature itself uses the principles of combinatorial synthesis to generate large and structurally diverse libraries by the combination of small biosynthetic building blocks (e.g., nucleic acids, amino acids, and other building blocks from primary metabolism like activated C_2-, C_3-, and C_4-carboxylic acids for polyketide-type assembling) [36].

Biomimetic combinatorial synthesis of polyketides starting from simple building blocks represent challenging targets for the production of compound libraries. First attempts towards the preparation of libraries have been reported [37, 38].

5.8   Conclusions and perspectives

If one considers the diversity of chemical structures found in nature with the narrow spectrum of structural variation of even the largest combinatorial library it can be expected that in drug discovery natural products will regain their importance. Mainly actinomycetes, fungi and higher plants have been proven to biosynthesize secondary metabolites of obviously unlimited structural diversity that can further be enlarged by structure modification applying strategies of combinatorial chemistry. Probably, a variety of novel concepts in natural products research is required to draw interest to incorporating natural sources into the HTS process.

Natural products libraries comprising only pure and structurally defined compounds will probably contribute to more successful competition of compounds from natural sources within the industrial drug discovery process. Only a minority of the natural products known so far has been biologically characterized in detail. Therefore, any novel target-directed screening assay may result in identification of a new lead structure even from sample collections comprising already described compounds. Today, selected targets of interest are transferred to HTS aiming at discovering a hit, and subsequently a lead structure within a 1 to 2-months period. After that time, in the HTS laboratory the corresponding assay system is replaced by a new one. Thus, rapid characterization and structure elucidation of the active principles of interest from natural sources are critical with respect to competing with synthetic libraries consisting of pure and structurally defined compounds. High-quality test sample preparation as well as elaboration of LC-MS and LC-NMR techniques to accelerate structure elucidation of bioactive principles from natural sources are currently underway. Combination of these techniques with databases comprising a maximum of known natural compounds will probably contribute substantially to reviving interest in natural products for application in drug discovery.

Indications that today only a small percentage of the organisms living in the biosphere is described, implies that there is an enormous reservoir of natural compounds which is still undiscovered. The United Nations Convention on Biological Diversity adopted in Rio de Janeiro in 1992 sets the basic principles of access to and exploitation of global biological sources in the future. As a result, the convention introduces national ownership of biological resources.

Implicating increasing efficiency and decreasing costs for HTS technologies, the basic limiting factor for finding new lead compounds will be the supply of structural diversity. The relevance of natural products in drug discovery consequently will highly depend on the efficiency and costs of access to compounds from natural origin compared to the supply from synthetic sources [39].

5.9   References

1 Shu Y-Z (1998) Recent natural products based drug development: a pharmaceutical industry perspective. *J Nat Prod* 61: 1053–1071
2 Grabley S, Thiericke R (eds) (1999) *Drug Discovery from Nature*. Springer Verlag, Berlin
3 Grabley S, Thiericke R (1999) Bioactive agents from natural sources: trends in discovery and application. *Adv Biochem Engin/Biotechnol* 64: 101–154
4 http://www.meb.uni-bonn.de/cancernet/600733.html
5 Reichenbach H, Höfle G (1999) Myxobacteria as producers of secondary metabolites. In: Grabley S,

Thiericke R (eds) *Drug Discovery from Nature*. Springer Verlag, Berlin, 149–179
6 http://cancertrials.nci.nih.gov/news/angio/table.html
7 Faulkner DJ (2000) Highlights of marine natural products chemistry (1972–1999). *Nat Prod Rep* 17: 1–6
8 Commission on Geosciences, Environment and Resources (1999) Marine-Derived Pharmaceuticals and Related Bioactive Agents. In*: From Monsoons to Microbes: Understanding the Ocean's Role in Human Health*. National Academy Press, 73–82
9 Rayl AJS (1999) Oceans: Medicine chests of the future. *The Scientist* 13: 1–5
10 http://www.pharmamar.es/english1.html
11 http:///www.phc.vcu.edu/feature/epi/index2.html
12 http://www.alzforum.org/members/research/drugs/HuperzineA.html
13 Omura S (ed) (1992) *The Search for Bioactive Compounds from Microorganisms*. Brock/Springer, New York
14 Gräfe U (1992) *Biochemie der Antibiotika*. Spektrum, Heidelberg
15 Grabley S, Thiericke R, Zeeck A (1994) Antibiotika und andere mikrobielle Wirkstoffe. In: Präve P, Faust U, Sittig W et al (eds): *Handbuch der Biotechnologie, 4th ed.* Oldenbourgverlag, München, 663–702
16 Kuhn W, Fiedler H-P (eds) (1995) Sekundärmetabolismus bei Mikroorganismen, Beiträge zur Forschung (engl). Attempto, Tübingen
17 Silverman RB (1992) *The Organic Chemistry of Drug Design and Drug Action*. Academic Press, London
18 Henkel T, Brunne RM, Müller H et al (1999) Statistical investigation into the structural complementarity of natural products and synthetic compounds. *Angew Chem Int Ed Engl* 111: 643–647
19 Luckas B (1996) Determination of algae toxins from seafood. *GIT* 4: 355–359
20 Ebert G (1999) Etablierung von Kultivierungs- und Bearbeitungstechniken zur Erschließung und Bewertung struktureller Diversität aus phototrophen Mikroorganismen für die Wirkstoffsuche. *PhD Thesis,* Technische Universität Berlin, Germany
21 Attaway DH, Zaborsky OR (eds) (1993*) Marine Biotechnology, Vol.1 Pharmaceutical and bioactive natural products*. Plenum Press, New York
22 König GM, Wright AD (1996) Marine natural products research: current directions and future potential. *Planta Medica* 62: 193–211
23 König GM, Wright AD (1999) Trends in Marine Biotechnology. In: Grabley S, Thiericke R (eds) *Drug Discovery from Nature*. Springer Verlag, Berlin, 180–187
24 Jensen PR, Fenical W (1994) Strategies for the discovery of secondary metabolites from marine bacteria: Ecological perspectives. *Annu Rev Microbiol* 48: 559–584
25 Gusovsky F, Daly JW (1990) Maitotoxin: a unique pharmacological tool for research on calcium-dependent mechanisms. *Biochem Pharmacol* 39: 1633–1639
26 Grabley S, Thiericke R, Zeeck A (1999) The Chemical Screening Approach. In: Grabley S, Thiericke R (eds) *Drug Discovery from Nature*. Springer Verlag, Berlin, 124–148
27 Grabley S, Thiericke R (1999) Access to structural diversity via chemical screening. In: Diederichsen U, Lindhorst TK, Westermann B et al (eds) *Bioorganic Chemistry*. Wiley-VCh, Weinheim, 409–417
28 Bindseil KU, God R, Gumm H et al (1997) Vollautomatische Isolierung von Naturstoffen. *GIT Spezial Chromatographie* 1: 19–21
29 Schmid I, Sattler I, Grabley S et al (1999) Natural products in high throughput screening: Automated high-quality sample preparation. *J Biomolecular Screening* 4: 15–25
30 Koch C, Neumann T, Thiericke R et al (1999) A central natural product pool – New approach in drug discovery strategies. In: Grabley S, Thiericke R (eds) *Drug Discovery from Nature*. Springer Verlag, Berlin, 51–55
31 Koch C, Neumann T, Thiericke R et al (1997) Der Naturstoff-Pool – Neuer Ansatz für die Wirkstoffsuche. *Nachr Chem Tech Lab* 45: 16–18
32 Koch C, Neumann T, Thiericke R et al (1997) Der Naturstoff-Pool – Ein neuartiges Konzept für die Wirkstoffsuche bringt Wirtschaft und Wissenschaft zusammen. *BIOspektrum* 3: 43–45
33 Bertels S, Frormann S, Jas G et al (1999) Synergistic use of combinatorial and natural product chemistry. In: Grabley S, Thiericke R (eds) *Drug Discovery from Nature*. Springer Verlag, Berlin, 72–105
34 Paululat T, Tang YQ, Grabley S et al. (1999) Combinatorial chemistry: The impact of natural products. *Chimica Oggi/Chemistry Today* May/June: 52–56
35 Sattler I, Grabley S, Thiericke R (1999) Structure modification via biological derivatization methods. In: Grabley S, Thiericke R (eds*) Drug Discovery from Nature*. Springer Verlag, Berlin, 191–214

36 Rohr J (1995) Combinatorial biosynthesis – an approach in the near future. *Angew Chem Int Ed Engl* 34: 881–885
37 Reggelin M, Brenig V, Welcker R (1998) Towards polyketide libraries-II: Synthesis of chiral aracemic di- and triketides on a solid support. *Tetrahedron Lett* 39: 4801–4804
38 Khosla C (1998) Combinatorial biosynthesis of "unnatural" natural products. In: EM Gordon, JF Kerwin (eds) *Combinatorial chemistry and molecular diversity in drug discovery.* Wiley-Liss, New York, 401–417
39 Beese K (1996) Pharmaceutical bioprospecting and synthetic molecular diversity. The convention on biological diversity and the value of natural products in the light of new technological developments. Institute for prospective technological studies – European Commission. Draft discussion paper, May 15: 1–43

Modern Methods of Drug Discovery
ed. by A. Hillisch and R. Hilgenfeld
© 2003 Birkhäuser Verlag/Switzerland

6 Combinatorial chemistry: Mixture-based combinatorial libraries of acyclic and heterocyclic compounds from amino acids and short peptides

Adel Nefzi, John M. Ostresh and Richard A. Houghten

Torrey Pines Institute for Molecular Studies, 3550 General Atomics Court, San Diego, CA 92121 USA

6.1   Introduction

Combinatorial chemistry is recognized worldwide as a powerful technology for drug discovery. This technology has gained wide acceptance by most pharmaceutical and biotechnology companies as well as academia [1–2]. The power of combinatorial chemistry lies in its ability to accelerate the drug discovery process through the rapid synthesis and subsequent screening of a larger number of compounds than previously possible. In a recent paper on combinatorial libraries, we reviewed the solid-phase chemistry used to prepare small molecule and heterocyclic mixture-based libraries [3]. Herein we provide a perspective on synthetic combinatorial approaches using mixture-based libraries, as well as an illustration of our work in the generation of libraries of acyclic and heterocyclic compounds from amino acids and/or short peptides.

6.2 Combinatorial chemistry techniques

Combinatorial techniques have their origins in parallel solid phase synthesis methods developed during the mid 1980s. These techniques were initially used for the synthesis of peptides, peptidomimetics, or oligonucleotides, and included the tea-bag [4], pin [5] and spot [6] approaches. Such approaches enabled hundreds of individual compounds to be prepared in a fraction of the time and cost previously required. The use of synthetic combinatorial library concepts and technologies for the synthesis of heterocycles and other small molecules was a logical continuation of this development, again with the aim of increasing the speed and throughput of synthesis and biological evaluation of compounds. Initially developed for peptides, the original combinatorial concept involved the generation of all possible sequence combinations for a peptide of a given length (e.g., $20^6 = 64$ million hexapeptides when the proteogenic amino acids are used), hence the origin of the term "combinatorial" [7–9]. Since the individual synthesis of the immense numbers of compounds required by true combinatorial libraries is unrealistic using current parallel synthetic methods, two approaches for the generation of very large compound mixtures have been developed and used successfully. First, libraries have been generated using recombinant DNA techniques in which large numbers of peptides can be expressed randomly in a fusion phage or other vector system [10]. This method has proved popular with laboratories already familiar with molecular biology techniques; however, it is restricted to the use of the 20 proteogenic amino acids as building blocks. In an article in *Nature Biotechnology*, a decapeptide that impedes both the growth and spread of cancerous tumors in animals has been identified following the screening of phage libraries [11]. In the second approach, combinatorial libraries have been produced by synthetic means [12–15]. For those laboratories having facilities for the generation of synthetic combinatorial libraries (SCLs), this approach allows for the introduction of any building block of interest, including D-amino acids [16, 17], non-proteogenic amino acids [18] and carboxylic acids [19].

Peptide combinatorial libraries have led to the identification of a wide range of bioactive peptides, including novel antibacterials [20], potent agonists and antagonists to opioid receptors [21–24], inhibitors of melittin's hemolytic activity [25], antigenic peptides recognized by monoclonal antibodies [26–28], and potent endothelin antagonists that were identified from a library of triamides [29]. Early work from this laboratory has shown the broad utility of mixture-based SCLs for the *de novo* identification of potent analgesics, highly active antimicrobial compounds and enzyme inhibitors, and highly specific antigenic determinants [13, 30]. Along with linear peptide sequences, our laboratory and other groups have also synthesized combinatorial libraries of cyclic compounds. Highly active chymotrypsin inhibitors have been identified upon screening of a cyclic peptide template combinatorial library [19, 31, 32]. A review providing a perspective on the synthesis and inherent strengths and weaknesses of the use of mixture-based combinatorial libraries, as well as the large number of successful applications of this technology has recently been published by our group [3].

6.3 Synthesis methods

Two synthetic approaches, involving either the mixing of multiple resins or the use of mixtures of incoming reagents, are now widely used to incorporate multiple functionalities at diverse positions within an SCL.

6.3.1 Resin mixtures

The "divide, couple, and recombine" (DCR) synthesis method [13], also know as the "split resin" method [14], was developed for use in the synthesis of peptide, acyclic and heterocyclic SCLs. This synthesis method, illustrated in Figure 1, involves the coupling of reactants to individual aliquots of resin followed by thorough mixing of the resin. This method allows the generation of approximately equimolar mixtures of compounds. Due to the statistical distribution of beads at each step, care should be used in determining the appropriate amount of resin to be used in the synthesis in order to ensure inclusion of all compounds in the library [33, 34]. An important aspect to the DCR approach is that, due to the physical nature of dividing and mixing the resin beads, each resin bead contains only one compound when the library is completed [14].

6.3.2 Reagent mixtures

A second synthetic approach, termed the "reagent mixture" method, generally uses a predefined ratio of reagents in excess to accomplish approximately equimolar incorporation of each reagent at a "position of diversity" [35]. This method offers the advantage that a mixture of reagents can be readily incorporated at any position in a sequence. A large excess of incoming reagents is used such that pseudo first order reaction kinetics are observed. It is important that the relative reaction rates of the incoming reagents are approximately equal and relatively independent of the resin-bound reagents (i.e., similar nucleophilicity, no significant steric hindrance, etc.). We have found that this concept applies equally well to mixtures of incoming reagents such as aldehydes, carboxylic acids, etc. (unpublished observation). We have recently applied the reagent mixture method to the synthesis of acyclic and heterocyclic compounds, such as polyamines [36, 37], cyclic ureas and cyclic thioureas [38, 39] and bicyclic guanidines [36].

A number of reports have been presented on the use of limiting reagents to accomplish the same result [40, 41]. This method relies on initially reacting equal amounts of all reagents, equimolar relative to the resin, in order to obtain a resin-bound mixture. The reaction is then repeated using excess reagents in order to drive the reaction of the remaining unreacted sites to completion. Reasonable results can be obtained for the addition of one position of diversity. However, a major disadvantage to this method is seen when incorporating more than one position of diversity. Incoming reagents can be preferentially consumed by particular resin-bound reactants even when there are small differences in reaction rates or relative reaction rates. Repetitive cycles using this method multiplies the problem, resulting in large deviations from equimolarity in the final products.

6.4   Deconvolution methods

Three approaches are generally used for the structural deconvolution of active compounds from assay data using nonsupport-bound SCLs: iterative deconvolution [13], positional scanning deconvolution [30], and tagging [42]. Each approach has been used to identify active individual compounds in a wide variety of SCLs and assays.

6.4.1   Iterative deconvolution

The iterative deconvolution method (Fig. 1) is illustrated with a generic heterocycle containing three positions of diversity, designated OXX (where O represents a defined position of diversity and X represents mixture positions) [13]. The SCL is first screened to identify active mixtures. Since for each mixture within the library one position of diversity is defined, active mixtures suggest the importance of the functionality at that position. The remaining two positions are then identified sequentially through an iterative process of synthesis and screening.

6.4.2   Positional scanning deconvolution

The positional scanning (PS) approach [30] is illustrated in the generic representation shown in Figure 2.

It involves the screening of separate, single defined position SCLs to individually identify the most important functionalities at each position of diversity within a library. A complete PS-SCL having three positions of diversity consists of three sublibraries (designated OXX, XOX, and XXO), each of which has a single defined functionality at one position and a mixture of functionalities at each of the other three positions. The structure of individual compounds can be determined from such a screening since each compound is present in only one mixture of each sublibrary. In theory, if only one compound was active in the library, activity corresponding to that compound would be found in the one mixture of each sublibrary containing that compound. When considered in concert, the

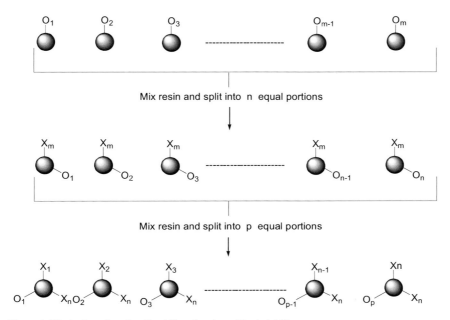

Figure 1. Illustration of an iterative trifunctional combinatorial library.

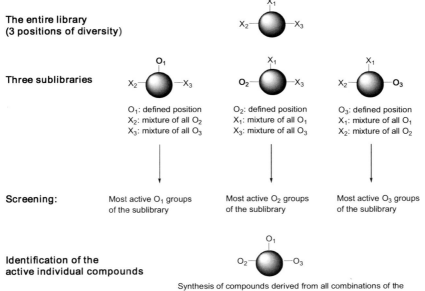

Figure 2. Illustration of a positional scanning trifunctional combinatorial library.

defined functionality in each mixture can then be used to identify the individual active compound responsible for the activity. In reality, the same result is seen, but the activity is generally due to the sum of activities of more than one compound. Anomalous results are seen if the activity is due to the sum of many weakly active compounds. PS-SCLs are generally prepared using the reagent mixture approach described above. Although the synthesis of PS-SCLs is theoretically possible with the DCR approach, in reality the labor involved makes the synthesis unfeasible. Freier and coworkers have performed a thorough examination of the theoretical and experimental aspects of iterative and positional scanning deconvolution [43, 44].

6.5   Solid phase synthesis of acyclic and heterocyclic compounds from amino acids and short peptides

Since peptides have limitations in their further development as pharmaceuticals due to their poor bioavailability and their rapid enzymatic degradation, the focus of combinatorial chemistry has shifted to libraries of small acyclic and heterocyclic compounds [45]. Due to their versatility, amino acids and short peptides have been extensively used for the synthesis of organic compounds. They possess a significant number of functional groups, which facilitates synthetic operations.

6.5.1   Peptidomimetics and acyclic compounds

We have developed an efficient method for the generation of peptidomimetic libraries by chemical transformation of an existing peptide library. As shown in Scheme 1, the peralkylation and/or the reduction of the amide bonds offer different classes of compounds, ranging from a peralkylated amide peptide to different shapes of polyamine compounds [46–52]. As an example, a soluble peptidomimetic combinatorial library of 57, 000 compounds having a dipeptide scaffold with each amide hydrogen replaced with different alkyl-groups has been reported [49].

In recent years, the focus of combinatorial chemistry has shifted to libraries of small acyclic and heterocyclic compounds. We review here part of our ongoing effort toward the synthesis of SCLs of small organic and heterocyclic compounds using the "libraries from libraries" approach [50].

A linear urea library has been prepared; the reaction of a resin-bound N-alkylated amino acid with an individual pre-formed isocyanate affords the linear urea in good yield. The isocyanate is generated by slowly adding an amine to a solution of triphosgene in anhydrous DCM in the presence of diisopropylethylamine (DIPEA). The condensation of the isocyanate with the resin-bound amino acid affords the linear urea (Scheme 2).

Following the individual synthesis of controls, a library of 125,000 linear N,N'-disubstituted ureas was prepared. This library has been tested for opioid activity

Scheme 1

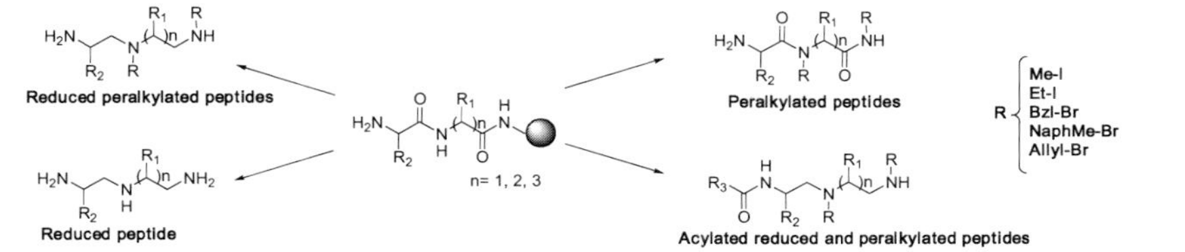

Reduced peralkylated peptides

Reduced peptide

Peralkylated peptides

Acylated reduced and peralkylated peptides

n= 1, 2, 3

R ⎰ Me-I
Et-I
Bzl-Br
NaphMe-Br
Allyl-Br

Scheme 2

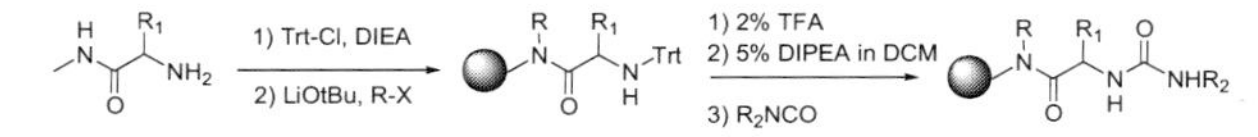

1) Trt-Cl, DIEA
2) LiOtBu, R-X

1) 2% TFA
2) 5% DIPEA in DCM
3) R₂NCO

Scheme 3

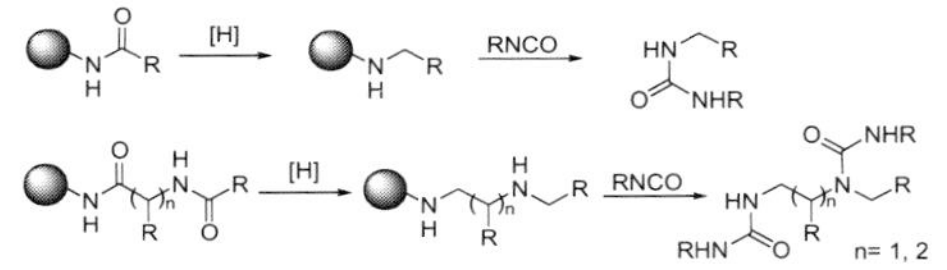

[H]   RNCO

[H]   RNCO

n= 1, 2

Scheme 4

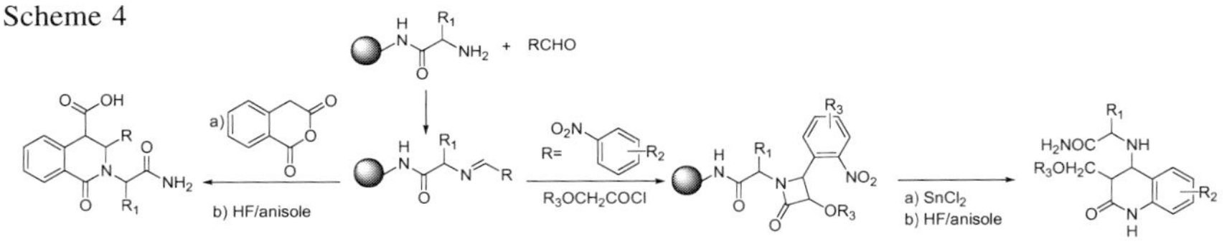

+ RCHO

a)

b) HF/anisole

R=

R₃OCH₂COCl

a) SnCl₂
b) HF/anisole

Scheme 5

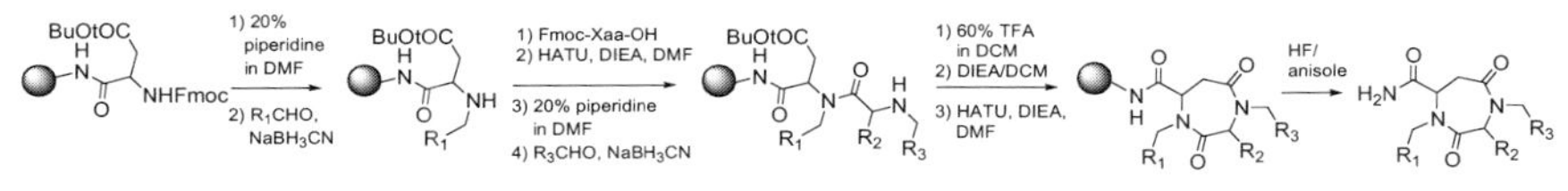

1) 20% piperidine in DMF
2) R₁CHO, NaBH₃CN

1) Fmoc-Xaa-OH
2) HATU, DIEA, DMF
3) 20% piperidine in DMF
4) R₃CHO, NaBH₃CN

1) 60% TFA in DCM
2) DIEA/DCM
3) HATU, DIEA, DMF

HF/anisole

Scheme 6

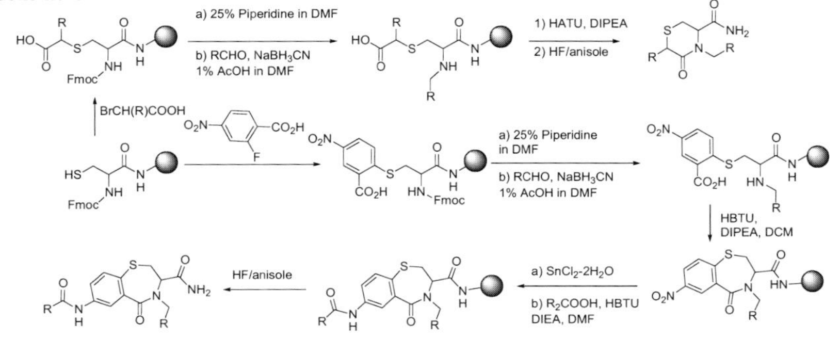

a) 25% Piperidine in DMF
b) RCHO, NaBH₃CN 1% AcOH in DMF

1) HATU, DIPEA
2) HF/anisole

BrCH(R)COOH

a) 25% Piperidine in DMF
b) RCHO, NaBH₃CN 1% AcOH in DMF

HBTU, DIPEA, DCM

HF/anisole

a) SnCl₂·2H₂O
b) R₂COOH, HBTU DIEA, DMF

at the mu, delta, kappa and sigma opioid receptors. Following deconvolution of this library, individual compounds having nM affinities at the mu or sigma receptors were found (unpublished observations).

We have also developed an efficient method for the solid phase synthesis of ureas, diureas and triureas from resin-bound monoamines, diamines and triamines. The exhaustive reduction of solid support-bound polyamides with borane in THF generated amines, which, following treatment with commercially available isocyanates and cleavage, provided the corresponding ureas in high purity and good yields (Scheme 3).

6.5.2   Heterocyclic compounds

Imines are often used as intermediates in organic synthesis and are the starting point for chemical reactions such as cycloadditions, condensation reactions and nucleophilic addition [7]. The formation of imines via condensation of amines with aldehydes has been extensively used for the synthesis of a variety of heterocyclic compounds.

The solid phase synthesis of a combinatorial library containing 43,000 tetrahydroisoquinolines has been reported by Griffith and coworkers [53]. The library was synthesized using a three-step procedure. An imine was formed by reacting a substituted benzaldehyde with an MBHA resin-bound amino acid. Imine formation was driven to completion using trimethylorthoformate as a dehydrating reagent. The treatment of the imine with homophthalic anhydride provided the desired tetrahydroisoquinoline (Scheme 4).

Isomerization to the more stable trans configuration was obtained following cleavage by treatment with 1N NaOH. The tetrahydroisoquinoline library was prepared using the DCR method with MBHA resin, 11 amino acid building blocks, 38 aldehydes and 51 amines. The library has been tested in a number of assays, including kappa and mu opioid radioreceptor binding assays and in a sigma radioreceptor binding assay [53]. Using the "libraries from libraries" concept [47, 50], a second copy of the isoquinoline library was reduced in the presence of borane in THF to generate a second library having different physical and chemical properties.

The solid phase synthesis of 4-amino-3,4-dihydro-2(1H)-quinolinones was developed by Pei and coworkers through the rearrangement of β-lactam intermediates on the solid phase [54]. The reaction of a resin-bound amino acid with ortho-nitrobenzaldehyde resulted in an imine that, following treatment with a ketene, undergoes a [2 + 2] cycloaddition to afford a four-member ring cis-β-lactam intermediate. The ketene was generated in situ from the corresponding phenoxyacetyl chloride in the presence of triethylamine. Following reduction of the nitro group with tin chloride, the β-lactam ring underwent an intramolecular rearrangement to afford the trans-3,4-dihydro-2(1H)-quinolinones through nucleophilic attack of the β-lactam moiety by the generated amine. Following HF cleavage and lyophilization, individual dihydroquinolinones were obtained in good yield. Using the tea-bag technology and in combination with the DCR synthesis method. A library of 2,070 dihydroquinolinones derived from (69 amino acids, 6 ortho-nitro-benzaldehydes and 5 acid chlorides) was produced.

6.5.3 Solid phase synthesis of heterocyclic compounds from C^α-functionalized amino acids

The solid phase synthesis of tetrasubstituted diazepine-2,5-diones from resin-bound t-butyl ester of aspartic acid was initiated by reductive alkylation with an aldehyde of the α-amino group of the p-methylbenzhydrylamine resin-bound aspartic acid. The coupling of an Fmoc amino acid to the resulting secondary amine does not readily go to completion. Satisfactory results were obtained using double coupling with HATU. This coupling step depends strongly on the incoming amino acid [55]. Good yields were obtained with Phe and Met(O), whereas low yields were obtained with hindered amino acids such as Val. Once the dipeptide was formed, the Fmoc protecting group was removed and a second reductive alkylation was carried out using the same conditions. Following tBu cleavage, the thermodynamically favorable coupling of the resulting secondary amine to the side chain of aspartic acid was readily accomplished in the presence of HATU (Scheme 5).

Starting from p-methylbenzhydrylamine (MBHA) resin-bound N-α-Fmoc-S-trityl-L-cysteine, and following cleavage of the trityl (Trt) group, the resin-bound Fmoc-cysteine was treated with a range of different α-bromo, α-alkyl acetic acid derivatives in DMF in the presence of N-methylmorpholine (NMM). Following the removal of the Fmoc protecting group with 20% piperidine in DMF, reductive alkylation of the free amine occurred in the presence of an aldehyde and sodium cyanoborohydride (NaBH₃CN). The formation of thiomorpholin-3-one occurred via intramolecular amidation using HATU as the coupling reagent (Scheme 6) [56].

Starting from the same resin and following the cleavage of the trityl (Trt) group, 2-fluoro-5-nitro-benzoic acid was added to the resin-bound Fmoc-cysteine. The Fmoc group was cleaved and the resulting free amine reductively alkylated with a variety of aldehydes in the presence of sodium cyanoborohydride. The resulting compound was treated with O-benzotriazolyl-N,N,N',N'-tetramethyluronium hexafluorophosphate (HBTU) in anhydrous DCM, which underwent intramolecular amide bond formation to afford the resin-bound nitro-benzothiazepine. The nitro group was reduced with tin chloride, followed by N-acylation, and following cleavage of the solid support yielded the desired product in good purity (Scheme 6).

Using 48 aldehydes and 95 carboxylic acids and in combination with the DCR synthesis method [13], a mixture-based library of 95 mixtures of 48 benzothiazepines was produced [57].

Scheme 7

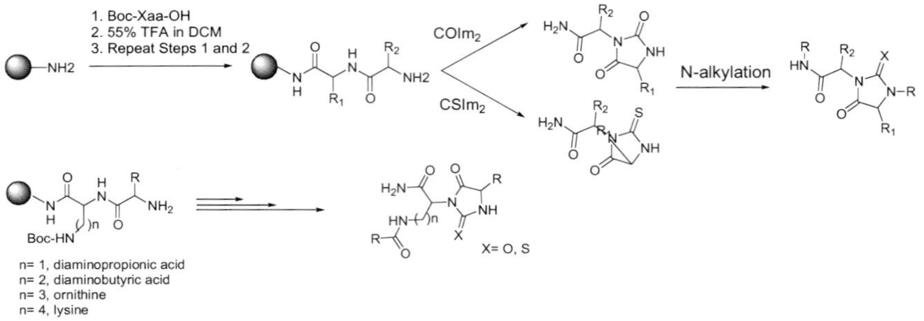

n= 1, diaminopropionic acid
n= 2, diaminobutyric acid
n= 3, ornithine
n= 4, lysine

Scheme 8

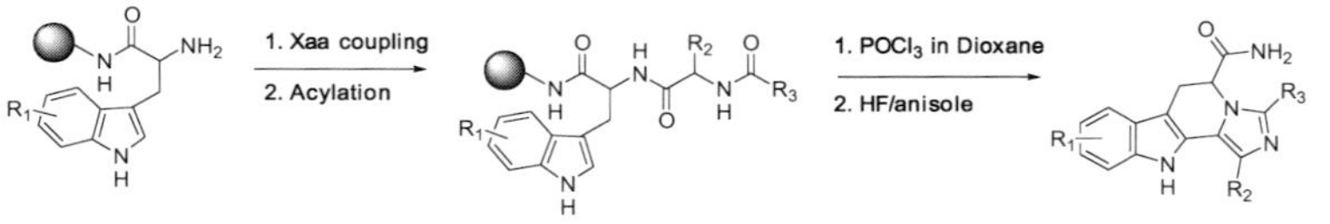

Scheme 9

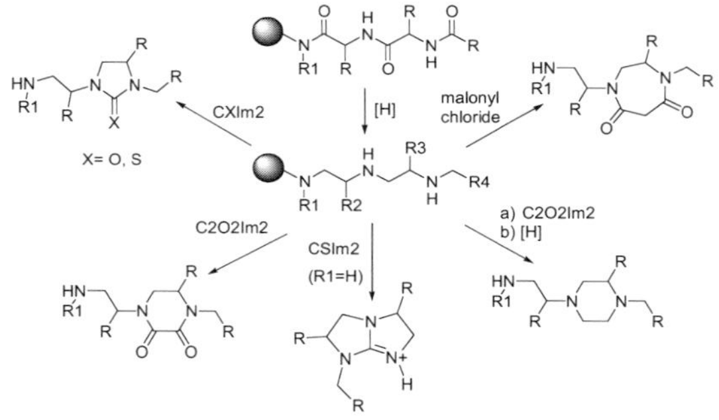

Scheme 10

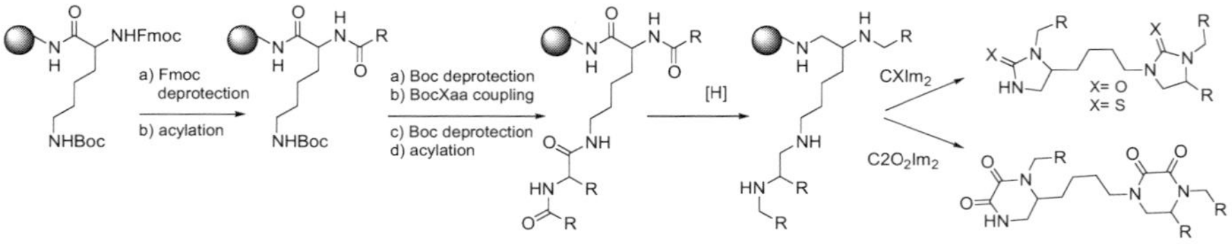

6.5.4   Solid phase synthesis of heterocyclic compounds from resin-bound dipeptides

We have developed a simple synthetic route to the solid phase synthesis of hydantoin and thiohydantoin SCLs from a resin-bound dipeptide SCL [58]. This involved the reaction of the N-terminal amino group of resin-bound dipeptide with phosgene or thiophosgene leading to the intermediate isocyanate or thioisocyanate that further reacted intra-molecularly to form the five member ring hydantoin or thiohydantoin. Using 54 amino acids for the first site of diversity and 60 amino acids for the second site of diversity, and using four different alkylating reagents a library of 38,880 compounds ($54 \times 60 \times 3 \times 4$) has been synthesized (Scheme 7).

In order to increase the number and class of available compounds, we selectively alkylated the resin-bound amide and then the remaining amides to generate a dialkylated hydantoin library (Scheme 7).

We also synthesized a different class of hydantoin compounds called branched hydantoins. Thus, starting from a resin-bound, orthogonally protected diamino acid (including diaminopropionic acid, diaminobutyric acid, ornithine or lysine), and following deprotection of the α-amino group and coupling of a second amino acid, the resin was treated with carbonyldiimidazole or triphosgene to afford the highly active intermediate isocyanate, which undergoes an intramolecular cyclization leading to the hydantoin. The diamino acid side chain is then deprotected and the free amine group is acylated with a range of carboxylic acids to yield the desired hydantoins following cleavage of the resin with hydrogen fluoride (Scheme 7) [59].

6.5.5 Solid phase synthesis of heterocyclic compounds from acylated dipeptides

We have also developed an efficient method for the solid-phase synthesis of imidazol-pyrido-indoles [60]. This class of compounds shows a broad range of biological activity. Starting from a resin-bound tryptophan or tryptophan analogs, a second amino acid was coupled and the resin bound dipeptide was acylated with a variety of carboxylic acids. This acylated dipeptide was subjected to Bischler-Napieralski condensation using phosphorus oxychloride in 1,4-dioxane at 85°C (Scheme 8).

Using 11 different functionalities derived from amino acids at R_1, 25 different functionalities derived from amino acids at R_2 and 92 different functionalities derived from carboxylic acids at R_3, we generated a library of 25,300 imidazol-pyrido-indole derivatives in the positional scanning format. Prior to the library synthesis, we synthesized individual control compounds in which the building blocks in each individual position were varied while the other three positions remained fixed. Those compounds were used to determine whether the individual building blocks could be successfully incorporated into the library. The complete PS-SCL was composed of three sublibraries, each containing the same 25,300 individual compounds ($11 \times 25 \times 92$) [60].

6.5.6 Solid phase synthesis of heterocyclic compounds from resin-bound reduced acylated dipeptides

Modified dipeptide SCLs having four positions of diversity were selected as starting materials for the solid phase synthesis of cyclic urea and thiourea libraries

[38]. The complete reduction of the amide carbonyls in a selectively N-alkylated acylated dipeptide SCL with borane in THF yielded a triamine SCL having two free secondary amines (Scheme 9).

Treatment of this triamine SCL with carbonyldiimidazole or thiocarbonyldiimidazole afforded the corresponding cyclic ureas and thioureas in good yield and high purity [38–39]. Using this approach and following the initial synthesis of individual control compounds, four PS-SCLs were generated from N-alkylated acylated dipeptide SCLs having either a methyl or benzyl group on the C-terminal amide.

The same template, a resin-bound polyamine library, when treated with oxalyldiimidazole led to the corresponding 2,3-diketopiperazines (following HF cleavage) in high purity [61]. Reduction on the solid support of the oxamide moiety with borane in THF afforded the corresponding trisubstituted piperazines following release from the solid support. Similarly, treatment of the reduced resin-bound acylated dipeptide with malonyl chloride, following HF cleavage, afforded the corresponding diazepinediones (manuscript submitted).

A similar strategy has been used for the solid phase synthesis of bicyclic guanidines from reduced N-acylated dipeptides having three available secondary amines (R_1 = H). Initially, cyclic thiourea was formed as previously described. In contrast to the previous synthetic route, the presence of three secondary amines allowed the reaction to proceed through the formation of a highly active intermediate that cyclized to afford a protonated bicyclic guanidine in good yield and high purity. Using 49 amino acids for the first site of diversity, 51 amino acids for the second site, and 41 carboxylic acids for the third site of diversity, a library of 102,459 ($49 \times 51 \times 41$) compounds was synthesized in the positional scanning format [36].

The bicyclic guanidine library was screened in a radioreceptor assay selective for the mu opioid receptor. A number of individual compounds showed binding affinities less than 200 nM. The most active individual bicyclic guanidine had an IC_{50} value of 37 nM [3]. In other studies, the cyclic urea and cyclic thiourea libraries were assayed for their ability to inhibit *Candida albicans* growth, which is one of the most common opportunistic fungi responsible for infections and the fungal infection most frequently associated with HIV-positive patients. Individual compounds were found to have MIC values ranging from 8 to 64 µg/ml [63].

6.5.7   Solid phase synthesis of bis heterocyclic compounds from resin-bound orthogonally protected lysine

Starting from a resin-bound orthogonally protected lysine, the N^α was deprotected and the free amine was acylated with carboxylic acids. The N^ξ was then deprotected and the generated amine was coupled to a protected amino acid. Following deprotection and acylation of the amine, the amide bonds were reduced with borane in THF to generate four secondary amines. The treatment of the resin-bound polyamines with oxalyldiimidazole, thiocarbonyldiimidazole and oxalyldiimidazole at low concentrations afforded the energetically favorable bis-cyclic ureas, bis-cyclic thioureas and bis-diketopiperazines, respectively (Scheme 10) [64].

6.6 Conclusions

Mixture-based synthetic combinatorial libraries offer a powerful advantage in that very large diversities can be synthesized and screened in a rapid and cost-efficient manner. Amino acids and short peptides are versatile precursors for the solid phase synthesis of acyclic and heterocyclic combinatorial libraries. Using the "libraries from libraries" concept, modified dipeptides have been successfully used for the solid phase synthesis of small molecule and heterocyclic combinatorial libraries.

6.7 Acknowledgments

Work carried out in the authors' laboratory was funded in part by National Cancer Institute Grant No. CA78040 (Houghten).

6.8 References

1 Anonymous (1998) Breakthrough of the year. *Science* 282: 2156–2161
2 Lebl M (1999) Parallel personal comments on "classical" papers in combinatorial chemistry. *J Comb Chem* 1: 3–24
3 Houghten RA, Pinilla C, Appel JR et al (1999) Mixture-based synthetic combinatorial libraries. *J Med Chem* 42: 3743–3778
4 Houghten RA (1985) General method for the rapid solid-phase synthesis of large numbers of peptides: Specificity of antigen-antibody interaction at the level of individual amino acids. *Proc Natl Acad Sci USA* 82: 5131–5135
5 Geysen HM, Meloen RH, Barteling SJ (1984) Use of a peptide synthesis to probe viral antigens for epitopes to a resolution of a single amino acid. *Proc Natl Acad Sci USA* 81: 3998–4002
6 Frank R, Doring R (1988) Simultaneous multiple peptide synthesis under continuous flow conditions on cellulose paper discs as segmental solid supports. *Tetrahedron* 44: 6031–6040
7 Nefzi A, Ostresh JM, Houghten RA (1997) The current status of heterocyclic combinatorial libraries. *Chem Rev* 97: 449–472
8 Fruchtel JS, Jung G (1996) Organic chemistry on solid supports. *Angew Chem Int Ed Engl* 35: 17–42
9 Fruchtel JS, Jung G (1996) Organic chemistry on solid supports – basic principles for combinatorial chemistry. In: Jung G (ed) *Combinatorial peptide and nonpeptide libraries*. Verlag Chemie, Weinheim, 19–78
10 Zwick MB, Shen JQ, Scott J (1998) Phage-displayed peptide libraries. *Curr Opin Biotechnol* 9: 427–436
11 Koivunen E, Arap W, Valtanen H et al (1999) Tumor targeting with a selective gelatinase inhibitor. *Nature Biotechnol* 17: 768–774
12 Geysen HM, Rodda SJ, Mason TJ (1986) A priori delineation of a peptide which mimics a discontinuous antigenic determinant. *Mol Immunol* 23: 709–715
13 Houghten RA, Pinilla C, Blondelle SE et al (1991) Generation and use of synthetic peptide combinatorial libraries for basic research and drug discovery. *Nature* 354: 84–86
14 Lam KS, Salmon SE, Hersh EM et al (1991) A new type of synthetic peptide library for identifying ligand-binding activity. *Nature* 354: 82–84

15 Furka A, Sebestyen F, Asgedom M et al (1991) General method for rapid synthesis of multicomponent peptide mixtures. *Int J Pept Protein Res* 37: 487–493

16 Dooley CT, Chung NN, Wilkes BC et al (1994) An all D-amino acid opioid peptide with central analgesic activity from a combinatorial library. *Science* 266: 2019–2022

17 Lam KS, Lebl M, Krchnak V et al (1993) Discovery of D-amino-acid-containing ligands with selectide technology. *Gene* 137: 13–16

18 Blondelle SE, Takahashi E, Weber PA et al (1994) Identification of antimicrobial peptides using combinatorial libraries made up of unnatural amino acids. *Antimicrob Agents Chemother* 38: 2280–2286

19 Eichler J, Lucka AW, Houghten RA (1994) Cyclic peptide template combinatorial libraries: Synthesis and identification of chymotrypsin inhibitors. *Pept Res* 7: 300–307

20 Blondelle SE, Pérez-Payá E, Houghten RA (1996) Synthetic combinatorial libraries: Novel discovery strategy for identification of antimicrobial agents. *Antimicrob Agents Chemother* 40: 1067–1071

21 Dooley CT, Chung, NN, Wilkes, BC et al (1994) An all D-amino acid opioid peptide with central analgesic activity from a combinatorial library. *Science* 266: 2019–2022

22 Dooley CT, Kaplan RA, Chung NN et al (1995) Six highly active mu-selective opioid peptides identified from two synthetic combinatorial libraries. *Pept Res* 8: 124–137

23 Kramer TH, Toth G, Haaseth RC et al (1991) Influence of peptidase inhibitors on the apparent agonist potency of delta selective opioid peptides *In vitro*. *Life Sci* 48: 882–886

24 Dooley CT, Houghten RA (1995) Identification of mu-selective polyamine antagonists from a synthetic combinatorial library. *Analgesia* 1: 400–404

25 Blondelle SE, Houghten RA, Pérez-Payá E (1996) Identification of inhibitors of melittin using non-support-bound combinatorial libraries. *J Biol Chem* 271: 4093–4099

26 Burton DR, Barbas CF, III, Persson MAA et al (1991) A large array of human monoclonal antibodies to type 1 human immunodeficiency virus from combinatorial libraries of asymptomatic seropositive individuals. *Proc Natl Acad Sci USA* 88: 10134–10137

27 Motti C, Nuzzo M, Meola A et al (1994) Recognition by human sera and immunogenicity of HBsAg mimotopes selected from an M13 phage display library. *Gene* 146: 191–198

28 Pinilla C, Appel J, Blondelle SE et al (1995) A review of the utility of peptide combinatorial libraries. *Biopolymers (Peptide Science)* 37: 221–240

29 Terret NK, Bojanic D, Brown D et al (1995) *BioMed Chem Lett* 5: 917–922

30 Pinilla C, Appel JR, Blanc P et al (1992) Rapid identification of high affinity peptide ligands using positional scanning synthetic peptide combinatorial libraries. *Biotechniques* 13: 901–905

31 Spatola AF, Crozet Y, DeWit D et al (1996) Rediscovering an endothelin antagonist (BQ-123): A self-deconvoluting cyclic pentapeptide library. *J Med Chem* 39: 3842–3846

32 Szymonifka MJ, Chapman KT (1995) Magnetically Manipulable Polymeric Supports for Solid Phase Organic Synthesis. *Tetrahedron Lett* 36: 1597–1600

33 Gallop MA, Barrett RW, Dower WJ et al (1994) Applications of combinatorial technologies to drug discovery. 1. Background and peptide combinatorial libraries. *J Med Chem* 37: 1233–1251

34 White SP, Scott DL, Otwinowski Z et al (1990) Crystal structure of cobra-venom phospholipase A_2 in a complex with a transition-state analogue. *Science* 250: 1560–1563

35 Ostresh JM, Winkle JH, Hamashin VT et al (1994) Peptide libraries: Determination of relative reaction rates of protected amino acids in competitive couplings. *Biopolymers* 34: 1681–1689

36 Ostresh JM, Schoner CC, Hamashin VT et al (1998) Solid phase synthesis of trisubstituted bicyclic guanidines via cyclization of reduced N-acylated dipeptides. *J Org Chem* 63: 8622–8623

37 Nefzi A, Ostresh JM, Houghten RA (1999) Parallel solid phase synthesis of tetrasubstituted diethylenetriamines via selective amide alkylation and exhaustive reduction of N-acylated dipeptides. *Tetrahedron* 55: 335–344

38 Nefzi A, Ostresh JM, Meyer J-P et al (1997) Solid phase synthesis of heterocyclic compounds from linear peptides: cyclic ureas and thioureas. *Tetrahedron Lett* 38: 931–934

39 Nefzi A, Ostresh JM, Giulianotti M et al (1999) Solid-phase synthesis of trisubstituted 2-imidazolidones and 2-imidazolidinethiones. *J Comb Chem* 1: 195–198

40 Udaka K, Wiesmüller K-H, Kienle S et al (1995) Decrypting the structure of major histocompatibility complex class I-restricted cytotoxic T lymphocyte epitopes with complex peptide libraries. *J Exp Med* 181: 2097–2108

41 Grass-Masse H, Ameisen JC, Boutillon C et al (1992) Synthetic vaccines and HIV-1 hypervariability: a mixotope approach. *Pept Res* 5: 211–216

42 Janda KD (1994) Tagged *versus* untagged libraries: Methods for the generation and screening of combinatorial chemical libraries. *Proc Natl Acad Sci USA* 91: 10779–10785

43 Konings DAM, Wyatt JR, Ecker DJ et al (1996) Deconvolution of combinatorial libraries for drug discovery: Theoretical comparison of pooling strategies. *J Med Chem* 39: 2710–2719

44 Wilson-Lingardo L, Davis PW, Ecker DJ et al (1996) Deconvolution of combinatorial libraries for drug discovery; experimental comparison of pooling strategies. *J Med Chem* 39: 2720–2726

45 Blomberg K, Granberg C, Hemmila I et al (1986) Europium0labelled target cells in an assay of natural killer cell activity. II. A novel non-radioactive method based on time-resolved fluorescence. Significance and specificity of the method. *J Immunol Meth* 92: 117–123

46 Dörner B, Ostresh JM, Blondelle SE et al (1997) Peptidomimetic synthetic combinatorial libraries. In: A Abell (ed): *Advances in Amino Acid Mimetics and Peptidomimetics Vol 1*. JAI Press, Greenwich, CT, 109–125

47 Ostresh JM, Dörner B, Houghten RA (1998) Peralkylation: "Libraries from libraries": Chemical transformation of synthetic combinatorial libraries. In: Cabilly S (ed): *Combinatorial peptide library protocols*. Humana Press, Totowa, New Jersey, 41–49

48 Ostresh JM, Dörner B, Blondelle SE, Houghten RA (1997) Soluble combinatorial libraries of peptides, peptidomimetics, and organics: fundamental tools for basic research and drug discovery. In: Wilson TR, Czarnik AW (eds): *Combinatorial Chemistry*. John Wiley & Son, Inc., New York, 225–240

49 Dörner B, Husar GM, Ostresh JM et al (1996) The synthesis of peptidomimetic combinatorial libraries through successive amide alkylation. *Bioorg Med Chem* 4: 709–715

50 Ostresh JM, Husar GM, Blondelle SE et al (1994) "Libraries from libraries": Chemical transformation of combinatorial libraries to extend the range and repertoire of chemical diversity. *Proc Natl Acad Sci USA* 91: 11138–11142

51 Ostresh JM, Blondelle SE, Dörner B et al (1996) Generation and use of nonsupport-bound peptide and peptidomimetic combinatorial libraries. *Methods Enzymol* 267: 220–234

52 Cuervo JH, Weitl F, Ostresh JM, Hamashin VT, Hannah AL, Houghten RA (1995) Polyalkylamine chemical combinatorial libraries. In: Maia HLS (ed): *Peptides 94: Proceedings of the 23rd European Peptide Symposium*. ESCOM, Leiden, 465–466

53 Griffith MC, Dooley CT, Houghten RA, Kiely JS (1996) Solid-phase synthesis, characterization, and screening of a 43,000-compound tetrahydroisoquinoline combinatorial library. In: Chaiken IM, Janda KD (eds): *Molecular Diversity and Combinatorial Chemistry: Libraries and Drug Discovery*. American Chemical Society, Washington, DC, 50–57

54 Pei Y, Houghten RA, Kiely JS (1997) Synthesis of 4-amino-3,4-dihydro-2(1H)-quinolines via B-lactam intermediates on the solid-phase. *Tetrahedron Lett* 38: 3349–3352

55 Nefzi A, Ostresh JM, Houghten RA (1997) Solid phase synthesis of 1,3,4,7-tetrasubstituted perhydro-1,4-diazepine-2,5-diones. *Tetrahedron Lett* 38: 4943–4946

56 Nefzi A, Giulianotti M, Houghten RA (1998) Solid phase synthesis of 2,4,5-trisubstituted thiomorpholin-3-ones. *Tetrahedron Letters* 39: 3671–3674

57 Nefzi A, Ong, NA, Giulianotti, MA et al (1999) Solid phase synthesis of 1,4-benzothiazepin-5-one derivatives. *Tetrahedron Lett* 40: 4939–4942

58 Meyer J-P, Ostresh JM, Houghten RA (1999) Combinatorial libraries of hydantoin and thiohydantoin derivatives, methods of making the libraries and compounds therein. U.S. Patent No. 5,859,190

59 Nefzi A, Ostresh JM, Giulianotti M et al (1998) Efficient solid phase synthesis of 3,5-disubstituted hydantoins. *Tetrahedron Lett* 39: 8199–8202

60 Ostresh JM, Houghten RA (1999) Combinatorial libraries of imidazol-pyrido-indole and imidazol-pyrido-benzothiophene derivatives, methods of making the libraries and compounds therein. U.S. Patent No. 5,856,107

61 Nefzi A, Giulianotti M, Houghten RA (1999) Solid phase synthesis of 1,6-disubstituted 2,3-diketopiperazines and 1,2-disubstituted piperazines from *N*-acylated amino acids. *Tetrahedron Lett* 40: 8539–8542

62 Nefzi A, Ostresh JM, Houghten RA (2001) Solid-phase synthesis of mixture-based acyclic and heterocyclic small molecule combinatorial libraries from resin-bound polyamides. *Biopolymers (Peptide Science)* 60: 212–219

63 Blondelle, SE, Crooks, E, Ostresh, JM et al (1999) Mixture-based heterocyclic combinatorial positional scanning libraries: discovery of bicyclic guanidines having potent antifungal activity against *Candida albicans* and *Cryptococcus neoformans*. *Antimicrob Agents & Chemother* 43: 106–114

64 Nefzi A, Giulianotti MA, Houghten RA (2001) Solid-phase synthesis of bis-heterocyclic compounds from resin-bound orthogonally protected lysine. *J Comb Chem* 3: 68–70

Modern Methods of Drug Discovery
ed. by A. Hillisch and R. Hilgenfeld
© 2003 Birkhäuser Verlag/Switzerland

7 Computational approaches towards the quantification of molecular diversity and design of compound libraries

Hans Matter

Aventis Pharma Deutschland GmbH, DI&A Chemistry, Molecular Modelling, Building G878, D-65926 Frankfurt am Main, Germany

7.1 Introduction

The concepts of molecular diversity and library design are increasingly important for the pharmaceutical industry due to the need to establish efficient technologies with potential for shortening the drug discovery process. This paradigm shift caused by cost pressure went along with the advent of high-throughput methods and produced an exponential increase of data from combinatorial chemistry [1] (see Chapter 6) and high-throughput screening (HTS, Chapter 4). Following this technological change, theoretical concepts for effective experimental design and data analysis were summarized under the term "*chemoinformatics*" [2], defined as "mixing of information resources to transform data into information and information into knowledge, for the intended purpose of making better decisions faster in the arena of drug lead identification and optimization".

In this Chapter computational approaches to diversity analysis, library design and comparison, compound selection, and prediction of drug-likeness will be reviewed. The similarity principle, which forms the conceptual basis for chemical diversity, is connected to molecular descriptors for an objective evaluation of diversity metrics. If valid descriptors are known, the next issue is related to the appropriate choice of a selection strategy to sample diverse or similar, but representative subsets from either existing or "virtual" compound libraries. The discussion of compound libraries then is focused on concepts and practical considerations for their design, assessment and comparison [3].

The ability to synthesize huge libraries in various formats is unprecedented in medicinal chemistry. Although combinatorial chemistry (see Chapter 6) has started with large libraries of mixtures followed by deconvolution strategies, the automated parallel or combinatorial synthesis of smaller, single compound libraries is today mainly applied and the following sections will refer to this library strategy [4]. For identification of novel bioactive compounds often a combination of historic compound collections and combinatorial libraries for high-throughput screening or focused biological testing are utilized [5]. Historic libraries are characterized by a limited number of templates from earlier projects. Hence, it is necessary to expand their structural diversity by directed acquisition of compound collection or synthesis [6].

Although combinatorial chemistry allows to synthesize large numbers of compounds, it requires rational design in order to concentrate on the "*best possible library*" in order to optimize resources for a successful discovery strategy. As experimental techniques are now sufficiently developed, it is possible to plan synthetic schemes for producing large populations of small, drug-like entities by systematic, repetitive connection of individual building blocks with varying chemical topology. Often more than 1000 reactants are commercially available for each randomization position, thus appropriate selection strategies must be applied. The identification of "redundant" compounds based on molecular diversity considerations is a key requirement, as any reduction of the number of compounds, while only reducing redundancy in a database, but not introducing any voids, should impact research efficiency [7]. Even with new miniaturization technologies (see Chapter 4), smaller subsets are essential to handle more assays in a given time frame.

However, it is not only molecular diversity, that makes a synthetically feasible combinatorial library a good choice. The integration of drug development activities into early stages of discovery [8] is increasingly important to rapidly flag those molecules, which are unlikely to become drugs. Hence, the incorporation of experimental knowledge into computational methods for filtering and prediction of "drug-likeness" is valuable permitting the rapid and cost-effective elimination of candidates prior to synthesis [9]. Here, approaches towards the prediction of drug-likeness in a general sense, as well as intestinal absorption through passive transport and the penetration of the blood-brain barrier will be presented, which are increasingly incorporated into library design, following the earlier interest in Lipinski's "rule-of-five" [10] (see also Chapter 12).

7.2 The similarity principle as basis for rational design

It was stated earlier that general screening libraries should *minimize redundancy* using reactants with *diverse structures*, while for *focused libraries* exploring the neighborhood of a lead *monomers with similar features* to individual building blocks should be employed [11]. This follows the *Similarity Principle* [12], stating that structurally similar molecules are likely to exhibit similar physicochemical and biological properties. Hence, it should be possible to predict biological properties of a structurally related compound given known activities for its nearest neighbors. Most approaches to quantify similarity [13] originate from methods like substructure and similarity searching [14], clustering [15] and quantitative structure-activity relationship (QSAR [16]). The use of very similar molecules does not enhance the probability to find novel structural classes for one target, while dissimilar molecules should enhance the probability for finding interesting leads on different targets. Such a diverse subset should cover a wide range of diverse, non-redundant compounds. Not only the risk of missing an interesting structural class should be low, but also the total number of compounds. When optimizing a hit on the other hand, the use of similar molecules is essential to establish a sound structure-activity relationship (SAR), which differs significantly from lead finding.

While this seems to be a clear concept, reality is more complicated, as there is no objective definition of "molecular diversity". One should be aware that all descriptors are derived *a priori* from molecular structures and two molecules might be similar given one descriptor, while there are dissimilar using another one. Similar molecules sometimes exhibit significant differences in biological activities due to differences in protein-ligand interactions, flexibility, binding mode, mechanism of action and others [17]. Hence, for any valid descriptor the number of exceptions from the *similarity principle* should be minimal by considering only relevant properties.

While historic databases are characterized by clusters of molecules, an optimal distribution of molecules should avoid redundancies for primary screening. An optimal selection would pick only dissimilar compounds outside of the smallest acceptable distance between two molecules (*similarity radius*), resulting in a diverse subset. Such a distribution may help to identify structurally different molecules being active on the same target, supposing that the interest lies in the differentiation of active *versus* inactive compounds regardless of a quantification of activity. In contrast, establishing a structure-activity relationship should be the subject of a lead optimization project starting from targeted libraries, enumerated within the similarity radius around known hits and enriched by known privileged structures.

7.3   Molecular descriptors to quantify diversity

For any computational application, chemical structure must be described in a relevant descriptor space for quantification of differences. Any valid descriptor must be able to group compounds with similar biological activities [18], following the similarity principle, and allow for a good coverage of the biological property space by selecting the most diverse range of structural classes following the previously identified and validated *similarity radius*. However, unlike QSAR descriptors, diversity descriptors do not relate to a common free-energy scale, which makes molecular comparisons non-intuitive [14, 19]. Selected descriptors and their validation are summarized below, while comprehensive reviews on this subject are found in the literature [20].

7.3.1   Two-dimensional descriptors

Rational synthesis planning assumes that a 2D structural diagram encodes molecular physicochemical properties and reactivity. Martin et al. demonstrated that 2D substructure descriptors indeed contain information about relevant protein-ligand interactions, like molecular shape, hydrophobicity, size, flexibility and hydrogen-bonding potential [21]. Many two-dimensional descriptors originated from 2D substructure searching systems [22], where characteristic fragments are used to rapidly screen out those candidates, which will never be able to match the 2D substructure query. As one requirement for effective substructure searching is that fragments should be statistically independent from each other and show a nearly equifrequent population, any optimal choice of descriptors is dependent on the database [23]. Following those studies, 2D fragment based descriptors were successfully applied for similarity searching and diversity studies, while similarity is quantified by evaluating the normalized number of fragments in common for two molecules. Every fragment is mapped unambiguously to a single bit within a bitstring for comparison. MACCS substructure keys [24] encode the presence of relevant 2D fragments like atom types, ring counts and augmented atoms, originally designed for substructure searching [25], they have been validated for quantification of diversity [21, 26, 27].

Although hashed 2D fingerprints [28, 29] also capture the presence of particular fragments, those are not unambiguously assigned to individual bits. For *Daylight* fingerprints all paths of predefined length in a compound are generated and mapped to several bit positions in a bitstring by a procedure known as "*hashing*," while they do not have specific bit ranges reserved for particular fragment lengths. This hashing is required due to the large number of possible fragments in a database, thus it is necessary to assign multiple fragments to each bit of the bitstring. This is achieved by a two-step procedure, as exemplified for *UNITY* fingerprints: First, each fragment is mapped to a unique integer, which then is projected to a size-limited bitstring by hashing, setting one or multiple bits to 1. For *UNITY* fingerprints, particular path lengths are assigned to separate bits in a bitstring, while they also denote the presence of predefined functional groups, rings and atoms in 60 of a total of 988 bits. In contrast to fragment fingerprints, hashed fingerprints describe structures with a similar quality independent from a predefined fragment dictionary, which might be focused towards pharma- or agro-relevant molecules. They have been originally combined with non-hierarchical clustering for similaritity-based compound evaluations [30] and in numerous diversity studies. Critical assessments of 2D fingerprint based similarity considerations are reported [31], while compact representations for storing chemical information (mini-fingerprints) have recently been applied [32].

Encoding molecular structures in the form of bitstrings allow to quantify molecular similarity using different functions, like the *Tanimoto* [33] or *Cosine* coefficients [34], while for non-bitstring descriptors the *Euclidean* distance is often used. Both coefficients are counting the number of bits in common set to 1 in slightly differing ways. The Tanimoto coefficient is widely used in database analysis, as it has properties making the work with larger datasets very efficient. A similarity coefficient of 0 means that both structures have no "1" bits in common.

Atom pair fingerprints as 2D descriptors count the shortest path of bonds between two atom types [35]. Each bit in a bitstring now corresponds to a pair of atom types separated by a fixed number of bonds. This concept has been extended to represent atoms by physicochemical properties rather then element types [36]. Based on these pharmacophoric atom types, a method for detecting meaningful common topological substructures among sets of active compounds was described based on a clique-based subgraph detection method to find the highest scoring common substructure for each pair of molecules [37]. Pharmacophoric definition groups rather than atom types have also been successfully used by others for similarity and diversity investigations [11, 38].

A conceptionally similar implementation of 2D pharmacophoric fingerprints within a 2D topological cross-correlation vector was successfully applied for identification of isofunctional molecules with significantly different backbones, thus demonstrating its potential for "*scaffold-hopping*" by topological pharmacophore searching [39]. The higher abstraction level, when introducing a pharmacophore concept into common descriptors, facilitates the identification of molecules with less similarity on the 2D structural diagram and 2D fingerprint level. Hence, for any objective assessment of a novel descriptor, not only the quantity of hits, but also their quality need to be assessed. A combination of 2D fragments with the medicinal chemistry concept of pharmacophoric groups should result in powerful and fast molecular descriptors.

2D or 3D autocorrelation vectors [40] have also been used as diversity descriptors, representing intramolecular 2D topologies or 3D distances. An autocorrelation coefficient is a sum over all atom pair properties separated by a predefined number of bonds (2D) or distance (3D), while the entire vector represents a series of coefficients for all topological or cartesian distances. This transformation converts a vector of variable length into information of fixed length for application of chemometrical comparison and evaluation approaches. Their drawback is that the transformation back into the original vector for chemical interpretation is not trivial. Atomic properties, which are mapped onto individual atoms involve hydrophobicity [41], partial atomic charges, hydrogen bonding properties and others. To reduce the number of variables and the model dimensionality prior to selection or analysis, often a principal component analysis (PCA) [42] is applied. Alternatively the autocorrelation vectors can be projected into 2D space using a Kohonen network [43] as nonlinear mapping approach. Three-dimensional autocorrelation vectors on properties based on distances calculated from 3D molecular surfaces [44] have also been applied to visually assess the diversity of different libraries [45].

BCUT values were developed by Pearlman [46] by generating an association matrix from the connection table of a particular molecule and then adding atomic charge, polarizability or hydrogen-bonding properties as diagonal elements, while non-diagonal elements encode intramolecular connectivities. For each matrix from each atomic property, the lowest and highest eigenvalue is extracted as descriptor, resulting in a low-dimensional chemistry space encompassing only six individual descriptors. The same approach has also been used to derive 3D descriptors by replacing connectivity by interatomic distance derived from a single 3D structure of a compound. These BCUT values are mainly used in combination with cell-based partitioning methods [46, 47] for compound selection and classification, while they were shown to be effective in QSAR studies [48].

Topological indices encode the molecular connectivity pattern [49] and quantify features like atom types, numbers of atoms and bonds, the extent and position of branchings and rings in a set of single-value real numbers. In addition, electro-topological state values [50], molecular shape indices [51] and topological symmetry indices are also used. A variety of different indices have been developed and successfully applied for QSAR studies, diversity studies and database comparisons [52]. Often the large number of correlated indices is condensed by principal component analysis to a few, orthogonal principal properties.

Similar to topological indices are global property molecular descriptors like size, computed logP, molecular refractivity, and key functional group counts. These descriptors characterize molecules using a single number representing the underlying physicochemical property. It was earlier reported that a basis set of six out of 49 2D descriptors are only weakly correlated to each other, suggesting its use to quantify diversity [47]. Using a genetic algorithm [53], it was possible to derive an effective set of descriptors from 111 possible ones computed from a molecule's 2D structure for classification of compounds to different biological activity classes. Interestingly it was reported that only four descriptors accounting for aromatic character, hydrogen bond acceptors, estimated polar vdW surface area, and a single structural key resulted in overall best performance, suggesting that only a few critical descriptors are preferred to partition compounds according to their activity. Those holistic descriptors are getting even more interest to be incorporated into library filter tools, as most of the properties captured within those descriptors correlate to pharmacokinetic observables or "drug-likeness" [54, 55].

7.3.2   Three-dimensional descriptors

The successful application of 3D-QSAR and protein-ligand docking for drug design suggests that biological activity and specificity is correlated with the ligand's 3D structure, as binding to a receptor or enzyme is a molecular recognition event driven by 3D complementarity (see Chapter 11). This led many groups to derive descriptors capturing the 3D nature of a ligand or the protein-ligand recognition process. There is an ongoing discussion of whether 2D or 3D descriptors are superior. Some explanation for the reported weakness of 3D descriptors is that 2D descriptors were more extensively developed. Other problems might relate to the inadequate handling of conformational flexibility, the choice of thresholds for conformations to be rejected and efficient, but fast searching strategies to evaluate conformational space [23]. Considering only a single conformation is inadequate for most classes of pharmaceutically relevant molecules. On the other hand, averaging over conformational space gives only a very rough view of what is intended to be described, namely the possible arrangement of pharmacophoric groups in cartesian space.

Three-dimensional screens, originally designed for 3D searching [22] encode interatomic distances of flexible molecules, in particular their ability to adopt a conformation with a specific spatial relationship between individual features. Distance ranges are predefined for each pair of features and each range is subdivided into a series of individual bins having a particular bin width (distance tolerance). If a distance in a particular conformation between predefined features matches to one of the bins, the corresponding bit is set to "1."

This concept was extended towards *pharmacophore keys* to consider pharmacophoric information for 3D searching [56]. Studies by Mason et al. [57], Martin et al. [21], Davies [58] and McGregor et al. [59] have reported interesting properties of pharmacophore keys. Those *"pharmacophore definition triplets"* refer to a reduction of molecules to pharmacophore groups and the recording of their spatial relationships in a set of triangles, i.e a set of three pharmacophoric points in a molecule, where multiple points can originate from similar features (e.g., a triangle acceptor-acceptor-hydrophobic). Each geometrically possible triangle with its point-to-point distance disregarding the order of specification is encoded in a fingerprint-like manner. Individual bits refer to different triangles, which can be formed between potential pharmacophoric points (e.g., acceptor-atoms, acceptor-sites, donor-atoms, donor-sites and hydrophobic centers [60]). Different approaches have in common that individual bits set to "1" encode a particular triangle geometry in pharmacophoric space. Those triplet fingerprints are typically computed for conformational ensembles to account for flexibility. Using 27 distance bins from 2.5 to 15 Å and five pharmacophoric types leads to ~300,000 bits for triangle geometries after correcting for symmetries and triangle inequalities. Often a single composite fingerprint is accumulated as union of all molecules in the database, although such a key can also be used to represent single molecules [61].

Successful applications of an extended four-point pharmacophore concept for molecular similarity and diversity have recently been reported [62]. In this approach up to seven features and 15 distance ranges are considered, leading to 350 million potential 4-point pharmacophores per molecule. The resulting pharmacophore fingerprint is reported as a powerful measure for diversity and similarity and provides a consistent frame of reference, independent from alignment of particular conformations, for comparing molecules, databases and evaluating the complementarity to a protein binding site.

Molecular steric fields were often applied as 3D shape descriptor, although a reliable superposition for the entire database is needed. Thus, topomeric fields as enhancement for comparing the diversity of a set of substituents at a common core have been proposed [63]. This descriptor is an implementation of the bioisosterism principle—similarly shaped molecules are more likely to share biological properties than other molecules—based on shape comparisons of a single rule-generated "topomer" conformation. Flexibility is accounted for depending on the shortest rotatable bond distance between any atom to the core structure. This descriptor is efficient for searching very large ($>10^{12}$) virtual libraries [64] of possible combinatorial reaction products. It is predictive for biological properties in retrospective studies, while recently a prospective trial of this descriptor was carried out by synthesis of a small, focused library resulting from searching a huge virtual library for antagonists of angiotensin II [65]. As expected the most similar compounds to the query were the most active ones.

WHIM descriptors [66] represent a different 3D approach to overcome the 3D alignment problem, as they are invariant to molecular rotations and translations. These indices capture global 3D chemical information at a molecular level in terms of size, shape, symmetry, atom distribution and electronic properties derived from cartesian coordinates. More recently these indices were enriched introducing new molecular surface properties related to hydrogen bonding capacity and hydrophobicity [67]. Many successful applications in QSAR [67, 68] and prediction of physicochemical and pharmacokinetic properties [69] have recently appeared.

7.3.3   Alternative approaches to quantify molecular similarity

The development of descriptors is still important towards an improved treatment of molecular similarity, while maintaining computation speed. Alternative approaches describing molecular similarity not on the 2D level are very valuable, even if they may show a lower performance in a quantitative evaluation, as novel structural motifs are likely to be found. Some of those approaches are summarized here.

Researchers at Telik [70] used a consistent set of ligands and determined experimental binding to a panel of proteins to describe similarity. The list of binding affinities per molecule is used as *affinity fingerprint*. When there is similarity between a target protein and proteins in the reference panel, binding affinities should also correlate due to common sub-pockets or pharmacophores. This concept was validated [71] by showing that binding data for human serum albumin correlate with three other proteins from the reference panel. Although those fingerprints can be used like conventional descriptors, they need additional experimental effort plus synthetic material for profiling. Dixon and Villar [71] used affinity fingerprints with Euclidean distances and a simple optimization algorithm to design diverse libraries from compound collections. They also show that a wide range of structurally diverse active molecules can be detected using the affinity data of a first screening experiment.

Starting from this biochemical concept, *in silicio* affinity fingerprints were computed [72] by estimating binding affinities for a series of molecules *versus* known 3D protein cavities in the reference panel using docking programs and scoring functions [73]. Although less performant than 2D techniques, this approach led to an enrichment of structurally diverse hits on the 2D level. For selection of proteins for the reference panel, optimization schemes were developed to increase predictivity of *in silicio* affinity fingerprints [74]. It was shown that this *Flexsim-X* approach is useful to detect molecules with similar biological activities from different classes without prior knowledge of the target protein structure.

Most descriptors represent 2D or 3D molecular features in a linear bitstring or vector, as such a representation allows for comparing two molecules using a simple, rapid mapping between their vectors using similarity coefficient (Tanimoto, Cosine, Euclidean distance). Hence, essential information about the topological arrangement of features is only partially conserved or lost. This led to the development of the *feature tree* descriptor [38], describing molecules by a tree structure. The nodes of this tree represent properties of individual molecular fragments, subdivided into steric and chemical features, while edges are connecting nodes in correspondence to fragments joined in the 2D chemical structure. For similarity evaluation, a mapping between parts of the trees must be computed, which is done by novel algorithms [38]. This descriptor introduces new concepts into descriptor technology. Due to its tree structure, the overall molecular structure is represented as trade-off between exactness and speed. In contrast to 2D fragment-based methods, feature trees detect similarities, if similar interaction pattern are possible, thus being a extension of the 2D topological pharmacophore approach discussed above [36, 39] with similar—and sometimes improved—performance compared to 2D fingerprints.

7.4   Subset selection approaches

7.4.1   Different types of libraries

There are several subset selection approaches from virtual libraries or compound collections, ranging from computationally demanding clustering [75] to fast maximum dissimilarity [18, 34, 76] and cell-based partitioning methods [46, 47]. A comparison of cell-based partitioning applications to similarity selections is given in [77] (see also Chapter 10), while reviews for this field have also appeared [3b,

19, 23]. Furthermore, stochastic methods like genetic algorithms and simulated annealing become increasingly popular, mainly for selection of focused libraries.

Selection strategies depend on the purpose of a library. For a novel, uncharacterized target, a generic screening library should encompass a larger number of molecules with many structural motifs. Such a generic screening library will also be used in general screening for other new targets in search for new leads. Much attention has been drawn to rational design for enhancing the structural variations in those libraries, while maintaining moderate library sizes [78]. Due to the lack of target classification, structural information and missing lead compounds, there is typically no information on constraints to be imposed during library design. Representative subsets should span the entire chemical property space of the parent library with a smaller number of preferably *drug-like* compounds. The intention to consider only *drug-like* might be valuable to introduce at least some constraints into the design procedure. Applying such a design strategy requires additional screening iterations test similar compounds (*analog libraries*) to identified hits for deriving structure-activity relationships. The ultimate goal of any rational design is to make sure that every new compound adds new *information* to the existing HTS compound collection [7]. For such databases an *optimally diverse* subset [7, 78] is designed, which include as few compounds as possible, but still is representative for the entire collection.

If some more information on the target is known, like its classification to a protein family, any design should bias the library towards structural target features and privileged structures from initial hits. When no structure-activity relationship can be generated, global similarity considerations during the selection from the underlying virtual library should be used, leading to *biased* or *targeted libraries*. Those targeted libraries might be directed towards particular protein families like metalloproteinases, kinases, ion-channels or G-protein coupled receptors (see also Chapter 10). Here, initial information could be collected from literature and corporate projects and subsequently turned into constraints plus privileged structures for library design.

With more information (3D-QSAR models [79], pharmacophore hypotheses [60], 3D protein structure), global similarity should be replaced by local similarity in the design, i.e., 3D regions or functional groups, which are correlated with biological activity should be considered. This leads to focused libraries, where available information on target and lead structures is high and thus constrains the structural space.

Subsequently, for every validated primary screening hit from rationally designed libraries, the nearest neighbors must be identified using similarity searching [80], synthesized, if not already available, and tested in a second iteration to develop an initial SAR (*structure-activity relationship*). If the nearest neighbor originates from a virtual library, there is a high probability that it can be made using the same, validated reaction protocol, which increases efficiency.

7.4.2 Maximum dissimilarity methods

For maximum dissimilarity methods new compounds are successively selected such that they are maximal dissimilar from the previously selected members of a subset [18]. There are several ways to determine the most dissimilar candidate to compounds already in the subset [81]. This process can be terminated either when a preset maximum number of compounds is selected or when no other molecules can be selected without being too similar to already selected molecules. There are several variations to this approximative technique to identify the most dissimilar subset [82]. To result in a quasi-deterministic procedure, one could select the most dissimilar compounds from all others as seed, or reject a couple of initially selected compounds after some selection cycles. A more flexible implementation of this method, the *OptiSim* algorithm, has recently been described [83]. This algorithm includes maximum and minimum dissimilarity-based selection as a special case. A parameter is used to adjust the balance between representativity and diversity in the selected compound subset [83, 84]. This approach is related to sphere exclusion algorithms, which operate by selecting a compound and excluding all other compounds more similar than a predefined threshold to that compound. This continues until all compounds within a database have been selected or rejected, which does not allow to predefine the number of compounds in a subset. There are different implementations of this concept [85], while a comparison of some maximum dissimilarity and sphere exclusion algorithms revealed the original maximum dissimilarity method [18] to be most effective in selecting compounds covering a broad variety of biological activities [82].

7.4.3 Cluster analysis

Cluster analysis [75, 86] as alternative method offers more specific control by assigning every structure to a group, which exhibit a high degree of both intracluster similarity and intercluster dissimilarity. A variety of different clustering techniques have been critically evaluated in the literature [33, 80]. There are no *a priori* guidelines for what technique will be most appropriate for a dataset, although some methods perform better for grouping similar compounds [26].

Cluster methods can be divided into hierarchical and non-hierarchical methods. The idea of hierarchical agglomerative clustering is to start with singleton clusters and merge those two clusters with minimal inter-cluster distance sequentially. This approach does not require any assumption about the number of clusters to generate, but is the computationally most expensive technique. Different methods can be used to compute distances between clusters: the distance between the closest pair of data points in both clusters (single linkage clustering), the distance between the most distant pair of data points in both clusters (complete linkage clustering), the average of all pairwise data points between two clusters, and the distance between two cluster centroids. For compound selection typically one or more structures closest to the centers are chosen to form a representative set.

Divisive hierarchical clustering methods is top-down: it starts with all compounds being members of one cluster and recursively divides it into smaller clusters.

For non-hierarchical methods, the *Jarvis-Patrick* algorithm [87] is often used, as it is much faster for larger datasets than hierarchical methods, although it was reported to be less effective than hierarchical clustering for grouping compounds with similar activity [26]. Molecules are clustered together, if they have a minimum number of nearest neighbors in common, so that this method first builds nearest neighbor-lists for each compound in the dataset. An effective cascaded clustering approach based on the *Jarvis-Patrick* approach has recently been published to overcome some of its inherent limitations by keeping the maximum cluster size and the number of produced singletons at a lower level [88].

Clustering methods are independent of the dimensionality of the underlying descriptors and it is possible to apply reasonable fast algorithms, although hierarchical clustering still gives superior results. Another drawback of clustering is that for new compounds the entire analysis has to be redone. Thus any library comparison starts by combining the reference and candidate database and redoing the entire time-consuming clustering procedure. Finally, it is not possible without an external database to identify diversity voids in an existing collection. The recently introduced stochastic clustering [89] aims to overcome some of those problems by subdividing the database into an appropriate number of clusters by identification of probe structures that fall outside a defined similarity cutoff with respect to each another. The binning of the remaining database is then done using this list of probes. This allows to directly add new compounds to the analysis. Active compounds were grouped into clusters based on an active probe six to 10 times more often than the incidence of active compounds in the entire database.

7.4.4   Cell-based methods

Cell-based or partitioning methods [46, 47, 52] are tailored to work in low-dimensional descriptor space by defining a small number of bins on each descriptor axis. Those bins are used to partition a multidimensional space into smaller subspaces (cells). Each compound to be analyzed in a virtual or existing library will be assigned to one of the cells (cubes for three dimensions, hypercubes for higher dimension) according to its descriptor values, i.e., physicochemical properties. The underlying assumption following the similarity principle is that compounds in one cell share similar biological properties. Diverse subsets might be derived by choosing one or more representative molecules from each occupied cell, while for focused libraries all molecules from a cell with the hit motif plus neighboring cells should be selected. In a recent study [90], different binning schemes were compared in terms of their ability to provide an even distribution of molecules across the chemistry space and to maximize the number of active molecules for hypothetical assay data.

This approach allows to efficiently identify unoccupied regions in descriptor space and thus guide rational library acquisition, while this is not possible using clustering or direct dissimilarity-based selection without an external database as reference frame. If particular members of a candidate library now populate empty cells, those are good candidates to fill diversity voids in the entire corporate collection. The key to meaningful partitioning is the choice of appropriate descriptors, which are able to span a low-dimensional space and the selection of a useful number of bins for partitioning. Too many cells after partitioning certainly will never be filled, while a reasonable number of bins provides a good reference for library comparison and thus diversity-driven compound acquisition. Hence, when using BCUT descriptors [46] designed for low-dimensional chemistry space, the selection, which descriptors to use for partitioning, is based on which metrics gives the most uniform distribution of compounds in space. Pearlman et al. [46] introduced the χ-squared-based *auto-choose* algorithm to tailor a chemistry space in order to best represent the diversity of a given database by appropriate selection of low-dimensional space descriptors.

7.4.5   Other selection approaches

Selection techniques from statistical experimental design have also been used for subset selection, like D-optimal design [91] to maximize diversity in reactant collections for combinatorial chemistry. This approach allows to incorporate a variety of different descriptors into the design, while it is known for D-optimal design to select molecules at the edges of the multidimensional descriptor space. Another strategy by Wold et al. [92] uses experimental design (factorial design, fractional factorial design and D-optimal design [93]) as key component for selecting diverse building blocks, while their combination should result in a diverse product library. Relevant descriptors are chosen specifically for the problem. Then a PCA (principal component analysis) is applied to reduce the dimensionality of the descriptor space, followed by statistical molecular design for selecting those building blocks, which optimally represent the chemistry space defined by PCA scores (*principal properties*). Although this approach works on building blocks rather than enumerated products, it led to diverse libraries. No efficiency difference was found comparing building block and product space selections, when critically investigating the building block space and select an appropriate number of reactants.

However, it is still an issue whether a selection in reactant or product space lead to more diverse libraries [94]. Certainly the most diverse library can be obtained by an unconstrained diversity selection on the full product matrix, while this approach violates the combinatorial scheme and thus is not very efficient, when addressing real chemistry problems. Although parallel synthesis schemes are compatible with this *pure diversity selection* after solving the logistic problems [7], a more economical solution is preferable for most synthesis schemes. Contrasting to results from Wold et al. [92], Gillet et al. [94] showed that product based design using a genetic algorithm, combinatorial constraints and a fingerprint-based library diversity measure as fitness function results in more diverse libraries than a simple reactant-based design, while the maximum diversity can only be achieved by product-based pure diversity selection disregarding the combinatorial scheme (see Fig. 1). Those conflicting results can be understood, if looking at the different set of descriptors and validation approaches employed in both studies. Another study [95] comparing the efficiency of reagent-based selections *versus* product-based selections under combinatorial constraints revealed that the advantage of working in product space indeed depends on the descriptors. While for some descriptors, product-based approaches provide an advantage, whereas for others results from reactant pools are comparable, which suggests that for each descriptor the ability to relate diversity from reagent to product space should be evaluated.

Several stochastic optimization techniques have been applied for selecting appropriate subsets mainly for focused libraries towards a particular target. Genetic algorithms [96] were utilized to identify a diverse set of compounds using topological similarity to a known lead structure as fitness function [97], while a direct optimization of experimental biological activities using a genetic algorithm was iteratively used for identification of trypsin inhibitors in a virtual library of

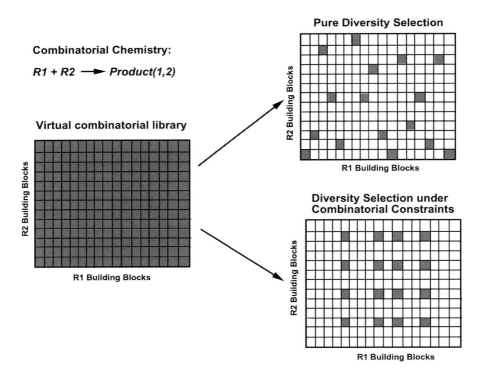

Figure 1. Comparison between pure diversity selection and selection under combinatorial constraints to obtain a representative subset [94]. Although pure diversity selection results in more diverse subsets, it is not economical, as most building blocks have to be used only once. With constrained selection a synthetically efficient and diverse combinatorial subset can be identified.

multi-component reaction products [98]. Another application is designed for efficient deconvolution of combinatorial libraries, which assumes a preclustered database [99]. The approach by Gillet et al. to select diverse combinatorial sublibraries in product space [94] was recently extended by the use of a multi-objective fitness function [100], allowing to include any number of rapidly available parameters in addition to a fingerprint-derived diversity measure. Simulated annealing as alternative stochastic approach for optimization [101] and neural networks [45] were also applied by several groups for library design.

7.5 Descriptor validation studies

This variety of methods immediately raises the question of what is useful for library design, comparison and subset selection? Any validation study to address this issue must evaluate, whether a method can discriminate between actives and inactives for a particular target. As one quantitative performance measure, the number of actives belonging to the same class than a reference compound could be used. Such an *enrichment factor* [38] is defined as number of hits in the first percent of the database sorted according to the similarity measure divided by the number of hits expected from a random selection. Intuitively, the enrichment factor describes how much is gained by using a similarity measure for selecting molecules compared to a random selection.

Several physicochemical descriptors were investigated to uncover the relationship between 2D/3D similarity and biological activity. The first study towards an objective validation of descriptors was done by Brown and Martin [26], who assessed the ability of 2D and 3D descriptors combined with hierarchical and nonhierarchical clustering techniques to separate active from inactive compounds. An *active cluster* contains at least one active compound for a particular assay, and the *active cluster subset* refers to the total number of molecules (actives and inactives) in all active clusters. Then the proportion of active structures in active clusters is computed and compared to the proportion of active structure in the entire dataset. Any increase in this proportion indicates a tendency to group active compounds. A superior behavior of 2D descriptors (MACCS keys, Daylight and UNITY fingerprints) compared to 3D descriptors was found, which partially was attributed to a lack in 3D descriptor development approaches and the problem of treating conformational flexibility. A following study demonstrated that a variety of 2D descriptors, MACCS keys in particular, implicitly contain information about physicochemical properties known to be relevant for stabilizing protein-ligand complexes [21], thus suggesting, why those simple descriptors might be successful in separating active from inactive compounds. The information content of each descriptor was assessed by its ability to accurately predict physicochemical properties from known values for related structures using similarity-based and cluster-based property prediction.

The correspondence between chemical and biological similarity was evaluated using structurally similar compounds known to act on different biological targets.

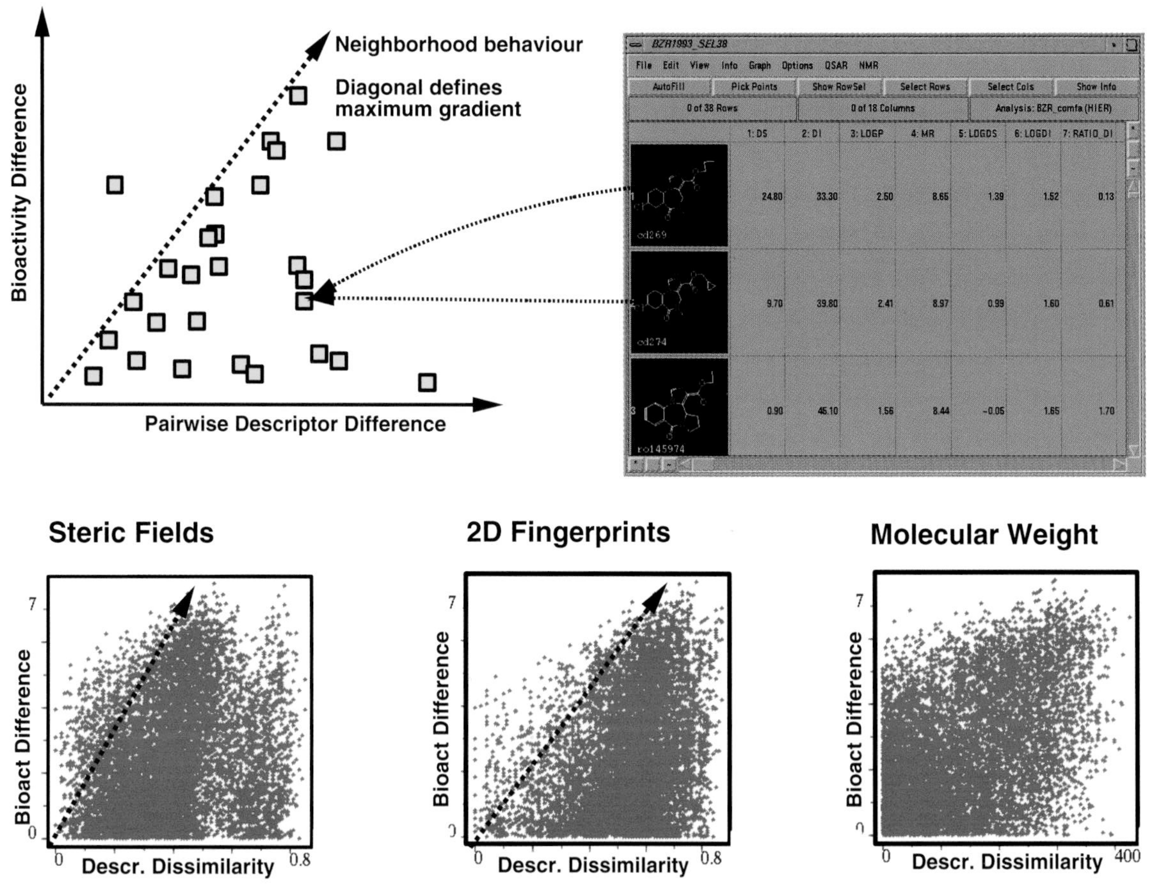

Figure 2. Generation of neighborhood plots for descriptor validation [102]. For each $n*(n-1)/2$ pairs of n molecules, the descriptor difference (1-Tanimoto coefficient) is plotted on the x-axis versus their difference in biological space $(Act_{(a)} - Act_{(b)})$. For a valid descriptor, a characteristic shape with only few outliers in the upper left triangle is observed. Examples for valid and invalid descriptors are plotted for 138 angiotensin-converting enzyme inhibitors. Left: steric fields; middle: 2D fingerprints; right: molecular weight [103].

Patterson et al. assessed the ability of 11 descriptors to group compounds from 20 representative QSAR datasets [102]. To be useful for design, a descriptor must exhibit a relationship to biological activity. Within a similarity radius of a bioactive compound there should be a high probability that any other compound belongs to the same activity class—otherwise the design is similar in result to random screening. Hence, a descriptor can be characterized using this similarity radius allowing to select minimally redundant compounds. The *neighborhood plot* as a useful graphical representation of this concept for validation was introduced, which compares differences in descriptor values with differences in biological activities for related compounds. Such a plot for a valid descriptor shows a characteristic shape with a filled lower right and an empty upper left triangle (Fig. 2). Their results show that CoMFA (Comparative Molecular Field Analysis) fields and 2D fingerprints for side chains outperform other descriptors.

In another study a broad range of 2D and 3D descriptors was validated for their ability to predict biological activity and to effectively sample structurally and biologically diverse datasets [103]. 2D fingerprints show the best performance using hierarchical clustering or maximum dissimilarity methods, while selection on clusters generated using atom-pair descriptors or a variety of 3D descriptors perform unsatisfactory. Again similarity radius for 2D fingerprints and molecular steric fields was estimated using an analysis of neighborhood plots for 10 diverse datasets. For each of the $n(n-1)/2$ pairs of n molecules per assay the pairwise differences for the descriptors and the biological differences were plotted, as shown in Figure 1 for 138 structurally diverse ACE inhibitors, revealing a characteristic shape for any valid metrics allowing to derive a maximum change for the biological activity per change in this descriptor. From the averaged gradients a similarity radius of 0.85 for 2D-fingerprints and 0.88 for steric fields was estimated, in agreement with other studies [102]. Thus two molecules with a Tanimoto coefficient larger than 0.85 should exhibit related biological activities. If a molecule A falls within the similarity radius of molecule B, it is sufficiently represented by B without loss of biological information for primary screening purposes. Of course, any hit follow-up needs to focus on neighboring compounds within the similarity radius to establish the underlying structure-activity relationship.

Similar validation studies were run to evaluate 3D PDT fingerprints [61] and MACCS substructure keys [27]. This comparison is based on their ability to cover representative biological classes from parent databases (*coverage analysis*) and the degree of separation between active and inactive compounds for a biological target from hierarchical clustering (*cluster separation analysis*). PDT fingerprints derived from a low number of conformers perform significantly better, but they are not comparable to 2D fingerprints or MACCS keys. When combining 2D and 3D descriptors with a weighting >0.5 for fingerprints, a significant improvement is observed.

The efficiency of rational design for sampling diverse subsets was compared to random selections using 2D and 3D PDT fingerprints [104]. In order to represent 90% of all biological classes from a biologically diverse dataset, 3.47 times more compounds must be randomly selected compared to the rational approach. Filling the gap between 90% and 100% biological representativity requires many more compounds. Remarkably, lower numbers of selected molecules led to almost similar coverage of biological classes for the rational and random approach.

Thus it was shown that designed subsets are superior in the sense of sampling more biological targets compared to randomly picked ensembles. Two-dimensional fingerprints as molecular descriptors were shown to be appropriate for designing subsets representing biological properties of parent databases. They perform better, when comparing the sampling properties of other descriptors carrying 2D or 3D molecular information (see Fig. 3). All studies reveal that 2D fingerprints alone or in combination with other metrics as primary descriptor allow to handle global diversity.

In a recent validation study for diversity selection approaches in combination with representativity techniques (selection of central compounds in a cluster, Kohonen neural networks, nonlinear scaling of descriptors) and descriptors (topological and 3D fingerprints), the differences between diversity and representativity was explored [105]. The authors report that only a cluster analysis operating on fingerprints or whole-molecule descriptors outperforms a random selection in a database of diverse drug compounds.

Another validation concept was introduced by Pearlman et al. [106], the *activity-seeded structure-based clustering,* which probes the assumption that active structures are grouped in chemistry space. The authors report that 74 active ACD inhibitors are found in only three clusters occupying less then 0.02% of the chemistry space of the entire MDDR database.

7.6   Comparing compound libraries

Those concepts for validating descriptors allow to apply them for design and comparison of compound libraries. Such a comparison should be based on the degree of complementarity to a *reference* database in a common reference frame, when adding a *candidate* database to enhance the number of structural motifs within a corporate collection. Such a candidate database might be an external collection of available compounds or a virtual library for combinatorial synthesis.

Database self-similarity and comparison plots based on a distribution of pairwise distances of validated descriptors between both datasets are useful tools to monitor their diversity, representativity and complementarity [27, 78]. In a typical application [78] the Tanimoto coefficient for every reference database structure to its nearest neighbor is computed using 2D fingerprints and used to generate a histogram, the maximum Tanimoto coefficient represents the closest pair of any two compounds in the dataset. For an idealized database designed using a Tanimoto coefficient of 0.85 as similarity radius (Fig. 4, far left panel) defining the boundary between similar and diverse, a self-similarity plot (Fig. 4, central left column) with a peak on the far left side is expected, while the comparison with the parent virtual library should not reveal any voids, leading to a histogram (Fig. 4, central right column) with a peak and a cut-off at on the right side of this similarity radius

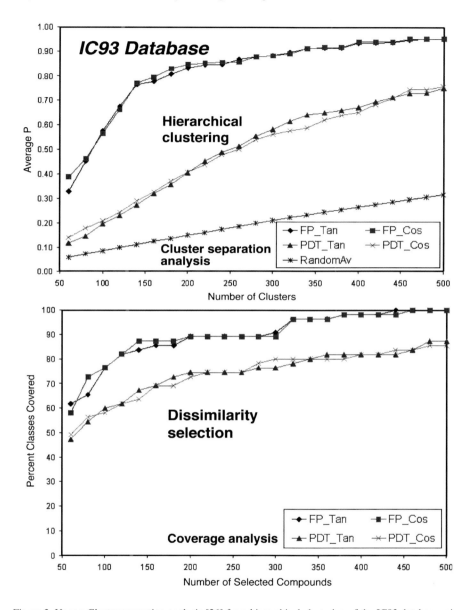

Figure 3. Upper: Cluster separation analysis [26] from hierarchical clustering of the IC93 database using 2D fingerprints (FP) and 3D pharmacophoric triplets (PDT) in comparison to an average of 10 analyses using random numbers (*RandomAv*) [61]. Tanimoto (TAN) and cosine coefficients (COS) are both evaluated. The average proportion *p* on the *y*-axis is plotted *versus* the number of clusters generated at different levels of the dendrogram. This analysis evaluates, whether a particular descriptor is able to group biologically similar compounds. Lower panel: Coverage analysis using the same database and descriptors [61]. This analysis evaluates the percent biological classes covered from the database in various subsets generated using maximum dissimilarity selection.

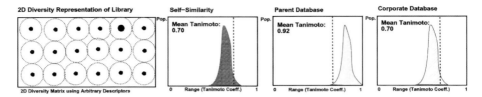

Figure 4. Database comparison histograms to illustrate a database selection generated using a sphere-exclusion algorithm with a predefined similarity radius [7, 78]. The left panel shows a simplified representation of this database as distribution of molecules (filled circles) in an arbitrary 2D molecular property space. In the middle left panel idealized self-similary histograms are shown, while the plots in the middle right panel show plots obtained by comparing the database subset to the entire database. The right column refers to plots obtained by comparison to a corporate database. Dotted lines indicate the similarity radius for exclusion.

(dotted vertical line). These tools allow to monitor complementarity to third-party databases when replacing the parent database by this additional database. A good fit is indicated by a peak of this distribution preferably on the right side of the dotted vertical line (see Fig. 4, far right column).

The mean pairwise dissimilarity as measure of similarity is also used to evaluate the diversity of databases. Although this method, based on a centroid algorithm for computing the intermolecular distances, allows to rapidly monitor the diversity change after merging two datasets, it does not provide an indication about overlap or complementarity. This measure is used for library design using 2D fingerprints [94, 100] and for comparison based on 3D pharmacophore bit-strings [107].

For information on the degree of complementarity, methods like cluster analysis of combined or separated databases can also be applied. One approach separately computes representative subsets from both databases using *Jarvis-Patrick* clustering and then combines both subsets for a second analysis. This allows to detect clusters, which are only populated by representative molecules from the candidate database, providing an indication of the overlap between both sets.

Other strategies involve principal component analysis to condense descriptors like topological indices [52] or 2D autocorrelation functions [40] into a set of characteristic principal properties for visualization and comparison of database overlap and complementarity. This allows to plot the scores from a candidate set in the PCA reference frame of a reference corporate database.

Any comparison of database could also be done on the level of fingerprints directly. There have been some reports on generating a database key using an appropriate logical combination of 2D [11] or 3D PDT fingerprints [61, 58]. Fingerprints representing different databases can be compared on the basis of similarity indices like the Tanimoto coefficient, while coverage and overlap could be directly assessed by bitwise comparison of the database keys. However, database fingerprints for any coverage analysis are only applicable in a meaningful way, if a single 2D fragment descriptor or 3D pharmacophore triangle geometry is mapped to only one bit.

Another approach is based on *spatial diversity* and evaluates the number of different functional groups within a certain 3D distance for a single or two databases [108]. Other ways for comparing databases and detecting complementary subsets involve profiling and categorizing based on a list of predefined 2D substructure features [109], ring systems, frameworks [110] or property distributions of relevant physicochemical descriptors, for which optimal ranges have been identified from statistical analysis of drug reference databases [55].

7.7   Design of combinatorial libraries

During the actual design, new compounds should be unique and thus add useful information to the library tailored for a particular purpose (primary screening library, targeted library or focused library). Such a library should exhibit a wide coverage of the physicochemical property space, any member should be *drug-like* to add biological relevance to the screening collection. For designing a primary screening library the use of two very similar compounds does not enhance the chance to find different types of biological activity or novel structural motifs related to biological activity for the same target. It would be advantageous to replace one with a more dissimilar candidate, which indeed is the motivating factor for the design of diverse compound libraries [7, 78] using dissimilarity selection, sphere exclusion, clustering, cell-based selection or other validated selection techniques. Hence, diversity-based selections are utilized to choose an optimal generic library from the possible virtual library, in some cases under combinatorial constraints to increase synthetic efficiency. In contrast, reverse selection strategies need to be utilized for targeted libraries, while similar descriptors and algorithms could be applied. For example, after clustering one would typically choose all compounds in the same cluster than an initial motif. Similarity searching around hits aids to design biased or focused libraries, which could be followed by a procedure to generate an optimal combinatorial library. The most detailed level of information can be incorporated by scoring a virtual library with 2D or 3D-QSAR models, pharmacophore models, or empirical protein-ligand scoring functions [111] (see Chapter 10), which is time-consuming and requires filtering strategy to reduce the virtual library size [112]. The fastest library design method today based on similarity searching of 3D shapes in combinatorial libraries has been reported [63–65]. Thus, as more information is known about a target, as more the choice of descriptors and selection techniques is shifted towards successful QSAR or structure-based design descriptors, techniques and methods for introducing a bias into the library.

Another point is to evaluate, which of many candidate library is more diverse or better applicable for a particular screening application. Here, validated quantitative measures of diversity are routinely applied to assess diversity, representativity plus complementarity to any corporate compound collection. It must be decided whether it is more effective to manufacture a sublibrary variation of an existing library or to establish a new synthesis scheme with a novel scaffold.

Library design typically starts with a new synthetic route or an accessible scaffold, followed by reaction validation to assess the compatibility of potential building blocks with the synthetic scheme. The undesirability of certain reagents, often from reactant databases like ACD (Available Chemical Directory), is defined by the presence of substructures that are known to have toxic side-effects and/or metabolically hazardous [113] as well as by the presence of groups that could interfere with or prevent the reaction. Reaction validation information has to be considered to exclude reaction specific unwanted substructures based on the likelihood to byproducts, their inertness, reactivity of non-reacting functional groups, stability of products, or solubility. Rejected and accepted compounds are stored in different files with physicochemical descriptions, such that a later selection is possible at every stage. This is useful, if at a later stage an educt is not longer commercially available. Such a refined assortment of reactive building blocks in accord with a reaction protocol now is the essential starting point for design, while the final acquisition and maintenance of a building block collection is a resource-intensive activity. Hence, such a collection will be used more than once by integrating them into the design of different libraries around alternative scaffolds with alternative reaction schemes to end up with a diverse portfolio of complementary libraries.

During the library planning phase one should decide whether to use commercially available reagents only or to incorporate proprietary building blocks from custom synthesis, which requires an initial effort to establish a tailored collection. Many researchers suggest to incorporate building blocks derived from historic medicinal chemistry knowledge (*privileged structures*) [114], which certainly depends on library type and motivation, the information about the biological target and the logistics and availability of suitable reactants.

If a product-based approach is used, acceptable reagents are combined to a virtual library. The theoretical number of individual compounds is given by the number of building blocks per step plus the number of steps. Efficient searching first rejects compounds using unwanted substructure queries and property ranges for molecular weight, computed logP, number of hydrogen bond donors, acceptors and others (see Chapter 12). Then an assessment of the drug-likeness of individual members could be applied, before diversity considerations are taken into account, again depending on the type of library to be generated.

7.8 Design of drug-like libraries

The consideration of drug-likeness in the library design process is gaining increasing importance and is likely to add more impetus to pure diversity based selections [8–10, 115, 116]. The rapid identification of candidates, which are unlikely to become drugs, would have a tremendous impact on research efficiency. It has been reported that only 10% of those compounds entering drug development finally become marketed drugs, while 40% fail due to their pharmacokinetic profile [117]. There is a high interest in computational prediction tools as complement to established *in vitro* ADME (**a**bsorption, **d**istribution, **m**etabolism, **e**xcretion)

assays like Caco-2 cell monolayers [118] for membrane permeability or hepatocytes or liver microsomes to predict possible metabolic instabilities [119], because
any *in vitro* profiling requires material, while *in-silico* methods allow the rapid filtering of *virtual* compound libraries prior to synthesis. For a review on computer-
aided prediction of drug toxicity and metabolism with respect to the design of
drug-like screening libraries see Chapter 13.

The importance of logP, the octanol/water partition coefficient, for distribution
of drugs and approaches towards the estimation of this parameter have recently
been reviewed [120], while a computational study [121] on the comparison of
logP prediction methods revealed the superiority of fragmental over atom-based
and molecular property approaches. This and other simple physicochemical
parameters were first used by Lipinski et al. [10] in a systematic comparison of
several orally absorbed drugs *versus* molecules, which are not expected to be
absorbed. A set of very simple filters to categorize compounds for their ability to
undergo possible transport only, known as *"rule-of-five,"* resulted from this study.
This rule to classify compounds for their ability to oral absorption had a major
impact on medicinal chemistry and library design (see Chapter 12).

Complementary approaches to a retrospective analysis of drugs involve the
generation of quantitative structure-property relationships (QSPR) using regression based statistical methods. For the majority of drugs the preferred route of
administration is oral absorption. Oral bioavailability refers to the fraction of the
oral dose, which reaches systemic circulation, influenced by intestinal absorption
plus metabolism in the membrane or liver. Those approaches utilize theoretical or
experimental descriptors capturing global properties related to lipophilicity, partition coefficients, surface areas, hydrogen bonding, pKa, electrostatics and others.

Recently, the importance of a simple, but physicochemically relevant parameter, the polar surface area (PSA), has been discovered in a variety of models to
predict *in vitro* membrane permeability, intestinal absorption or blood-brain barrier penetration. The PSA is computed based on the sum of van-der-Waals or sol-
vent-accessible surface areas of oxygen and nitrogen atoms including polar hydrogens, it correlates to molecular hydrogen bonding and donating properties
[122–125]. A first study [123] on 17 diverse drugs resulted in a regression model
correlating Caco-2 membrane permeabilities with molecular weight and PSA
only, showing that an increase in PSA is detrimental for absorption. A following
study [126] using MolSurf descriptors from quantum chemical calculations in
combination with PLS also suggests that a reduced hydrogen bonding potential is
favorable for absorption.

An extension led to the dynamic PSA_d [122, 127], where multiple conformations of one molecule are averaged during the PSA calculations. However, single
conformer derived PSA were able to give almost similar results, which allows to
apply this rapid descriptor for library filtering [125, 124]. Another study by Palm
et al. [127] evaluated human absorption data for a set of 20 passively transported
drugs covering a diverse range of human fractional absorption data (Fig. 6). A
strong sigmoidal correlation between the dynamic PSA and the human fractional
absorption was found. From this and another study [124], an upper PSA threshold

Nordiazepam

99.0 % mean fractional absorption

Lactulose

0.6 % mean fractional absorption

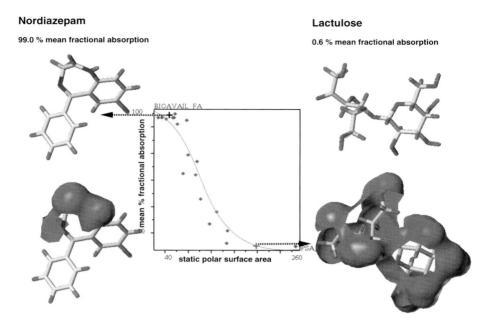

Figure 5. Sigmoidal correlation between human fractional absorption and dynamic polar surface area for 20 diverse drugs (data taken from [127]). The influence of the polar surface area on fractional absorption is illustrated plotting the small static PSA for the oral available Nordiazepam (left) *versus* the large area for the non-available Lactulose (right).

of 140 A^2 for oral absorption was proposed, while compounds with a PSA lower than 60 A^2 are found to favorable absorption. Other studies using PSA involve the successful prediction of blood-brain barrier penetration [128], 125] and the use of the non-PSA in addition to establish predictive models [129]. Those studies led to the conclusion that the polar molecular surface is the dominating determinant for oral absorption and brain penetration of drugs that are transported via the transcellular route. Some library design strategies now include filtering using PSA descriptors to enhance the hit-to-lead properties for focused libraries [130].

Recently, a novel set of alignment-free 3D descriptors named *VolSurf* was developed, referring to several calculated molecular properties extracted from 3D grid maps of molecular interaction energies with particular probe atoms [131]. These descriptors have been successfully correlated with bioavailability, blood-brain partitioning, membrane transport and other properties [132, 133]. The general conclusions of these models for factors influencing permeability and absorption are in agreement with those reported above. However, due to the nature of the descriptors this approach allows to better understand the underlying physico-chemical requirements for a pharmacokinetic effect and use it for further design, for example the balance between lipophilic and hydrophilic parts in combination with size, volume and other effects.

Another study on a larger set of 86 structurally diverse drugs by Wessel et al. [134] uses six 3D descriptors in combination with three topological descriptors to predict the fractional absorption of molecules. The 3D descriptors involved the charged partial surface area (CPSA). One caveat here is that the dataset encompasses drugs being transported actively and passively, while it is unlikely that one model is useful for those fundamentally different processes.

The quantum chemically derived MolSurf descriptors have also been correlated to human intestinal absorption [135] and blood-brain partitioning [136]. Other approaches towards the prediction of human fractional absorption have been published [137–139], while only a limited amount of experimental data is available today. Hence, it remains to be investigated whether those conclusions can be extended to structurally more diverse compounds, while they are certainly useful during lead optimization and focused library design.

There have been some interesting approaches to predict the *drug-likeness* in a more general sense either by an empirically compiled list of undesirable fragments [113], a detailed analysis of the importance of simple global molecular properties by means of a genetic algorithm [54], the existence of frequently populated molecular frameworks in drug databases [110], or the use of scoring schemes to discriminate between drugs and non-drugs. Two groups at BASF [140] and Vertex [141] independently reported neural networks to account for nonlinearities to recognize drug-like molecules based on existing drug databases like CMC or WDI. Using global descriptors combined with MACCS keys [24] a Bayesian neural network was trained, which is able to classify 90% CMC and 89% ACD (Available Chemical Directory) compounds correctly as drugs or non-drugs, respectively [141]. In addition, simple structural parameters like extended atom types [41] have been used to train a feed-forward neural network, which classified 77% of the WDI (World Drug Index) and 83% of the ACD correctly, respectively. In a subsequent study, decision trees in combination with extended atom types have also been successfully applied to discriminate between drugs (WDI) and non-drugs (ACD) [142]. One advantage of the latter study is that the structural origin for classification was identified. The authors report that just by evaluating the presence of hydroxyl, tertiary or secondary amines, carboxyl, phenol or enol groups almost 75% of all drugs were classified correctly, while non-drugs were characterized by their aromatic nature in combination with a general low content of functional groups except halogens.

Another approach towards designing drug-like libraries is based on the concept of multilevel chemical compatibility between a candidate molecule and an entire drug database as a measure of drug-likeness for that candidate molecule [143]. A systematic comparison of the atomic environment in a molecule to those of existing drugs is used as basis for discrimination and for library filtering. Interestingly, a convergent number of unique structural types were found in the analysis of libraries of existing drugs, suggesting that the discovery of a drug with a novel atomic local environment is not very likely.

All these approaches have shown to be effective in providing a reasonable ranking of compounds in a virtual or existing library, so that synthetic and screening resources can be focused on compound subsets of general biological interest. However, the definition of the term *drug-likeness* is not clear and there are certainly many compounds, which could not easily be assigned to either of both categories. Furthermore those classifications are merely reflecting the characteristics of existing drugs, which might impede the discovery of novel structural motifs.

Thus, alternative approaches to the empirical definition of *drug-like* molecules were investigated using database profiling approach [55] with respect to their pharmacologically relevant physicochemical properties, global molecular descriptors and particular chemical functionalities. Acceptable ranges for several of properties could be extracted, covering, for example, more than 80% of drugs in CMC, thus showing that acceptable ranges for *drug-like* molecules indeed can be defined. Those ranges are intended to be used in future design of combinatorial libraries. A more comprehensive analysis of a broader variety of drug databases like MDDR, Current Patents Fast-alert, CMC, Physician Desk Reference, New Chemical Entities and ACD as control was recently performed [55], showing that although the *rule-of-five* is not able to sufficiently discriminate between drugs and non-drugs, it was possible to extract acceptable ranges for 70% of all drugs for number of hydrogen bond donors (0–2), acceptors (2–9), rotatable bonds (2–8) and rings (1–4), while other ranges have been found for ACD database members. Those investigations also demonstrate that a reference frame together with ranges for acceptable compounds can be defined and certainly will influence library design and compound selection in a similar way as the well-known *rule-of-five*.

7.9   Conclusions

There is growing interest to augment the structural and biological diversity in databases and combinatorial libraries in order to optimize resources in today's discovery strategies. Many approaches towards the design and analysis of combinatorial libraries based on molecular diversity selection emerged, utilizing validated descriptors and selection techniques. However, one should be aware that similarity and dissimilarity strongly depends on the descriptors for a particular task. Hence, any conclusion about similarity is only valid for a particular descriptor space. For some applications, 2D similarity defined by the presence of functional groups according to a 2D structural drawing is sufficient, while for others, related motifs in 3D should be identified. 2D substructure information is important for valid descriptors, while it was shown that design outperforms a random selection. The question 2D *versus* 3D descriptors has not been answered yet, while the development of 3D descriptors is an active research field. Hence, objective descriptor validation and comparison methods are needed for evaluating different approaches. Some were highlighted: the sampling of biological classes for diverse library design, enrichment factors and neighborhood plots. This implicitly means that it is not possible to define a universal set of representative molecules working for all biological targets and that it is difficult to derive any "optimal" number of diverse compounds for successful high-throughput screening strategies.

This chapter summarized approaches in the field of primary screening library design and designing a collection to explore a target or optimize some hits. Practicability and automation of the design process are other key issues for a successful use of combinatorial chemistry. However, finding a potent inhibitor is only the first step, as there are more properties expected for a viable drug candidate. Many researchers became aware that the early screening for those properties in combination with computational prediction of parameters related to pharmacokinetics is crucial to lower the rate of compounds, which are unlikely to survive later development stages due to their pharmacokinetic profile. This paradigm shift of activities towards earlier discovery phases should have a critical impact on research efficiency.

As many research groups today have gained experience with library design and production, it becomes possible to produce tailored libraries for lead identification and optimization directed towards biological targets using a broad variety of approaches. The critical importance of computational chemistry and chemoinformatic approaches here is without question. Now the high-quality analysis of a huge number of hopefully high-quality datapoints generated by today's screening technologies remains a challenge for the next years to extract valuable knowledge for future design.

7.10 Acknowledgements

The author thanks K.H. Baringhaus, C. Giegerich, T. Naumann (Aventis), T. Pötter (Bayer), R. Vaz (Aventis), W. Guba (Hoffmann-La Roche), B. Wendt, R.D. Cramer, D.E. Patterson (Tripos), G. Cruciani (Univ. Perugia) and M. Rarey (Univ. Hamburg) for many stimulating discussions on various issues covered in this chapter.

7.11 References

1 Gordon EM, Kerwin JF (eds) (1998) *Combinatorial chemistry and molecular diversity in drug discovery.* Wiley, New York
2 Brown FK (1998) Chemoinformatics: what is it and how does it impact drug discovery. *Annu Rep Med Chem* 33: 375–384
3 Martin EJ, Critchlow RE, Spellmeyer DC et al (1998) Diverse approaches to combinatorial library design. In: van der Goot H (ed): *Trends in drug research II.* Elsevier, 133–146
4 Kubinyi H (1998) Combinatorial and computational approaches in structure-based design. *Curr Opin Drug Disc Dev* 1: 16–27
5 Ash JE, Warr WA, Willett P (eds) (1997) *Chemical information systems.* Ellis Horwood, Chichester
6 Warr WA (1997) Combinatorial chemistry and molecular diversity. An overview. *J Chem Inf Comput Sci* 37: 134–140
7 Ferguson AM, Patterson DE, Garr C et al (1996) Designing chemical libraries for lead discovery. *J Biomol Screen* 1: 65–73
8 Smith DA, van de Waterbeemd H (1999) Pharmacokinetics and metabolism in early drug discovery. *Curr Opin Chem Biol* 3: 373–378
9 Clark DE, Pickett SD (2000) Computational methods for the prediction of "drug-likeness". *Drug Disc Today* 5: 49–58
10 Lipinski CA, Lombardo F, Dominy BW et al (1997) Experimental and computational approaches to

estimate solubility and permeability in drug discovery and development settings. *Adv Drug Delivery Rev* 23: 3–25

11 Martin EJ, Blaney JM, Siani MA, et al (1995) Measuring diversity: experimental design of combinatorial libraries for drug discovery. *J Med Chem* 38: 1431–1436

12 Maggiora GM, Johnson MA (eds) (1990) *Concepts and applications of molecular similarity.* Wiley, New York

13 Bures MG, MartinYC (1998) Computational methods in molecular diversity and combinatorial chemistry. *Curr Opin Chem Biol* 2: 376–380

14 Willett P, Barnard JM, Downs GM (1998) Chemical similarity searching. *J Chem Inf Comput Sci* 38: 983–996

15 Willett P (ed) (1987) *Similarity and clustering in chemical information systems.* Letchworth, Research Studies Press

16 Hansch C, Leo A (eds) (1995) *Exploring QSAR: Fundamentals and applications in chemistry and biology.* Am Chem Soc, Washington, DC

17 Kubinyi H (1998) Similarity and dissimilarity: a medicinal chemist's view. In: Kubinyi H, Folkers G, Martin YC (eds): *3D-QSAR in drug design* Vol 2. Kluwer, Dordrecht, 225–252

18 Lajiness MS (1997) Dissimilarity-based compound selection techniques. *Perspect Drug Disc Design* 7/8: 65–84

19 Martin YC, Brown RD, Bures MG (1998) Quantifying diversity. In: Gordon EM, Kerwin JF (eds): *Combinatorial chemistry and molecular diversity in drug discovery.* Wiley, New York, 369–388

20 Brown RD (1997) Descriptors for diversity analysis. *Perspect Drug Disc Design* 7/8: 31–49

21 Brown RD, Martin YC (1997) The information content of 2D and 3D structural descriptors relevant to ligand-receptor binding. *J Chem Inf Comput Sci*, 37: 1–9

22 Barnard JM (1993) Substructure searching methods – old and new. *J Chem Inf Comput Sci* 33: 572–584

23 Gillet VJ (1999) Computational aspects of combinatorial chemistry. In: Miertus S, Fassina G (eds) *Comb Chem Technol.* Dekker, 251–274

24 ISIS/Base 2.1.3., Molecular Design Ltd, 14600 Catalina Street, San Leandro, CA 94577

25 Adamson GW, Cowell J, Lynch MF, et al (1973) Strategic considerations in the design of a screening system for substructure searches of chemical structure files. *J Chem Doc* 13: 153–157

26 Brown RD, Martin YC (1996) Use of structure-activity data to compare structure-based clustering methods and descriptors for use in compound selection. *J Chem Inf Comput Sci* 36: 572–584

27 Matter H, Rarey M (1999) Design and diversity analysis of compound libraries for lead discovery. In: Jung G (ed): *Combinatorial chemistry.* Wiley-VCH, Weinheim, 409–439

28 UNITY Chemical Information Software, Tripos Inc, 1699 S Hanley Road, St Louis, MO 63144, USA

29 Daylight Chemical Information Systems, Inc, 3951 Claremont Street, Irvine, CA 92714

30 Willett P, Winterman V, Bawden D (1986) Implementation of non-hierarchical cluster analysis methods in chemical information systems: Selection of compounds for biological testing and clustering of substructure search output. *J Chem Inf Comput Sci* 26: 109–118

31 Godden JW, Xue L, Bajorath J (2000) Combinatorial preferences affect molecular similarity/diversity calculations using binary fingerprints and Tanimoto coefficients. *J Chem Inf Comput Sci* 40: 163–166

32 Xue L, Godden JW, Bajorath J (1999) Database searching for compounds with similar biological activity using short binary bit string representations of molecules. *J Chem Inf Comput Sci* 39: 881–886

33 Willett P, Winterman V (1986) Comparison of some measures for the determination of intermolecular structural similarity. *Quant Struct-Act Relat* 5: 18–25

34 Holliday JD, Ranade SS, Willett P (1995) A fast algorithm for selecting sets of dissimilar structures from large chemical databases. *Quant Struct-Act Relat* 14: 501–506

35 Sheridan RP, Nachbar RB, Bush BL (1994) Extending the trend vector: The trend matrix and sample-based partial least squares. *J Comput Aided Mol Des* 8: 323–340

36 Sheridan RP, Miller MD, Underwood DJ et al (1996) Chemical similarity using geometrical atom pair descriptors. *J Chem Inf Comput Sci* 36: 128–136

37 Sheridan RP, Miller MD (1998) A Method for visualizing recurrent topological substructures in sets of active molecules. *J Chem Inf Comput Sci* 38: 915–924

38 Rarey M, Dixon JS (1998) Feature trees: a new molecular similarity measure based on tree matching. *J Comput Aided Mol Des* 12: 471–490

39 Schneider G, Neidhart W, Giller T et al (1999) "Scaffold-Hopping" by topological pharmacophore search: a contribution to virtual screening. *Angew Chem* 111: 3068–3070

40 Moreau G, Turpin C (1996) Use of similarity analysis to reduce large molecular libraries to smaller

sets of representative molecules. *Analusis* 24: M17-M22

41 Ghose A, Crippen G (1986) Atomic physicochemical parameters for three-dimensional structure-directed quantitative structure-activity relationships. 1. Partition coefficients as a measure of hydrophobicity. *J Comp Chem* 7: 565–577

42 Wold S, Albano C, Dunn WJ et al (1984) Multivariate data analysis in chemistry. In: Kowalski BR (ed): *Chemometrics: mathematics and statistics in chemistry*. NATO, ISI Series C 138, Reidel Publ Co, Dordrecht, 17–96

43 Kohonen T (ed) (1989) *Self-organization and associative memory*. Springer, Berlin

44 Wagener M, Sadowski J, Gasteiger J (1995) Autocorrelation of molecular surface properties for molecular corticosteroid binding globulin and cytosolic Ah receptor activity by neural networks. *J Am Chem Soc* 117: 7769–7775

45 Sadowski J, Wagener M, Gasteiger J (1996) Assessing similarity and diversity of combinatorial libraries by spatial autocorrelation functions and neural networks. *Angew Chem* 34: 23–24

46 Pearlman RS, Smith KM (1998) Novel software tools for chemical diversity. *Perspect Drug Disc Des* 9: 339–353

47 Lewis RA, Mason JS, McLay IM (1997) Similarity measures for rational set selection and analysis of combinatorial libraries: The diverse property-derived (DPD) approach. *J Chem Inf Comput Sci* 37: 599–614

48 Stanton DT (1999) Evaluation and use of BCUT descriptors in QSAR and QSPR studies. *J Chem Inf Comput Sci* 39: 11–20

49 Kier LB, Hall LH (eds) (1976) *Molecular connectivity and drug research*. Academic Press, New York

50 Hall LH, Mohney B, Kier LB (1991) The electrotopological state: structure information at the atomic level for molecular graphs. *J Chem Inf Comput Sci* 31: 76–82

51 Gombar VK, Jain DVS (1987) Quantification of molecular shape and its correlation with physico-chemical properties. *Ind J Chem* 26A: 554–555

52 Cummins DJ, Andrews CW, Bentley JA et al (1996) Molecular diversity in chemical databases: comparison of medicinal chemistry knowledge bases and databases of commercially available compounds. *J Chem Inf Comput Sci* 36: 750–763

53 Xue L, Bajorath J (2000) Molecular descriptors for effective classification of biologically active compounds based on principal component analysis identified by a genetic algorithm. *J Chem Inf Comput Sci* 40: 801–809

54 Gillet VJ, Willett P, Bradshaw J (1998) Identification of biological activity profiles using substructural analysis and genetic algorithms. *J Chem Inf Comput Sci* 38: 165–179

55 Oprea TI (2000) Property distribution of drug-related chemical databases. *J Comp-Aided Mol Des* 14: 251–264

56 Sheridan RP, Nilikantan R, Rusinko A et al (1989) 3DSEARCH: a system for three-dimensional substructure searching. *J Chem Inf Comput Sci* 29: 255–260

57 Pickett SD, Mason JS, McLay IM (1996) Diversity profiling and design using 3D pharmacophores: Pharmacophore-derived queries (PDQ). *J Chem Inf Comput Sci* 36: 1214–1223

58 Davies K (1996) Using pharmacophore diversity to select molecules to test from commercial catalogues. In: Chaiken IM, Janda KD (eds): *Molecular diversity and combinatorial chemistry: Libraries and drug discovery*. Am Chem Soc, Washington DC, 309–316

59 McGregor MJ, Muskal SM (1999) Pharmacophore fingerprinting. 1. Application to QSAR and focused library design. *J Chem Inf Comput Sci* 39: 569–574

60 Martin YC, Bures MG, Danaher EA et al (1993) A fast new approach to pharmacophore mapping and its application to dopaminergic and benzodiazepine agonists. *J Comput Aided Mol Des* 7: 83–102

61 Matter H, Pötter T (1999) Comparing 3D pharmacophore triplets and 2D fingerprints for selecting diverse compound subsets. *J Chem Inf Comput Sci* 39: 1211–1225

62 Mason JS, Morize I, Menard PR et al (1999) New 4-point pharmacophore method for molecular similarity and diversity applications: Overview of the method and applications, including a novel approach to the design of combinatorial libraries containing privileged substructures. *J Med Chem* 42: 3251–3264

63 Cramer RD, Clark RD, Patterson DE et al (1996) Bioisosterism as a molecular diversity descriptor: Steric fields of single "topomeric" conformers. *J Med Chem* 39: 3060–3069

64 Cramer RD, Patterson DE, Clark RD et al (1998) Virtual compound libraries: A new approach in decision making in molecular discovery research. *J Chem Inf Comput Sci*, 38: 1010–1023

65 Cramer RD, Poss MA, Hermsmeier MA et al (1999) Prospective identification of biologically active structures by topomeric shape similarity searching. *J Med Chem* 42: 3919–3933

66 Todeschini R, Gramatica P (1998) New 3D molecular descriptors. The WHIM theory and QSAR applications. *Perspect Drug Discovery Des* 9/10/11: 355–380

67 Bravi G, Wikel JH (2000) Application of MS-WHIM descriptors: 1. Introduction of new molecular surface properties and 2. Prediction of binding affinity data. *Quant Struct-Act Relat* 19: 29–38

68 Ekins S, Bravi G, Binkley S et al (1999) Three and four dimensional-quantitative structure activity relationship (3D/4D-QSAR) analyses of CYP2D6 inhibitors. *Pharmacogenetics* 9: 477–489

69 Bravi G, Wikel JH (2000) Application of MS-WHIM descriptors: 3. Prediction of molecular properties. *Quant Struct-Act Relat* 19: 39–49

70 Kauvar LM, Higgins DL, Villar HO et al (1995) Predicting ligand-binding to proteins by affinity fingerprinting. *Chemistry & Biology* 2: 107–118

71 Dixon SL, Villar HO (1998) Bioactive diversity and screening library selection via affinity fingerprinting. *J Chem Inf Comput Sci* 38: 1192–1203

72 Briem H, Kuntz ID (1996) Molecular similarity based on dock-generated fingerprints. *J Med Chem* 39: 3401–3408

73 Böhm H-J (1994) The development of a simple empirical scoring function to estimate the binding constant for a protein-ligand complex of known three-dimensional structure. *J Comput Aided Mol Des* 8: 243–256

74 Lessel UF, Briem H (2000) Flexsim-X: A method for the detection of molecules with similar biological activity. *J Chem Inf Comput Sci* 40: 246–253

75 Barnard JM, Downs GM (1992) Clustering of chemical structures on the basis of two-dimensional similarity measures. *J Chem Inf Comput Sci* 32: 644–649

76 Lajiness M, Johnson MA, Maggiora GM (1989) Implementing drug screening programs by using molecular similarity methods. In: Fauchere, JL (ed): *QSAR: Quantitative structure-activity relationships in drug design*. Alan R Liss Inc, New York, 173–176

77 Van Drie JH, Lajiness MS (1998) Approaches to virtual library design. *Drug Disc Today* 3: 274–283

78 Matter H, Lassen D (1996) Compound libraries for lead discovery. *Chim Oggi* 14: 9–15

79 Kubinyi H (ed) (1993) *3D-QSAR in drug design theory, methods and applications*. ESCOM, Leiden

80 Willett P, Barnard JM, Downs GM (1998) Chemical similarity searching. *J Chem Inf Comput Sci* 38: 983–996

81 Holliday JD, Willett P (1996) Definitions of "disimilarity" for dissimilarity-based compound selection. *J Biomolecul Screen* 1: 145–151

82 Snarey M, Terrett NK, Willett P et al (1997) Comparison of algorithms for dissimilarity-based compound selection. *J Mol Graphics* 15: 372–385

83 Clark RD (1997) OptiSim: an extended dissimilarity selection method for finding diverse representative subsets. *J Chem Inf Comput Sci* 37: 1181–1188

84 Clark RD, Langton WJ (1998) Balancing representativeness against diversity using optimizable K-dissimilarity and hierarchical clustering. *J Chem Inf Comput Sci* 38: 1079–1086

85 Hudson BD, Hyde RM, Rahr E et al (1996) Parameter based methods for compound selection from chemical databases. *Quant Struct-Act Relat* 15: 285–289

86 Downs GM, Willett P (1994) Clustering of chemical structure databases for compound selection. In: Van de Waterbeemd H (ed): *Advanced computer-assisted techniques in drug discovery* Vol 3. VCH, Weinheim, 111–130

87 Jarvis RA, Patrick EA (1973) Clustering using a similarity measure based on shared nearest neighbours. *IEEE Trans Comput*, C22: 1025–1033

88 Menard PR, Lewis RA, Mason JS (1998) Rational screening set design and compound selection: Cascaded clustering. *J Chem Inf Comput Sci* 38: 497–505

89 Reynolds CH, Druker R, Pfahler LB (1998) Lead discovery using stochastic cluster analysis (SCA): A new method for clustering structurally similar compounds. *J Chem Inf Comput Sci* 38: 305–312

90 Bayley MJ, Willett P (1999) Binning schemes for partition-based compound selection. *J Mol Graphics Modell* 17: 10–18

91 Mitchell TJ (1974) An algorithm for the construction of "D-optimal" experimental designs. *Technometrics* 16: 203–210

92 Linusson A, Gottfries J, Lindgren F et al (2000) Statistical molecular design of building blocks for combinatorial chemistry. *J Med Chem* 43: 1320–1328

93 Box GEP, Hunter WG, Hunter JS (eds) (1978) *Statistics for experimenters* Wiley, New York

94 Gillet VJ, Willett P, Bradshaw J (1997) The effectiveness of reactant pools for generating structurally diverse combinatorial libraries. *J Chem Inf Comput Sci* 37: 731–740

95 Jamois EA, Hassan M, Waldman M (2000) Evaluation of reagent-based and product-based strategies

in the design of combinatorial library subsets. *J Chem Inf Comput Sci* 40: 63–70

96 Weber L (1998) Applications of genetic algorithms in molecular diversity. *Drug Disc Today* 3: 379–385

97 Sheridan RP, Kearsley SK (1995) Using a genetic algorithm to suggest combinatorial libraries. *J Chem Inf Comput Sci* 35: 310–320

98 Weber L, Wallbaum S, Broger C et al (1995) Optimization of the biological activity of combinatorial libraroes by a genetic algorithm. *Angew Chem* 34: 2281–2282

99 Brown RD, Martin YC (1997) Designing combinatorial library mixtures using a genetic algorithm. *J Med Chem* 40: 2304–2313

100 Gillet VJ, Willett P, Bradshaw J et al (1999) Selecting combinatorial libraries to optimize diversity and physical properties. *J Chem Inf Comput Sci* 39: 169–177

101 Agrafiotis DK (1997) Stochastic algorithm for molecular diversity. *J Chem Inf Comput Sci* 37: 841–851

102 Patterson DE, Cramer RD, Ferguson AM et al (1996) A useful concept for validation of molecular diversity descriptors. *J Med Chem* 39: 3049–3059

103 Matter H (1997) Selecting optimally diverse compounds from structure databases: A validation study of 2D and 3D molecular descriptors. *J Med Chem* 40: 1219–1229

104 Pötter T, Matter H (1998) Random or rational design? Evaluation of diverse compound subsets from chemical structure databases. *J Med Chem* 41: 478–488

105 Bayada DM, Hamersma H, van Geerestein VJ (1999) Molecular diversity and representativity in chemical databases. *J Chem Inf Comput Sci* 39: 1–10

106 Pearlman RS, Smith KM (1999) Metric validation and the receptor-relevant subspace concept. *J Chem Inf Comput Sci* 39 28–35

107 Pickett SD, Luttmann C, Guerin V et al (1998) DIVSEL und COMPLIB – Strategies for the design and comparison of combinatorial libraries using pharmacophoric descriptors. *J Chem Inf Comput Sci* 38: 144–150

108 Boyd SM, Beverley M, Norskov L et al (1995) Characterizing the geometric diversity of functional groups in chemical databases. *J Comput Aided Mol Des* 9: 417–424

109 Bemis GW, Murcko MA (1999) Properties of known drugs. 2.Side chains. *J Med Chem* 42: 5095–5099

110 Murcko MA, Bemis GA (1996) Properties of known drugs. 1. Molecular frameworks. *J Med Chem* 39: 2887–2893

111 Hirst JD (1998) Predicting ligand binding energies. *Curr Opin Drug Disc Dev* 1: 28–33

112 Walters WP, Stahl MT, Murcko MA (1998) Virtual screening – an overview *Drug Disc Today* 3: 160–178

113 Rishton GM (1997) Reactive compounds and *in vitro* false positives in HTS. *Drug Disc Today* 2: 382–384

114 Lewell XQ, Judd DB, Watson SP et al (1998) RECAP-retrosynthetic combinatorial analysis procedure: A powerful new technique for identifying privileged molecular fragments with useful applications in combinatorial chemistry. *J Chem Inf Comput Sci* 38: 511–522

115 Walters WP, Ajay Murcko MA (1999) Recognizing molecules with drug-like properties. *Curr Opin Chem Biol* 3: 384–387

116 Blake JF (2000) Chemoinformatics - predicting the physicochemical properties of "drug-like" molecules. *Curr Opin Biotechnol* 11: 104–107

117 Prentis RA, Lis Y, Walker SR (1988) Pharmaceutical innovation by the seven UK-owned pharmaceutical companies (1964–1985). *Br J Clin Pharmacol* 25: 387–396

118 Delie F, Rubas WA (1997) A human colonic cell line sharing similarities with enterocytes as a model to examine oral absorption: Advantages and limitations of the Caco-2 model. *Crit Rev Ther Drug Carrier Syst* 14: 221–286

119 Eddershaw PJ, Dickens M (1999) Advances in *in vitro* drug metabolism screening. *Pharm Sci Technol Today* 2: 13–19

120 Buchwald P, Bodor N (1998) Octanol-water partition: searching for predictive models. *Curr Med Chem* 5: 353–380

121 Mannhold R, Cruciani G, Dross K et al (1998) Multivariate analysis of experimental and computational descriptors of molecular lipophilicity. *J Comput Aided Mol Des* 12: 573–581

122 Palm K, Luthman K, Ungell A-L et al (1998) Evaluation of dynamic molecular surface area as predictor of drug absorption: Comparison of other computational and experimental predictors. *J Med Chem* 41: 5382–5392

123 van de Waterbeemd H, Camenisch G, Folkers G et al (1996) Estimation of Caco-2 cell permeability using calculated molecular descriptors. *Quant Struct-Act Relat* 15: 480–490

124 Clark DE (1999) Rapid calculation of polar molecular surface area and its application to the prediction of transport phenomena. 1. Prediction of intestinal absorption. *J Pharm Sci* 88: 807–814

125 Kelder J, Grootenhuis PDJ, Bayada DM et al (1999) Polar molecular surface as a dominating determinant for oral absorption and brain penetration of drugs. *Pharm Res* 16: 1514–1519

126 Norinder U, Osterberg T, Artursson P (1997) Theoretical calculation and prediction of Caco-2 cell permeability using MolSurf parametrization and PLS statistics. *Pharm Res* 14: 1786–1791

127 Palm K, Stenberg P, Luthman K et al (1997) Polar molecular surface properties predict the intestinal absorption of drugs in humans. *Pharm Res* 14: 568–571

128 Clark DE (1999) Rapid calculation of polar molecular surface area and its application to the prediction of transport phenomena. 2. Prediction of blood-brain penetration. *J Pharm Sci* 88: 815–821

129 Stenberg P, Luthman K, Ellens H et al (1999) Prediction of intestinal absorption of endothelin receptor antagonits using three theoretical methods of increasing complexity. *Pharm Res* 16: 1520–1526

130 Pickett SD, McLay IM, Clark DE (2000) Enhancing the hit-to-lead properties of lead optimization libraries. *J Chem Inf Comput Sci* 40: 263–272

131 Cruciani G, Pastor M, Clementi S (2000) Handling information from 3D grid maps for QSAR studies. In: Gundertofte K, Jorgensen FS (eds): *Molecular modelling and prediction of bioactivity.* Plenum Press, New York

132 Guba W, Cruciani G (2000) Molecular field-derived descriptors for the multivariate modelling of pharmacokinetic data. In: Gundertofte K, Jorgensen FS (eds): *Molecular modelling and prediction of bioactivity.* Plenum Press, New York

133 Alifrangis LH, Christensen IT, Berglund A, et al (2000) Structure-property model for membrane partitioning of oligopeptides. *J Med Chem* 43: 103–113

134 Wessel MD, Jurs PC, Tolan JW et al (1998) Prediction of human intestinal absorption of drug compounds from molecular structure. *J Chem Inf Comput Sci* 38: 726–735

135 Norinder U, Österberg T, Artursson P (1999) Theoretical calculation and prediction of intestinal absorption of drugs in humans using MolSurf parametrization and PLS statistics. *Eur J Pharm Sci* 8: 49–56

136 Norinder U, Sjöberg P, Österberg T (1998) Theoretical calculation and prediction of brain-blood partitioning of organic solutes using MolSurf parametrization and PLS statistics. *J Pharm Sci* 87: 952–959

137 Sugawara M, Takekuma Y, Yamada H et al (1998) A general approach for the prediction of the intestinal absorption of drugs: Regression analysis using the physicochemical properties and drug-membrane electrostatic interactions. *J Pharm Sci* 87: 960–966

138 Winiwarter S, Bonham NM, Ax F, et al (1998) Correlation of human jejunal permeability (*in vivo*) of drugs with experimentally and theoretically derived parameters. A multivariate data analysis approach. *J Med Chem* 41: 4939–4949

139 Ghuloum AM, Sage CR, Jain AN (1999) Molecular Hashkeys: A novel method for molecular characterization and its application for predicting important pharmaceutical properties of molecules. *J Med Chem* 42: 1739–1748

140 Sadowski J, Kubinyi H (1998) A scoring scheme for discriminating between drugs and nondrugs. *J Med Chem* 41: 3325–3329

141 Ajay Walters WP, Murcko MA (1998) Can we learn to distinguish between "drug-like" and "nondruglike" molecules? *J Med Chem* 18: 3314–3324

142 Wagener M, van Geerestein VJ (2000) Potential drugs and nondrugs: prediction and identification of important structural features. *J Chem Inf Comput Sci* 40: 280–292

143 Wang J, Ramnarayan K (1999) Toward designing drug-like libraries: A novel computational approach for prediction of drug feasibility of compounds. *J Comb Chem* 1: 524–533

Modern Methods of Drug Discovery
ed. by A. Hillisch and R. Hilgenfeld
© 2003 Birkhäuser Verlag/Switzerland

8   The role of protein 3D-structures in the drug discovery process

Alexander Hillisch[1] and Rolf Hilgenfeld[2]

[1] *EnTec GmbH, Adolf Reichwein Str. 20, D-07745 Jena, Germany*
[2] *Institut für Molekulare Biotechnologie e.V., Beutenbergstr. 11, D-07745 Jena, Germany*

8.1   Introduction

The majority of drugs available today were discovered either by chance observations or by screening synthetic or natural products libraries. In many cases, a trial-and-error based approach of chemical modification of lead compounds led to an improvement with respect to potency and reduced toxicity. Since this approach is labor and time-intense, researchers in the pharmaceutical industry are constantly developing methods to increase the efficiency of the drug finding process. Simply put, two directions have evolved from these efforts. The "random" approach involves the development of high-throughput screening assays and the testing of a large number of compounds. Combinatorial chemistry is used to satisfy the need for huge substance libraries. The "rational," structure-based approach relies on an iterative procedure of structure determination of the target protein, prediction of hypothetical ligands by molecular modeling, specific chemical synthesis and biological testing of compounds (the structure-based drug design cycle). It is becoming evident, that the future of drug discovery does not lie in one of these approaches solely, but rather an intelligent combination. In this chapter, we will concentrate on the protein structure-based drug discovery approach and discuss possible overlaps with complementary technologies.

8.2   Techniques for protein structure determination and prediction

The three methods contributing most to experimental as well as theoretical protein structure information are x-ray crystallography, nuclear magnetic resonance (NMR) spectroscopy and comparative modeling. The following part will provide a short overview of the underlying principles, their particular strengths and weaknesses.

8.2.1   X-ray crystallography

The vast majority of experimentally determined protein structures were solved using x-ray crystallography. The method allows to obtain images of molecules embedded in a crystal lattice environment (see Fig. 1a). A prerequisite for the application of x-ray crystallography is the availability of a pure, homogenous protein preparation in multi-milligram quantities. Crystals of at least 0.15 to 0.2 mm in size are then grown by applying micro-dialysis, or vapor diffusion techniques such as the hanging or sitting drop methods. Since the conditions under which crystallization occurs cannot be predicted *a priori,* several parameters like the precipitating agent, pH, ionic strength, protein concentration, temperature etc. must be varied in order to produce crystals. This proverbial search for the needle in the haystack often is the rate-limiting step in structure determination by x-ray crystallography. However, crystals are essential since single molecules can neither be "seen" by x-rays nor handled. A significant amplification of the diffraction image is obtained only if molecules are oriented in the three-dimensional periodic array of the crystal. After characterization of the crystal quality and symmetry, diffraction data are collected from the crystal. To this end, the crystal is exposed to a collimated beam of monochromatic x-rays and the resulting diffraction pattern is recorded by a recording device, such as an image plate, a multiwire area detector, or a charge-coupled device (CCD) camera. The entire diffraction data set obtained consists of multiple images with x-ray intensities that are diffracted at various angles from the crystal. When both the amplitude and the phases of the diffracted waves are available, it is possible to compute a three-dimensional electron density map of the molecule under study, by applying the Fourier transformation. The amplitudes can be calculated directly from the obtained diffraction intensities whereas the phases cannot directly be measured. These are derived by one of three methods: i) isomorphous replacement: heavy atom compounds (mer-

---

Figure 1. a) Schematic representation of the process of protein structure determination by x-ray crystallography. The structure elucidation of A21Gly-B31,B32Arg2 insulin (HOE 901, Lantus®), a long-acting insulin derivative, is depicted [161]. b) Schematic representation of the process of protein structure elucidation by multidimensional NMR spectroscopy. The determination of the solution structure of the plasminogen-activator protein staphylokinase is shown as an example [162]. Two NMR-active nuclei (A, red and B, blue) are depicted in the 3D-spectrum, with the set of distance restraints (green lines) and the final structure. c) Schematic representation of the process of protein structure prediction by comparative modeling. Structure modeling of the bacterial transcriptional repressor CopR is shown [163]. Although the model is based on a fairly low-sequence identity of only 13.8% to the P22 c2 repressor, several experimental methods support this prediction.

a) X-ray crystallography

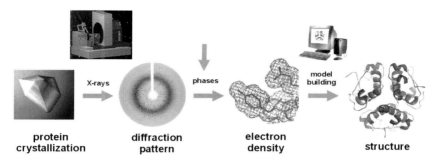

b) NMR spectroscopy

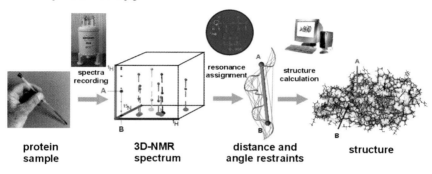

c) Comparative modeling

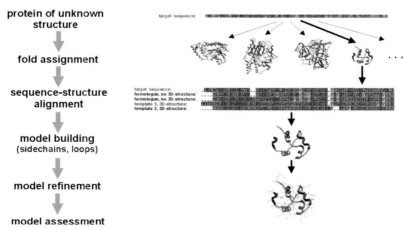

cury, gold, or other salts) are soaked into the crystal lattice; these heavy atom derivatives of the native protein crystal give rise to specific changes in the intensities of the diffracted x-rays, from which the phases can be estimated, ii) multi-wavelength anomalous dispersion (MAD) phasing: the so-called anomalous diffraction of atoms such as selenium or bromine is explored for phase determination by collecting diffraction data at different wavelengths at a synchrotron; these atoms are incorporated into the native protein by replacing methionine by selenomethionine, or simply by soaking potassium bromide into the crystals; iii) molecular replacement: the atomic coordinates of a homologous protein with a known three-dimensional structure are used to derive the orientation of the molecule under study in the unit cell of the crystal. This latter approach is the method of choice for determining the structure of a protein with a series of different ligands, frequently used in structure-based drug design.

After successful phase determination, the polypeptide chain is modeled into the electron density according to the known amino acid sequence of the protein. An iterative process of adjusting the model and recomputing the phases leads to a gradual improvement of the agreement between the observed intensities and those calculated from the model. This process is called refinement and leads to the final structural model of the biopolymer.

In contrast to other methods of structure determination, diffraction techniques such as x-ray crystallography provide a direct image of the molecule under study, in an exact mathematical process where the Fourier transformation can be compared to the lens of a microscope. The major disadvantage of the method is the need for crystals, which can be sometimes difficult or even impossible to obtain. In the past, there have occasionally been concerns that contacts between neighboring molecules in the crystal might influence the conformation of proteins and lead to artifacts. However, meanwhile we know that this is extremely rare, and if it occurs, this phenomenon is restricted to the chain termini or to flexible loops on the surface of proteins, which are typically remote from the active sites. [1, 2]. The reason is that protein crystals typically contain 35–70% solvent (mostly water) and molecules in the crystal are highly hydrated, behaving as if in aqueous solution and maintaining the catalytic activity. Comparisons between proteins determined with x-ray crystallography and NMR-spectroscopy generally show good agreement [3-5]. A disadvantage of the method is, that unless the resolution (smallest interplanar spacing of diffraction pattern for which data have been collected) is less than 1.2 Å, the positions of hydrogen atoms cannot be resolved by x-ray crystallography and must be modeled. Also the exact orientation of Asn, Gln and His is usually only inferred from the protein environment of the side chain (i.e., hydrogen bonds to neighboring amino acids).

8.2.2   NMR spectroscopy

NMR spectroscopy is a technique that allows structures and dynamics of proteins to be determined in aqueous solution (for reviews see [6]). Intramolecular distances and torsion angles in molecules are measured and translated into structure information using computer programs (see Fig. 1b). As with x-ray crystallography, multi-milligram amounts of pure, homogenous protein are needed. In addition, ^{13}C and ^{15}N labeled protein [7] should be available and the protein must be stable up to 30–40°C and soluble to a concentration of ~1 mM.

The principle of NMR is based on the quantum mechanical property of nuclear spin. The magnetic moments of certain nuclei occurring in biomacromolecules such as ^{1}H, ^{31}P and, after labeling, ^{13}C, ^{15}N, adopt a parallel or anti-parallel orientation when placed in an external magnetic field provided by an NMR spectrometer. These nuclei precess about the external magnetic field with the so-called Larmor frequency. The different orientations (parallel *versus* anti-parallel) of these "small bar magnets" in the external field correspond to different energy states, which are unequally distributed according to Boltzmann's distribution law. Spins with parallel orientation are of slightly lower energy than those with the anti-parallel one, leading to a net magnetization in the external magnetic field. This net magnetization can be measured in an NMR experiment. In order to induce transitions between the different energy states and hence a net magnetization, the spins in an NMR sample are irradiated with a radio-frequency that corresponds to the Larmor frequency. The resulting magnetization, inducing an electric current in the detection coil of the NMR spectrometer, returns to its equilibrium condition. This signal is called *free induction decay* (FID) and can be recorded as a function of time. The time-domain NMR signal is converted into the frequency domain by applying a Fourier transformation (FT). A one-dimensional NMR spectrum is obtained by plotting the absorbed irradiation *versus* the frequency of the irradiation. The utility of NMR originates from the fact that the frequency at which absorption occurs is not only dependent on the type of nucleus, but also on the molecular environment.

The protein structure determination with NMR spectroscopy typically involves four steps: 1) recording of NMR spectra, 2) assigning of all resonances in the spectrum (which means establishing correlations between atoms in the protein and resonance peaks in the NMR spectrum), 3) determination of distances and angles between atoms in the protein (restraints) and 4) calculation of the three-dimensional structure from a set of distance and angle restraints (see Fig. 1b).

The major difficulty in studying the three-dimensional structures of biopolymers by means of NMR spectroscopy is spectral overlap (respectively resolution). A peptide of 50 amino acids has approximately 300 protons which can result in a one-dimensional proton NMR spectrum with more than 600 peaks. Assignment of one-dimensional spectra becomes impossible in this case. Solutions to this problem are using higher magnetic field strength (up to 900 MHz proton resonance frequencies) and spreading the spectrum into two or more dimensions. A two-dimensional NMR spectrum can be obtained by applying, for example, two radio-frequency pulses temporally separated by a certain interval. The magnetization of a nucleus, resulting from the first pulse, is transferred to a second nucleus and is again allowed to oscillate in the static magnetic field at a frequency characteristic of the second nuclear spin. The signal is recorded as a function of two time domains. A two-dimensional absorption nuclear spin *versus* frequency spectrum is obtained by applying a two-dimensional Fourier transformation. This concept has been extended by dispersing the NMR signals along three and four frequency axes (three- and four dimensional NMR spectroscopy) using additional signals from isotopically labeled NMR active nuclei such as ^{13}C and ^{15}N.

Structure information originates mainly from two different types of 2D NMR experiments:

NOESY (nuclear Overhauser enhancement spectroscopy) is used to detect cross-relaxation processes between protons. Three radio frequency pulses are applied in these experiments. During the second and third pulse, the so-called mixing time, magnetization is transferred between different spins by cross-relaxation. For this dipole-dipole interaction, the amount of transferred magnetization is dependent on the sixth power of the distance between the nuclei. Distances between protons of up to 5 Å can be measured by integrating the volume of the corresponding cross-peak in the spectrum. This distance information between as much as possible protons is used to generate sets of NOE-restraints necessary for 3D structure determination.

COSY (correlation spectroscopy) is a 2D NMR experiment in which cross-peaks arise between protons correlated through scalar J-coupling interactions. These "through-bond" interactions occur between nuclei separated by one or more covalent bonds. At least two radio-frequency pulses are applied. Nuclei that are J-coupled exchange magnetization. This can be expressed in terms of coupling constants, where the magnitude of the coupling constant varies sinusoidally with the torsion angle between the two protons. An empirical equation, the so-called Karplus relation [8] can be employed to convert coupling constants into torsion angle information. Sets of torsion angle restraints, aiding to determine the three-dimensional structure of molecules, can be achieved with these NMR experiments.

In multidimensional NMR-spectroscopy NOE experiments are combined with correlation experiments between ^1H, ^{13}C and ^{15}N nuclei. Dispersing the NMR signals along three and four frequency axes leads to better resolved signals and enables to investigate even larger proteins.

Restraints on proton-proton distances from NOEs and torsion angles from scalar couplings are used along with structural information (sequence, standard bond lengths and angles) to calculate the three-dimensional structure of biological macromolecules. A combination of distance geometry with subsequent restraint molecular mechanics or molecular dynamics calculations are used to obtain several similar conformers that represent the solution structure of the molecule under study (for reviews see [9, 10]).

In contrast to x-ray crystallography, NMR does not require crystals to be grown and is thus applicable to flexible molecules as well. Dynamic processes occurring over a wide range of time scales (spanning 10 orders of magnitude) can be examined in solution. This includes protein folding and dynamics. Therefore, NMR spectroscopy and x-ray crystallography are not competing techniques, but rather complement each other [11]. A certain drawback of NMR is the limitation concerning the size of molecules that can be studied. *De novo* protein structure determination with NMR is possible today for proteins with molecular weights up to about 30 kDa. However, recent developments such as TROSY (transverse relaxation-optimized spectroscopy) [12] and CRINEPT (cross-correlated relaxation-enhanced polarization transfer) [13, 14] have proven useful for studying multimeric proteins with molecular weight >100 kDa. Progress in the development of spectrometers with higher magnetic field strengths of up to 21 Tesla (corresponding to 900 MHz proton resonance frequencies) will also contribute to NMR structure determination.

8.2.3 Comparative modeling

Comparative or homology modeling utilizes experimentally determined protein structures to predict the conformation of another protein with similar amino acid sequence. The method relies on the observation that in nature, structure is more conserved than sequence and small or medium changes in the sequence normally result only in small changes of the three-dimensional structure [15]. Homologuous proteins are proteins related by evolutionary processes of divergence from a common ancestor.

Model building
Generally, homology modeling involves four steps: fold assignment, sequence alignment, model building and model refinement (see Fig. 1c). The first step aims at identifying proteins with a known three-dimensional structure (so called template structures) which are related to the polypeptide sequence with unknown structure (the target sequence). A sequence database of proteins with known structures (e.g., the PDB-sequence database) is searched with the target sequence using sequence similarity search algorithms. Programs like BLAST [16] and FASTA [17] are mainly used for these purposes (for a description of sequence searches see Chapter 3). These sequence-based methods are complemented with so-called fold recognition or threading methods that rely explicitly on the known structures of the template proteins (see also Chapter 3). Once a clear similarity (>30% sequence identity) of the target protein with one of the proteins with known 3D structure is identified, both sequences are aligned. This is to establish the best possible correspondence between residues in the template and the target sequence. Extending this pairwise alignment by considering further homologues in a multiple sequence alignment normally improves the quality of the final structure prediction. In the third step a model of the target protein is built by exchanging amino acids in the 3D structure of the template protein and introducing insertions and/or deletions according to the sequence alignment.

Loop modeling
Loops at the surface of proteins are generally less conserved and can adopt different conformations in even closely related homologues. The goal of loop mod-

eling is to predict the conformation of the backbone fixed at both ends of structurally conserved regions. Two general approaches exist. *Ab initio* methods
involve the generation of a large number of possible loop conformers and the subsequent evaluation with respect to energetic and other criteria [18-20]. In the second knowledge-based approach, a database of loop fragments in experimentally
determined protein structures (subset of PDB, see below) is searched to fit on the
fixed backbone endpoints. The fragments are identified either on the basis of
sequence relationship or of geometric criteria such as the distance between the
amino and carboxyl termini of the conserved regions flanking the loop in the
model [21-24].

Modeling of amino acid side-chains

Side-chain conformations of the modeled protein are predicted either from similar structures, from proteins in general, or from steric or energetic considerations
(for a review on various approaches to side-chain modeling see [25]). Most programs for side-chain prediction are based on rotamer libraries, which are obtained
by statistical analysis of side-chain torsional angles for preferred backbone conformations of particular amino acid side-chains using many high-resolution x-ray
structures. It has been observed that side-chain dihedral angles are dependent on
the protein backbone conformation [26]. Dunbrack and colleagues have constructed a library giving the probabilities for side-chain rotamers that depend on
the main-chain (Φ, Ψ) values and the type of amino acid [27]. This widely used
library is implemented in the program SCWRL [27, 28]. Using a large data set of
high-resolution protein structures, an improved rotamer library was constructed
allowing the elimination of an increased number of higher energy rotameric states
[29]. Another type of library is characterized by the inclusion of a third angle (χ_1,
the torsion angle over the C_α–C_β bond) derived from the analysis of known protein structures, and resulted in trivariate (Φ, Ψ, χ_1) angle distributions for each
amino acid type [30]. A Monte Carlo approach for sampling conformational space
of protein side-chains at a fixed backbone using a trivariate rotamer library is
described by Shenking et al. [31]. Lovell et al. published a rotamer library with
improved detection of internal van der Waals clashes between backbone and side-
chains, increasing the accuracy of predictions [32].

Generally, the side-chain prediction accuracy depends on the degree to which
the protein backbone is known. Thus core residues (conserved backbone) are in
general better predicted than residues at the surface or in loop regions. This has
some implications for drug design, since many ligand binding pockets occur in
regions of protein families with conserved sequence and conserved structure. The
prediction of exact side-chain conformations can be crucial for homology model-
based drug design and the design of selective ligands.

Model refinement

After loop modeling and adjusting the side-chains, the resulting homology model
is subjected to an energy minimization protocol using force field methods to
remove remaining steric clashes. Molecular dynamic simulations using explicit

solvent environments are also applied for these purposes and can help characterize the plasticity of binding pockets. Finally, the model is evaluated by considering stereochemical (e.g., Ramachandran plot, χ dihedral angles, H-bond geometries) and energetic (e.g., van der Waals clashes) criteria. Several programs, such
as PROCHECK [33] and WHATCHECK [34], are used for these purposes (see
also [35]). Model refinement and evaluation are often applied iteratively to obtain
the final model.

In addition, model evaluation helps to decide what structure information can be
drawn from the model. For drug design purposes, the exact positions and rotamers
of the amino acids defining the binding pocket are often crucial. The necessary
accuracy can be achieved with current homology modeling programs for target
sequences showing >50% sequence identity to a template protein for which an
experimental high-resolution structure is available. However, even models based
on 30–50% sequence identity can yield important hints for structure-activity-relationships of ligands, if the binding pocket is structurally conserved. Caution
should be paid with models based on lower sequence similarity. Due to the inherent uncertainties, such models can be misleading with respect to ligand design.

The particular advantage of homology models is the short time needed to construct such models. Preliminary models may be built and "refined" within a few
hours. Of course, and this is the major disadvantage, one is restricted to proteins
which show relatively high sequence identity to experimental structures, and sur-
prises with respect to conformational changes with even small sequence differences cannot be excluded.

8.3   Sources for protein 3D-structure information

The development of effective protein expression systems and major progress in
instrumentation for structure determination have contributed to an exponential
growth in the number of experimental protein 3D structures (see Fig. 2a). By
October 2001 the Protein Data Bank (PDB) contained about 14,700 experimental
protein structures for approximately 6200 different proteins (proteins with less
than 95% sequence identity). A recent analysis of all protein chains in the PDB
shows that these proteins can be grouped into 1699 protein families comprising
648 unique protein folds [36] (and updates at the internet page: http://scop.mrc-
lmb.cam.ac.uk). More than seven new entries are added to the database daily,
leading to an annual growth of about 25%. Most of the structures (82%) in the
PDB are determined by x-ray crystallography and 15% by NMR spectroscopy
Other techniques like theoretical modeling (2%) cryo electron microscopy (11
structures), neutron diffraction (nine structures), electron diffraction (eight structures) and fluorescence resonance energy transfer (one structure) contribute as
well. In this database 89% of the structures are proteins or peptides (46%
enzymes), 4% are protein/nucleic acid complexes, 6% nucleic acids and only 18
structures (~0.1%) of larger carbohydrates are known. Experimental information
on 3500 unique ligands (small organic molecules and ions) bound to more than

a)

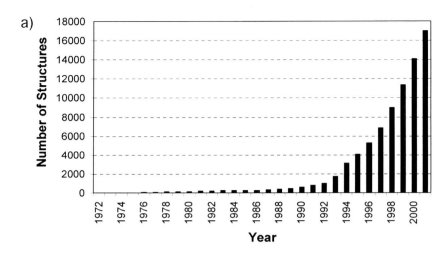

b)

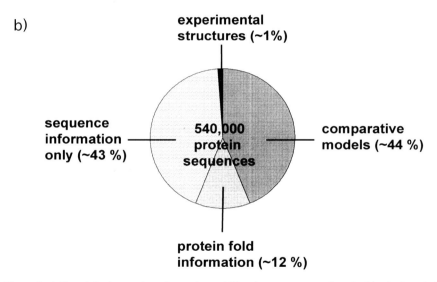

Figure 2. a) Growth in the number of experimental biopolymer structures deposited in the Protein Data Bank (PDB). By early 2002, about 17,000 structures of biopolymers were publically accessible in the PDB. b) Structure information on 540,000 protein sequences in the Swiss-Prot and TrEMBL databases. For about 1% of these sequences (~6200 different proteins), experimental structure information (x-ray, NMR) is available. Comparative models can be built for up to 44% (237,000) and protein fold information can be obtained for an additional ~12% of the sequences. The remaining 43% of the proteins actually cannot be assigned any fold or other structure information yet.

50,000 different binding sites are covered by this database and can be analyzed using, for example, programs like ReliBase [37] and PDBsum [38] or the HIV Protease Database [39].

Although the experimental structure database is growing rapidly, there is a huge gap between the number of highly annotated sequences (540,000 different sequences in Swiss-Prot (release 39) and TrEMBL (release 15)) and known structures (6200 proteins, see above). The number of known sequences is about 15 Mio. (Gen Bank, NCBI, see chapters 1 and 3) This gap can partly be filled with homology models. For example, the querieable database MODBASE provides access to an enormous number of annotated comparative protein structure models [40]. The program PSI-BLAST (see Chapter 3) was used to assign protein folds to all 540,000 unique sequence entries in Swiss-Prot and TrEMBL (see above). For 44% of these sequences, comparative models could be built using the program MODELLER [41]. Thus, by early 2002, 237,000 3D-structure models of proteins are accessible via the internet page http://pipe.rockefeller.edu. The models are predicted to have at least 30% of their Cα atoms superimposed within 3.5 Å of their correct positions. Similar results were obtained within the 3DCrunch Project [42]; 211,000 sequences were submitted for homology modeling using the automated comparative protein modeling server SWISS-MODEL and led to 85,000 structure models (~40%). This database of annotated homology models is accessible at http://www.expasy.org/swissmod. It should be realized that the level of accuracy, especially for the "low sequence identity" models is, in many cases, not sufficient for a detailed structure-based ligand design.

Thus, for 40–44% of all highly annotated sequences homology models with varying quality can be generated (see Fig. 2b). These comparative models have already led, and will continue to do so, to an enormous increase in structure information for drug target proteins.

8.4 Structural genomics

It is highly likely that structure determination of target proteins will undergo the same development that DNA sequencing and chemical synthesis and testing of compounds (see Chapters 4 and 6) underwent several years ago. The linear approach to structure elucidation will in many cases be replaced by massively parallel efforts for protein production [43], purification, crystallization [44], crystal handling, data collection and analysis [45], spectra interpretation and model generation [46]. These structural genomics efforts, which aim at solving or modeling all (or a large fraction) of the soluble proteins in an organism [47], will lead to an enormous increase in structure information for drug target proteins. Several companies and publicly funded projects which focus on structural genomics have been initiated during recent years [48]. In many of these projects, it is planned to select the proteins for experimental structure determination based on sequence analysis [49]. Those proteins which are predicted to have sufficiently novel structures from sequence analyses are preferred targets for protein expression, purification and structure determination. Remaining proteins not covered by experimental structure determination methods will be predicted by homology modeling. The experimental techniques will solve new structures while homology modeling fills the gaps in structure space. Comparative models which are built on an overall sequence identity of >30% tend to have >85% of the C_α atoms within 3.5 Å of their correct position [50]. Considering the stronger structural conservation in active-site or binding-niche regions, even some of these models can be suitable for predicting ligand binding modes [51], for virtual screening [52] and for the design of site-directed mutations in order to probe ligand-binding characteristics [53].

8.5 Success stories of protein structure-based drug design

8.5.1 X-ray structure-based drug design

Generally, the success rate of a drug discovery method is often measured by the number of compounds that reach the market. Based on current drug development times of about 12–15 years (see Chapter 1) and the enormous increase in available protein 3D structures at the beginning of the 1990s, we will certainly see more drugs in the future where protein structure-based design will have played a more or less important part in the lead discovery and/or optimization process. Figure 3 shows the structures of such compounds. The first success story in protein structure-based drug design was Captopril, an angiotensin-converting enzyme (ACE) inhibitor. ACE inhibitors block the conversion of the decapeptide angiotensin I to the potent pressor substance angiotensin II (an octapeptide), leading to decreased blood pressure. Initial lead compounds were discovered in the 1960s by screening mixtures of peptides isolated from the venom of the snake *Bothrops jararaca* (see Chapter 5). Based on these observations, the nonapeptide teprotide was synthesized and tested in humans with i.v. administration. It was found to lower blood pressure in patients with essential hypertension. The search for orally active ACE inhibitors was dominated by a rational, structure-based strategy. Cushman et al. [54] from the Squibb Institute for Medical Research designed several series of orally active nonpeptidic ACE inhibitors. Based on the predicted similarity of the structure of the zinc proteinase carboxypeptidase A [55] with ACE, they derived a binding site model of the ACE active site. Since it was known that D-benzylsuccinic acid inhibited carboxypeptidase A, they hypothesized that ACE could be blocked by succinyl amino acids that correspond in lengths and shape to the dipeptide cleaved by ACE. This hypothesis proved to be true and led to the synthesis of carboxy alkanoyl and mercapto alkanoyl derivatives [56]. The ACE inhibitor with the lowest IC_{50} in this series was developed as captopril and is marketed as Lopirin®/Capoten®. Fifteen additional ACE inhibitors with different pharmacodynamic and pharmacokinetic properties have been developed since and are employed worldwide now [57].

The carbonic anhydrase (CA) inhibitor dorzolamid is a further example for the successful application of the structure-based drug design. Carbonic anhydrases are ubiquitous zinc enzymes that convert carbon dioxide into bicarbonate. Fourteen isozymes are presently known in higher vertebrates. In the ciliary processes of the eye, carbonic anhydrases mediate the formation of large amounts of bicarbonate in aqueous humor. Inhibition of these enzymes (especially type II and type VI carbonic anhydrases) leads to reduced formation of aqueous humor thereby reducing intraocular pressure. These inhibitors are valuable remedies for treatment of glaucoma. The best studied compounds, acetazolamide, metazolamide and dichlorophenamide (aryl- and heteroaryl-sulfonamides), are given systemically in high doses, unfortunately leading to numerous side-effects. Therefore the topic route of administration is preferred for the treatment of glaucoma. Since the compounds mentioned above were not effective with topical administration, carbonic anhydrase inhibitors having a proper balance between aqueous solubility and lipophilicity were desired. Merck & Co. succeeded with a series of thienothiopyran sulfonamides. An iterative approach of synthesis, biological testing, x-ray crystallographic analysis and molecular modeling led to the compound later marketed as dorzolamide (Trusopt®) (see Fig. 3), the first topical CA inhibitor for the treatment of glaucoma. The x-ray structure of carbonic anhydrase II [58] guided the design of this compound [59]. *Ab initio* quantum mechanical calculations were used to predict substitution patterns that stabilize a certain conformation, necessary for good steric complementarity of the inhibitors and the active site. *In vitro* potency and affinity of inhibitors was optimized by introducing substituents that form polar interactions with the protein, displace ordered water molecules, and show a large lipophilic contact surface. A balanced pharmacokinetic behavior with regard to aqueous solubility and lipophilicity was taken into account during the entire design process. Brinzolamide (Azopt®), a close analogue of dorzolamide, was recently developed by Alcon Laboratories, Inc.

Probably the most prominent examples of structure-based drug design are HIV-protease inhibitors. HIV-1 protease is a virally-encoded enzyme that makes a number of specific cleavages within the gag and gag-pol polyproteins, producing nucleocapsid proteins and viral enzymes. In 1988, it was observed that mutations in the HIV protease gene lead to noninfectious, immature virus particles. This experiment was the basis for the concept of HIV-protease inhibitors, which should block the enzyme active site, thereby inhibiting virus assembly and maturation. First inhibitors were discovered due to the fact that HIV protease and renin belong to the same class of proteases, the aspartic proteases. Several companies shifted their activities from research on renin to HIV protease. The first inhibitors resembled the peptide sequence of the substrates, with the only difference that the scissile amide bond was replaced by statin or other non-cleavable moieties, a proven concept adopted from research on renin inhibition. In 1989, the x-ray crystal structure of HIV-1 protease was solved [60], initiating the structure-based design of second-generation HIV-protease inhibitors. These compounds are peptidomimetics with a central hydroxyl group and larger lipophilic substituents filling the S1, S1', S2, and S2' pockets on both sides of the catalytic Asp 25, Asp 25' pair of residues. The number of amide bonds is in the ranges of two or three in these inhibitors. The potency and/or affinity of most of these compounds for HIV protease was then optimized using several cycles of chemical synthesis, biological testing, x-ray crystallographic analysis and molecular modeling. In this process, the structure of HIV protease provided essential information, especially concerning the properties (size, lipophilicity) of the substituents filling the lipophilic pockets of the substrate-binding site. In 1996, the first HIV protease inhibitor reached the market and was followed by further four compounds (see Fig. 3). Another non-peptidic HIV-protease inhibitor, tipranavir [61], is likely to enter the market soon. The availability of HIV-protease inhibitors revolutionized the therapy of AIDS. Given in combination with HIV reverse transcriptase inhibitors, the viral load of patients can be reduced to undetectable limits, prolonging the lives

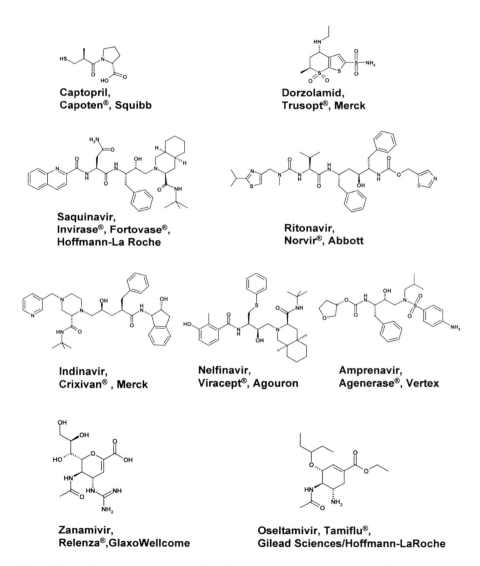

Figure 3. Examples of compounds designed on the basis of protein structure that reached the market.

of these individuals, and partially restoring immune functions. However, the current therapy is far from achieving viral eradication.

A further example of direct structure-based design are influenza neuraminidase inhibitors. Sialidases or neuraminidases catalyze the removal of sialic acid (several analogues of N-acetyl neuraminic acid) from various glycoconjugates. In animals, these enzymes fulfil several roles in the immune system and in apoptosis by regulating the surface sialic acid profile of cells. The location of sialic acids at the termini of various carbohydrates associated with animal cells is exploited by a number of pathogenic viruses such as e.g., influenza virus. These viruses have proteins that recognize sialic acid for cell attachment, and many have acquired sialidases to aid in their pathogenesis.

Influenza virus neuraminidase is a viral coat protein that catalyzes the hydrolysis of the linkage joining a terminal sialic acid residue to a D-galactose or a D-galactosamine residue. The role of neuraminidase is not completely understood, but it is thought to facilitate the liberation of progeny virions from the cell surface and to assist the virus in penetrating through mucin of the infection site. Since the inhibition of neuraminidase leads to aggregation of virus particles at the cell surface [62], it was hypothesized that blocking this enzyme could be a promising strategy against influenza infection. Synthesis of transition state analogues led to the discovery of DANA (Neu5Ac2en). The fluorinated compound FANA showed only little improvement. A crucial breakthrough in the discovery of potent inhibitors was the determination of the x-ray crystal structure of influenza virus neuraminidase in 1983 [63]. The analysis of the interactions of DANA in the substrate binding pocket revealed a negatively charged sub-pocket (involving Glu119 and Glu227) which could be filled with positively charged moieties attached to DANA. The resulting compound 4-guanidino-Neu5Ac2en designed and synthesized by von Itzstein et al. [64] showed a k_i of ~0.1 nM, about four orders of magnitude lower than Neu5Ac2en, and a high selectivity for influenza A and B. This compound was developed under the name zanamivir by GlaxoWellcome Inc. and introduced to market as Relenza® in 1999. The drug is delivered using a dry powder inhaler. Structure-based design at Gilead Sciences Inc. resulted in another potent neuraminidase inhibitor, a carbocyclic transition state analogue with a lipophilic sidechain replacing the glycerol moiety in DANA [65] (see Fig. 3). In contrast to zanamivir, this compound is orally bioavailable. Oseltamivir was approved for the oral prevention and treatment of influenza A and B in 2000 and is marketed as Tamiflu® by Hoffmann-La Roche.

In many other drug discovery projects the structure of target proteins proved to be useful in the discovery and optimization of lead compounds. A comprehensive list of target proteins with known 3D structure and examples of successful application of structure-based drug design is given in Table 1. Of course, this list does not lay claim to completeness. In this table, review articles are listed along with recent examples of structure-based design.

Table 1. List of drug target proteins with known 3D structure and examples of successful application of structure-based drug design

Target protein	Reference
β-lactamase	[66, 67]
acetylcholinesterase	[68]
aldose reductase	[69, 70]
angiotensin converting enzyme	[71]
carbonic anhydrase type II	[59, 72]
catechol O-methyltransferase	[73, 74]
cathepsins	[75, 76]
cyclooxygenases	[77]
cytochromes P450	[78]
dihydrofolate reductases	[79–81]
elastase	[82, 83]
estrogen receptors	[84, 85]
factor Xa	[86–88]
farnesyltransferase	[89]
FK506-binding protein	[90, 91]
G protein-coupled receptors	[92, 93]
hepatitis virus C protease	[94]
HIV gp41	[95]
HIV integrase	[96]
HIV protease	[59, 61, 97, 98]
HIV reverse transcriptase	[99, 100]
influenza virus neuraminidase	[64, 65, 101]
interleukin-1beta converting enzyme	[102]
matrix metalloproteases	[103–106]
nitric oxide synthases	[107]
peroxisome proliferator-activated receptors	[108]
phospolipase A2	[109, 110]
progesterone receptor	[111]
protein-tyrosine kinases	[112–116]
protein-tyrosine phosphatase 1B	[117–120]
purine nucleoside phosphorylase	[121, 122]
renin	[123–125]
retinoic acid receptors	[52, 126]
rhinoviral coat protein	[127]
rhinovirus 3C protease	[128, 129]
telomerase	[130]
thrombin	[131, 132]
thymidylate synthase	[59, 133]
thyroid hormone receptor	[134]
trypanosomatid glyceraldehyde-3-phosphate dehydrogenase	[135, 136]
tubulin	[137]
urokinase	[138, 139]

8.5.2 Homology model-based drug design

There are numerous examples where protein homology models have supported the discovery and the optimization of lead compounds. A homology model of a cysteine proteinase of the malaria parasite *Plasmodium falciparum* lead to the discovery of an inhibitor (IC_{50} = 6 μM) using 3D-database searches [140] (see also Chapter 10). The predicted binding mode is supported by the fact that the lead compound could be improved (IC_{50} = 0.2 μM) by homology model-based design and subsequent synthesis [141].

Several compounds that inhibit trypanothione reductase were designed on the basis of a homology model of the target enzyme [142]. From a homology model of malarial dihydrofolate reductase/thymidylate synthase mutant protein, inhibitors were designed and subsequently synthesized that overcome pyrimethamine resistance [143]. Homology model-based drug design has been applied to numerous kinases such as herpes simplex virus type 1 thymidine kinase [144], EGF-receptor protein tyrosine kinase [145], Bruton's tyrosine kinase [146], Janus kinase 3 [115] and cyclin-dependent kinase 1 [147]. Using a muscarinic M1 receptor homology model, a potent selective muscarinic M1 agonist was designed and subsequently synthesized [148]. Structure activity relationships in a series of hydroxyflutamid-like compounds could be explained with a homology of the androgen receptor [149].

8.6 Applications of protein structure information in the drug discovery process

It is clear that protein structure information (e.g., generated within the frame of structural genomics initiatives) will contribute to many stages of drug discovery process (see Fig. 4). Since structure is associated with protein function [150], the hope is that high-throughput structure analysis will reveal functions of proteins relevant for therapeutic intervention, thus providing novel drug targets. There are successful examples of inferring function from structure [150-153], but the utility of this approach for target identification has yet to be proven.

The analysis of ligand binding sites on the surface of proteins will probably contribute to drug target validation. If, for example, inhibition of protein-protein interactions is regarded a valuable therapeutic principle, proteins with deeply buried interaction sites are clearly superior to those with shallow hydrophilic interfaces. In addition, the design of ligands for proteins of known/predicted structure but unknown function could also contribute to target validation.

Protein structure information is frequently applied to identify lead compounds through virtual screening, 3D-database searching and interactive ligand design (see above). These methods also support the optimization of lead compounds with regard to potency. Rapidly growing structure information on entire target protein families will also enable the design of selective ligands. Once the structure of many or all members of a target protein are known, differences and common fea-

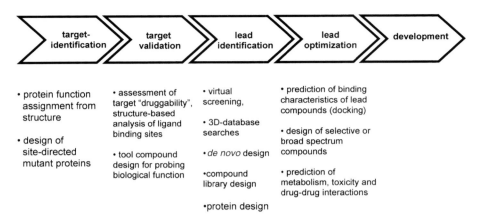

Figure 4. Applications of protein structure information in the drug discovery process. The enormous pool of protein structure data currently available may not only support lead compound identification and optimization, but also many other steps in the drug discovery process, such as target identification, target validation, selection of appropriate animal test models and prediction of metabolic stability or drug-drug interactions.

tures of the binding pockets can be identified and compounds can be designed that contact differing amino acids in the binding niche. If the paraloguous target proteins (paralogues are evolutionary related proteins that perform different but related functions within one organism) show sufficiently different tissue distribution or expression characteristics, drugs with increased selectivity and/or reduced side-effects may be designed [77]. This is especially true for the design of antiparasitic drugs, which should not interact with human orthologues [79–81, 142], but show broad spectrum effects against similar microorganisms. Detailed structural knowledge of ligand binding sites of target proteins could also facilitate the selection of animal models for *in vivo* tests. Animals which possess target proteins with significantly different binding sites compared to human orthologues could be excluded as test models.

A further application for structural genomics in lead optimization is the prediction of drug metabolism (see also Chapter 13). Structure determination in combination with comparative modeling has shed light on the structure of many cytochrome P450 enzymes and other drug metabolizing enzymes. Virtual screening and docking of compounds to these enzymes provide a rational basis for the prediction of metabolic stability and drug-drug interactions [78, 154].

8.7 Conclusions

Numerous examples for the successful application of protein structure-based drug design are described here. Several drugs discovered and optimized with at least some help from protein structure information have made their way into the market. We conclude that today the question is not if protein structure-based design works, but if protein structures are available at the beginning of a drug discovery project in order to support medicinal chemistry. Decades of protein structure elucidation, recent structural genomics initiatives and advances in homology modeling techniques have already generated a wealth of structure information. Today, homology models with varying quality for up to 40% of all sequences can be generated. The 3D-structures of exciting drug targets such as ion channels (a potassium channel [155, 156]) and membrane receptors (nicotinic receptors [157]), a G protein-coupled receptor [158]), the HMG-CoA reductase [159] and a bacterial ribosome [160] have recently been elucidated. This entire pool of structure data may not only support lead compound identification or optimization, but also many other steps in the drug discovery process, such as target identification, target validation, selection of appropriate animal test models and prediction of metabolic stability, side-effects or drug-drug interactions.

8.8 Acknowledgements

The authors thank Oliver Ohlenschläger (IMB-Jena), Tom Sicker (IMB-Jena) and Ursula Kaulmann (HKI-Jena) for their contributions to Figure 1, and Walter Elger (EnTec GmbH) for stimulating discussions.

8.9   References

1 Zimmerle CT, Alter GM (1983) Crystallization-induced modification of cytoplasmic malate dehydrogenase structure and function. *Biochemistry* 22: 6273–6281
2 Tanenbaum DM, Wang Y, Williams SP et al (1998) Crystallographic comparison of the estrogen and progesterone receptor's ligand binding domains. *Proc Natl Acad Sci USA* 95: 5998–6003
3 Lu J, Lin CL, Tang C et al (1999) The structure and dynamics of rat apo-cellular retinol-binding protein II in solution: comparison with the X-ray structure. *J Mol Biol* 286: 1179–1195
4 Billeter M, Kline AD, Braun W et al (1989) Comparison of the high-resolution structures of the alpha-amylase inhibitor tendamistat determined by nuclear magnetic resonance in solution and by X-ray diffraction in single crystals. *J Mol Biol* 206: 677–687
5 Billeter M (1992) Comparison of protein structures determined by NMR in solution and by X-ray diffraction in single crystals. *Q Rev Biophys* 25: 325–377
6 Wuthrich K (1989) Determination of three-dimensional protein structures in solution by nuclear magnetic resonance: an overview. *Methods Enzymol* 177: 125–131
7 McIntosh LP, Dahlquist FW (1990) Biosynthetic incorporation of 15N and 13C for assignment and interpretation of nuclear magnetic resonance spectra of proteins. *Q Rev Biophys* 23: 1–38
8 Karplus M (1963) Vicinal proton coupling in nuclear magnetic resonance. *J Am Chem Soc* 85: 2870–2871
9 Kuntz ID, Thomason JF, Oshiro CM (1989) Distance geometry. *Methods Enzymol* 177: 159–204
10 Scheek RM, van GW, Kaptein R (1989) Molecular dynamics simulation techniques for determination of molecular structures from nuclear magnetic resonance data. *Methods Enzymol* 177: 204–218
11 Brunger AT (1997) X-ray crystallography and NMR reveal complementary views of structure and dynamics. *Nat Struct Biol* 4 Suppl: 862–865
12 Pervushin K, Riek R, Wider G et al (1997) Attenuated T2 relaxation by mutual cancellation of dipole-dipole coupling and chemical shift anisotropy indicates an avenue to NMR structures of very large biological macromolecules in solution. *Proc Natl Acad Sci USA* 94: 12366–12371
13 Riek R, Wider G, Pervushin K et al (1999) Polarization transfer by cross-correlated relaxation in solution NMR with very large molecules. *Proc Natl Acad Sci USA* 96: 4918–4923

14 Riek R, Pervushin K, Wuthrich K (2000) TROSY and CRINEPT: NMR with large molecular and supramolecular structures in solution. *Trends Biochem Sci* 25: 462–468

15 Lesk AM, Chothia CH (1986) The response of protein structures to amino-acid sequence changes. *Philos Trans R Soc London Ser B* 317: 345–356

16 Altschul SF, Gish W, Miller W et al (1990) Basic local alignment search tool. *J Mol Biol* 215: 403–410

17 Pearson WR, Lipman DJ (1988) Improved tools for biological sequence comparison. *Proc Natl Acad Sci USA* 85: 2444–2448

18 Zheng Q, Kyle DJ (1996) Accuracy and reliability of the scaling-relaxation method for loop closure: an evaluation based on extensive and multiple copy conformational samplings. *Proteins* 24: 209–217

19 Rapp CS, Friesner RA (1999) Prediction of loop geometries using a generalized born model of solvation effects. *Proteins* 35: 173–183

20 Fiser A, Do RK, Sali A (2000) Modeling of loops in protein structures. *Protein Sci* 9: 1753–1773

21 Jones TA, Thirup S (1986) Using known substructures in protein model building and crystallography. *EMBO J* 5: 819–822

22 Claessens M, Van Cutsem E, Lasters I et al (1989) Modelling the polypeptide backbone with 'spare parts' from known protein structures. *Protein Eng* 2: 335–345

23 Li W, Liu Z, Lai L (1999) Protein loops on structurally similar scaffolds: database and conformational analysis. *Biopolymers* 49: 481–495

24 Wojcik J, Mornon JP, Chomilier J (1999) New efficient statistical sequence-dependent structure prediction of short to medium-sized protein loops based on an exhaustive loop classification. *J Mol Biol* 289: 1469–1490

25 Vasquez M (1996) Modeling side-chain conformation. *Curr Opin Struct Biol* 6: 217–221

26 Dunbrack RLJ, Karplus M (1994) Conformational analysis of the backbone-dependent rotamer preferences of protein sidechains. *Nat Struct Biol* 1: 334–340

27 Bower MJ, Cohen FE, Dunbrack RLJ (1997) Prediction of protein side-chain rotamers from a backbone-dependent rotamer library: a new homology modeling tool. *J Mol Biol* 267: 1268–1282

28 Dunbrack RLJ (1999) Comparative modeling of CASP3 targets using PSI-BLAST and SCWRL. *Proteins* Suppl 3: 81–87

29 De Maeyer M, Desmet J, Lasters I (1997) All in one: a highly detailed rotamer library improves both accuracy and speed in the modelling of sidechains by dead-end elimination. *Fold Des* 2: 53–66

30 Cheng B, Nayeem A, Scheraga HA (1996) From secondary structure to three-dimensional structure: Improved dihedral angle probability distribution function for use with energy searches for native structures of polypeptides and proteins. *J Comp Chem* 17: 1453–1480

31 Shenkin PS, Farid H, Fetrow JS (1996) Prediction and evaluation of side-chain conformations for protein backbone structures. *Proteins* 26: 323–352

32 Lovell SC, Word JM, Richardson JS et al (2000) The penultimate rotamer library. *Proteins* 40: 389–408

33 Laskowski RA, MacArthur MW, Moss DS et al (1993) PROCHECK: a program to check the stereochemical quality of protein structures. *J Appl Cryst* 26: 283–291

34 Hooft RW, Vriend G, Sander C et al (1996) Errors in protein structures. *Nature* 381: 272

35 EU 3-D Validation Network (1998) Who checks the checkers? Four validation tools applied to eight atomic resolution structures. *J Mol Biol* 276: 417–436

36 Murzin AG, Brenner SE, Hubbard T et al (1995) SCOP: a structural classification of proteins database for the investigation of sequences and structures. *J Mol Biol* 247: 536–540

37 Hendlich M (1998) Databases for protein-ligand complexes. *Acta Crystallogr D Biol Crystallogr* 54: 1178–1182

38 Laskowski RA (2001) PDBsum: summaries and analyses of PDB structures. *Nucleic Acids Res* 29: 221–222

39 Vondrasek J, van Buskirk CP, Wlodawer A (1997) Database of three-dimensional structures of HIV proteinases. *Nat Struct Biol* 4: 8

40 Sanchez R, Pieper U, Mirkovic N et al (2000) MODBASE, a database of annotated comparative protein structure models. *Nucleic Acids Res* 28: 250–253

41 Sali A, Blundell TL (1993) Comparative protein modelling by satisfaction of spatial restraints. *J Mol Biol* 234: 779–815

42 Peitsch MC, Schwede T, Guex N (2000) Automated protein modelling—the proteome in 3D. *Pharmacogenomics* 1: 257–266

43 Edwards AM, Arrowsmith CH, Christendat D et al (2000) Protein production: feeding the crystallographers and NMR spectroscopists. *Nat Struct Biol* 7 Suppl: 970–972

44 Abola E, Kuhn P, Earnest T et al (2000) Automation of x-ray crystallography. *Nat Struct Biol* 7 Suppl: 973–977
45 Lamzin VS, Perrakis A (2000) Current state of automated crystallographic data analysis. *Nat Struct Biol* 7 Suppl: 978–981
46 Montelione GT, Zheng D, Huang YJ et al (2000) Protein NMR spectroscopy in structural genomics. *Nat Struct Biol* 7 Suppl: 982–985
47 Burley SK (2000) An overview of structural genomics. *Nat Struct Biol* 7 Suppl: 932–934
48 Service RF (2000) Structural genomics offers high-speed look at proteins. *Science* 287: 1954–1956
49 Brenner SE (2000) Target selection for structural genomics. *Nat Struct Biol* 7 Suppl: 967–969
50 Marti-Renom MA, Stuart AC, Fiser A et al (2000) Comparative protein structure modeling of genes and genomes. *Annu Rev Biophys Biomol Struct* 29: 291–325
51 Schafferhans A, Klebe G (2001) Docking Ligands onto Binding Site Representations Derived from Proteins built by Homology Modelling. *J Mol Biol* 307: 407–427
52 Schapira M, Raaka BM, Samuels HH et al (2000) Rational discovery of novel nuclear hormone receptor antagonists. *Proc Natl Acad Sci USA* 97: 1008–1013
53 Spencer TA, Li D, Russel JS et al (2001) Pharmacophore analysis of the nuclear oxysterol receptor LXRalpha. *J Med Chem* 44: 886–897
54 Cushman DW, Cheung HS, Sabo EF et al (1977) Design of potent competitive inhibitors of angiotensin-converting enzyme. Carboxyalkanoyl and mercaptoalkanoyl amino acids. *Biochemistry* 16: 5484–5491
55 Lipscomb WN, Reeke GNJ, Hartsuck JA et al (1970) The structure of carboxypeptidase A. 8. Atomic interpretation at 0.2 nm resolution, a new study of the complex of glycyl-L-tyrosine with CPA, and mechanistic deductions. *Philos Trans R Soc Lond B Biol Sci* 257: 177–214
56 Petrillo EWJ, Ondetti MA (1982) Angiotensin-converting enzyme inhibitors: medicinal chemistry and biological actions. *Med Res Rev* 2: 1–41
57 Goodman LS, Gilman A (1996) *Goodman & Gilman's The pharmakological basis of therapeutics.* McGraw-Hill, New York
58 Eriksson AE, Jones TA, Liljas A (1988) Refined structure of human carbonic anhydrase II at 2.0 A resolution. *Proteins* 4: 274–282
59 Greer J, Erickson JW, Baldwin JJ et al (1994) Application of the three-dimensional structures of protein target molecules in structure-based drug design. *J Med Chem* 37: 1035–1054
60 Wlodawer A, Miller M, Jaskolski M et al (1989) Conserved folding in retroviral proteases: crystal structure of a synthetic HIV-1 protease. *Science* 245: 616–621
61 Thaisrivongs S, Strohbach JW (1999) Structure-based discovery of Tipranavir disodium (PNU-140690E): a potent, orally bioavailable, nonpeptidic HIV protease inhibitor. *Biopolymers* 51: 51–58
62 Palese P, Schulman JL, Bodo G et al (1974) Inhibition of influenza and parainfluenza virus replication in tissue culture by 2-deoxy-2,3-dehydro-N-trifluoroacetylneuraminic acid (FANA). *Virology* 59: 490–498
63 Varghese JN, Laver WG, Colman PM (1983) Structure of the influenza virus glycoprotein antigen neuraminidase at 2.9 A resolution. *Nature* 303: 35–40
64 von Itzstein M, Wu WY, Kok GB et al (1993) Rational design of potent sialidase-based inhibitors of influenza virus replication. *Nature* 363: 418–423
65 Kim CU, Lew W, Williams MA et al (1997) Influenza neuraminidase inhibitors possessing a novel hydrophobic interaction in the enzyme active site: Design, synthesis, and structural analysis of carbocyclic sialic acid analogues with potent anti-influenza activity. *J Am Chem Soc* 119: 681–690
66 Mascaretti OA, Danelon GO, Laborde M et al (1999) Recent advances in the chemistry of beta-lactam compounds as selected active-site serine beta-lactamase inhibitors. *Curr Pharm Des* 5: 939–953
67 Heinze-Krauss I, Angehrn P, Charnas RL et al (1998) Structure-based design of beta-lactamase inhibitors. 1. Synthesis and evaluation of bridged monobactams. *J Med Chem* 41: 3961–3971
68 Kryger G, Silman I, Sussman JL (1999) Structure of acetylcholinesterase complexed with E2020 (Aricept): implications for the design of new anti-Alzheimer drugs. *Structure Fold Des* 7: 297–307
69 Wilson DK, Petrash JM, Quiocho FA (1997) Structural studies of aldose reductase inhibition. In: *Structure-based drug design*, edited by P. Veerapandian, pp. 229–246. Marcel Dekker, Inc., New York, Basel, Hong Kong
70 Iwata Y, Arisawa M, Hamada R et al (2001) Discovery of novel aldose reductase inhibitors using a protein structure-based approach: 3D-database search followed by design and synthesis. *J Med Chem* 44: 1718–1728
71 Ondetti MA, Rubin B, Cushman DW (1977) Design of specific inhibitors of angiotensin-converting

enzyme: new class of orally active antihypertensive agents. *Science* 196: 441–444

72 Bohacek RS, McMartin C, Guida WC (1996) The art and practice of structure-based drug design: a molecular modeling perspective. *Med Res Rev* 16: 3–50

73 Vidgren J (1998) X-ray crystallography of catechol O-methyltransferase: perspectives for target-based drug development. *Adv Pharmacol* 42: 328–331

74 Masjost B, Ballmer P, Borroni E et al (2000) Structure-based design, synthesis, and *in vitro* evaluation of bisubstrate inhibitors for catechol O-methyltransferase (COMT). *Chemistry* 6: 971–982

75 Katunuma N, Murata E, Kakegawa H et al (1999) Structure based development of novel specific inhibitors for cathepsin L and cathepsin S *in vitro* and *in vivo*. *FEBS Lett* 458: 6–10

76 Katunuma N, Matsui A, Inubushi T et al (2000) Structure-based development of pyridoxal propionate derivatives as specific inhibitors of cathepsin K *in vitro* and *in vivo*. *Biochem Biophys Res Commun* 267: 850–854

77 Bayly CI, Black WC, Leger S et al (1999) Structure-based design of COX-2 selectivity into flurbiprofen. *Bioorg Med Chem Lett* 9: 307–312

78 Szklarz GD, Halpert JR (1998) Molecular basis of P450 inhibition and activation: implications for drug development and drug therapy. *Drug Metab Dispos* 26: 1179–1184

79 Gschwend DA, Sirawaraporn W, Santi DV et al (1997) Specificity in structure-based drug design: identification of a novel, selective inhibitor of Pneumocystis carinii dihydrofolate reductase. *Proteins* 29: 59–67

80 Zuccotto F, Brun R, Gonzalez PD et al (1999) The structure-based design and synthesis of selective inhibitors of Trypanosoma cruzi dihydrofolate reductase. *Bioorg Med Chem Lett* 9: 1463–1468

81 Rosowsky A, Cody V, Galitsky N et al (1999) Structure-based design of selective inhibitors of dihydrofolate reductase: synthesis and antiparasitic activity of 2, 4-diaminopteridine analogues with a bridged diarylamine side chain. *J Med Chem* 42: 4853–4860

82 Cregge RJ, Durham SL, Farr RA et al (1998) Inhibition of human neutrophil elastase. 4. Design, synthesis, X-ray crystallographic analysis, and structure-activity relationships for a series of P2-modified, orally active peptidyl pentafluoroethyl ketones. *J Med Chem* 41: 2461–2480

83 Filippusson H, Erlendsson LS, Lowe CR (2000) Design, synthesis and evaluation of biomimetic affinity ligands for elastases. *J Mol Recognit* 13: 370–381

84 Tedesco R, Thomas JA, Katzenellenbogen BS et al (2001) The estrogen receptor: a structure-based approach to the design of new specific hormone-receptor combinations. *Chem Biol* 8: 277–287

85 Stauffer SR, Coletta CJ, Tedesco R et al (2000) Pyrazole ligands: structure-affinity/activity relationships and estrogen receptor-alpha-selective agonists. *J Med Chem* 43: 4934–4947

86 Phillips G, Davey DD, Eagen KA et al (1999) Design, synthesis, and activity of 2,6-diphenoxypyridine-derived factor Xa inhibitors. *J Med Chem* 42: 1749–1756

87 Arnaiz DO, Zhao Z, Liang A et al (2000) Design, synthesis, and *in vitro* biological activity of indole-based factor Xa inhibitors. *Bioorg Med Chem Lett* 10: 957–961

88 Han Q, Dominguez C, Stouten PF et al (2000) Design, synthesis, and biological evaluation of potent and selective amidino bicyclic factor Xa inhibitors. *J Med Chem* 43: 4398–4415

89 Kaminski JJ, Rane DF, Snow ME et al (1997) Identification of novel farnesyl protein transferase inhibitors using three-dimensional database searching methods. *J Med Chem* 40: 4103–4112

90 Navia MA (1996) Protein-drug complexes important for immunoregulation and organ transplantation. *Curr Opin Struct Biol* 6: 838–847

91 Burkhard P, Hommel U, Sanner M et al (1999) The discovery of steroids and other novel FKBP inhibitors using a molecular docking program. *J Mol Biol* 287: 853–858

92 van Neuren AS, Muller G, Klebe G et al (1999) Molecular modelling studies on G protein-coupled receptors: from sequence to structure? *J Recept Signal Transduct Res* 19: 341–353

93 Müller G (2000) Towards 3D structures of G protein-coupled receptors: a multidisciplinary approach. *Curr Med Chem* 7: 861–888

94 Martin F, Dimasi N, Volpari C et al (1998) Design of selective eglin inhibitors of HCV NS3 proteinase. *Biochemistry* 37: 11459–11468

95 Debnath AK, Radigan L, Jiang S (1999) Structure-based identification of small molecule antiviral compounds targeted to the gp41 core structure of the human immunodeficiency virus type 1. *J Med Chem* 42: 3203–3209

96 Hong H, Neamati N, Wang S et al (1997) Discovery of HIV-1 integrase inhibitors by pharmacophore searching. *J Med Chem* 40: 930–936

97 Wlodawer A, Vondrasek J (1998) Inhibitors of HIV-1 protease: a major success of structure-assisted drug design. *Annu Rev Biophys Biomol Struct* 27: 249–284

98 De Lucca GV, Erickson-Viitanen S, Lam PY (1997) Cyclic HIV protease inhibitors capable of displacing the active site structural water molecule. *Drug Discov Today* 2: 6–18

99 Arnold E, Das K, Ding J et al (1996) Targeting HIV reverse transcriptase for anti-AIDS drug design: structural and biological considerations for chemotherapeutic strategies. *Drug Des Discov* 13: 29–47

100 Mao C, Sudbeck EA, Venkatachalam TK et al (2000) Structure-based drug design of non-nucleoside inhibitors for wild-type and drug-resistant HIV reverse transcriptase. *Biochem Pharmacol* 60: 1251–1265

101 Wade RC (1997) 'Flu' and structure-based drug design. *Structure* 5: 1139–1145

102 Shahripour AB, Plummer MS, Lunney EA et al (2002) Structure-based design of nonpeptide inhibitors of interleukin-1beta converting enzyme (ICE, caspase-1). *Bioorg Med Chem* 10: 31–40

103 Zask A, Levin JI, Killar LM et al (1996) Inhibition of matrix metalloproteinases: structure based design. *Curr Pharm Des* 2: 624–661

104 Brown PD (1998) Matrix metalloproteinase inhibitors. *Breast Cancer Res Treat* 52: 125–136

105 Matter H, Schwab W, Barbier D et al (1999) Quantitative structure-activity relationship of human neutrophil collagenase (MMP-8) inhibitors using comparative molecular field analysis and X-ray structure analysis. *J Med Chem* 42: 1908–1920

106 Cheng M, De B, Pikul S et al (2000) Design and synthesis of piperazine-based matrix metalloproteinase inhibitors. *J Med Chem* 43: 369–380

107 Huang H, Martasek P, Roman LJ et al (2000) Synthesis and evaluation of peptidomimetics as selective inhibitors and active site probes of nitric oxide synthases. *J Med Chem* 43: 2938–2945

108 Oliver WRJ, Shenk JL, Snaith MR et al (2001) A selective peroxisome proliferator-activated receptor delta agonist promotes reverse cholesterol transport. *Proc Natl Acad Sci USA* 98: 5306–5311

109 Schevitz RW, Bach NJ, Carlson DG et al (1995) Structure-based design of the first potent and selective inhibitor of human non-pancreatic secretory phospholipase A2. *Nat Struct Biol* 2: 458–465

110 Mihelich ED, Schevitz RW (1999) Structure-based design of a new class of anti-inflammatory drugs: secretory phospholipase A(2) inhibitors, SPI. *Biochim Biophys Acta* 1441: 223–228

111 Bursi R, Groen MB (2000) Application of (quantitative) structure-activity relationships to progestagens: from serendipity to structure-based design. *Eur J Med Chem* 35: 787–796

112 al-Obeidi FA, Wu JJ, Lam KS (1998) Protein tyrosine kinases: structure, substrate specificity, and drug discovery. *Biopolymers* 47: 197–223

113 Shakespeare W, Yang M, Bohacek R et al (2000) Structure-based design of an osteoclast-selective, nonpeptide src homology 2 inhibitor with *in vivo* antiresorptive activity. *Proc Natl Acad Sci USA* 97: 9373–9378

114 Toledo LM, Lydon NB, Elbaum D (1999) The structure-based design of ATP-site directed protein kinase inhibitors. *Curr Med Chem* 6: 775–805

115 Sudbeck EA, Liu XP, Narla RK et al (1999) Structure-based design of specific inhibitors of Janus kinase 3 as apoptosis-inducing antileukemic agents. *Clin Cancer Res* 5: 1569–1582

116 Furet P, Garcia-Echeverria C, Gay B et al (1999) Structure-based design, synthesis, and X-ray crystallography of a high-affinity antagonist of the Grb2-SH2 domain containing an asparagine mimetic. *J Med Chem* 42: 2358–2363

117 Burke TRJ, Zhang ZY (1998) Protein-tyrosine phosphatases: structure, mechanism, and inhibitor discovery. *Biopolymers* 47: 225–241

118 Doman TN, McGovern SL, Witherbee BJ et al (2002) Molecular docking and high-throughput screening for novel inhibitors of protein tyrosine phosphatase-1B. *J Med Chem* 45: 2213–2221

119 Yao ZJ, Ye B, Wu XW et al (1998) Structure-based design and synthesis of small molecule protein-tyrosine phosphatase 1B inhibitors. *Bioorg Med Chem* 6: 1799–1810

120 Iversen LF, Andersen HS, Branner S et al (2000) Structure-based design of a low molecular weight, nonphosphorus, nonpeptide, and highly selective inhibitor of protein-tyrosine phosphatase 1B. *J Biol Chem* 275: 10300–10307

121 Ealick SE, Babu YS, Bugg CE et al (1991) Application of crystallographic and modeling methods in the design of purine nucleoside phosphorylase inhibitors. *Proc Natl Acad Sci USA* 88: 11540–11544

122 Montgomery JA (1994) Structure-based drug design: inhibitors of purine nucleoside phosphorylase. *Drug Des Discov* 11: 289–305

123 Dhanaraj V, Cooper JB (1997)Rational design of renin inhibitors. In: *Structure-based drug design*, edited by P. Veerapandian, pp. 321–342. Marcel Dekker, Inc., New York, Basel, Hong Kong

124 Hutchins C, Greer J (1991) Comparative modeling of proteins in the design of novel renin inhibitors. *Crit Rev Biochem Mol Biol* 26: 77–127

125 Rahuel J, Rasetti V, Maibaum J et al (2000) Structure-based drug design: the discovery of novel non-

peptide orally active inhibitors of human renin. *Chem Biol* 7: 493–504
126 Nagpal S, Chandraratna RA (2000) Recent developments in receptor-selective retinoids. *Curr Pharm Des* 6: 919–931
127 Giranda VL (1994) Structure-based drug design of antirhinoviral compounds. *Structure* 2: 695–698
128 Matthews DA, Dragovich PS, Webber SE et al (1999) Structure-assisted design of mechanism-based irreversible inhibitors of human rhinovirus 3C protease with potent antiviral activity against multiple rhinovirus serotypes. *Proc Natl Acad Sci USA* 96: 11000–11007
129 Reich SH, Johnson T, Wallace MB et al (2000) Substituted benzamide inhibitors of human rhinovirus 3C protease: structure-based design, synthesis, and biological evaluation. *J Med Chem* 43: 1670–1683
130 Read M, Harrison RJ, Romagnoli B et al (2001) Structure-based design of selective and potent G quadruplex-mediated telomerase inhibitors. *Proc Natl Acad Sci USA* 98: 4844–4849
131 Sanderson PE, Naylor-Olsen AM (1998) Thrombin inhibitor design. *Curr Med Chem* 5: 289–304
132 Stubbs MT, Bode W (1993) A player of many parts: the spotlight falls on thrombin's structure. *Thromb Res* 69: 1–58
133 Costi MP (1998) Thymidylate synthase inhibition: a structure-based rationale for drug design. *Med Res Rev* 18: 21–42
134 Greenidge PA, Carlsson B, Bladh LG et al (1998) Pharmacophores incorporating numerous excluded volumes defined by X-ray crystallographic structure in three-dimensional database searching: application to the thyroid hormone receptor. *J Med Chem* 41: 2503–2512
135 Callens M, Hannaert V (1995) The rational design of trypanocidal drugs: selective inhibition of the glyceraldehyde-3-phosphate dehydrogenase in Trypanosomatidae. *Ann Trop Med Parasitol* 89 Suppl 1: 23–30
136 Aronov AM, Suresh S, Buckner FS et al (1999) Structure-based design of submicromolar, biologically active inhibitors of trypanosomatid glyceraldehyde-3-phosphate dehydrogenase. *Proc Natl Acad Sci USA* 96: 4273–4278
137 Uckun FM, Mao C, Vassilev AO et al (2000) Structure-based design of a novel synthetic spiroketal pyran as a pharmacophore for the marine natural product spongistatin 1. *Bioorg Med Chem Lett* 10: 541–545
138 Zeslawska E, Schweinitz A, Karcher A et al (2000) Crystals of the urokinase type plasminogen activator variant beta(c)-uPAin complex with small molecule inhibitors open the way towards structure-based drug design. *J Mol Biol* 301: 465–475
139 Nienaber VL, Davidson D, Edalji R et al (2000) Structure-directed discovery of potent non-peptidic inhibitors of human urokinase that access a novel binding subsite. *Structure Fold Des* 8: 553–563
140 Ring CS, Sun E, McKerrow JH et al (1993) Structure-based inhibitor design by using protein models for the development of antiparasitic agents. *Proc Natl Acad Sci USA* 90: 3583–3587
141 Li Z, Chen X, Davidson E et al (1994) Anti-malarial drug development using models of enzyme structure. *Chem Biol* 1: 31–37
142 Garforth J, Yin H, McKie JH et al (1997) Rational design of selective ligands for trypanothione reductase from Trypanosoma cruzi. Structural effects on the inhibition by dibenzazepines based on imipramine. *J Enzyme Inhib* 12: 161–173
143 McKie JH, Douglas KT, Chan C et al (1998) Rational drug design approach for overcoming drug resistance: application to pyrimethamine resistance in malaria. *J Med Chem* 41: 1367–1370
144 Folkers G, Alber F, Amrhein I et al (1997) Integrated homology modelling and X-ray study of herpes simplex virus I thymidine kinase: a case study. *J Recept Signal Transduct Res* 17: 475–494
145 Traxler P, Furet P, Mett H et al (1997) Design and synthesis of novel tyrosine kinase inhibitors using a pharmacophore model of the ATP-binding site of the EGF-R. *J Pharm Belg* 52: 88–96
146 Mahajan S, Ghosh S, Sudbeck EA et al (1999) Rational design and synthesis of a novel anti-leukemic agent targeting Bruton's tyrosine kinase (BTK), LFM-A13 [alpha-cyano-beta-hydroxy-beta-methyl-N-(2, 5-dibromophenyl)propenamide]. *J Biol Chem* 274: 9587–9599
147 Gussio R, Zaharevitz DW, McGrath CF et al (2000) Structure-based design modifications of the paullone molecular scaffold for cyclin-dependent kinase inhibition. *Anticancer Drug Des* 15: 53–66
148 Sabb AL, Husbands GM, Tokolics J et al (1999) Discovery of a highly potent, functionally-selective muscarinic M1 agonist, WAY-132983 using rational drug design and receptor modelling. *Bioorg Med Chem Lett* 9: 1895–1900
149 Marhefka CA, Moore BM, Bishop TC et al (2001) Homology modeling using multiple molecular dynamics simulations and docking studies of the human androgen receptor ligand binding domain bound to testosterone and nonsteroidal ligands. *J Med Chem* 44: 1729–1740

150 Thornton JM, Todd AE, Milburn D et al (2000) From structure to function: approaches and limitations. *Nat Struct Biol* 7 Suppl: 991–994

151 Zarembinski TI, Hung LW, Mueller-Dieckmann HJ et al (1998) Structure-based assignment of the biochemical function of a hypothetical protein: a test case of structural genomics. *Proc Natl Acad Sci USA* 95: 15189–15193

152 Kleywegt GJ (1999) Recognition of spatial motifs in protein structures. *J Mol Biol* 285: 1887–1897

153 Xu LZ, Sanchez R, Sali A et al (1996) Ligand specificity of brain lipid-binding protein. *J Biol Chem* 271: 24711–24719

154 Lewis DF (2000) Modelling human cytochromes P450 for evaluating drug metabolism: an update. *Drug Metabol Drug Interact* 16: 307–324

155 Doyle DA, Morais CJ, Pfuetzner RA et al (1998) The structure of the potassium channel: molecular basis of K$^+$ conduction and selectivity. *Science* 280: 69–77

156 Armstrong N, Sun Y, Chen GQ et al (1998) Structure of a glutamate-receptor ligand-binding core in complex with kainate. *Nature* 395: 913–917

157 Brejc K, van Dijk WJ, Klaassen RV et al (2001) Crystal structure of an ACh-binding protein reveals the ligand-binding domain of nicotinic receptors. *Nature* 411: 269–276

158 Palczewski K, Kumasaka T, Hori T et al (2000) Crystal structure of rhodopsin: A G protein-coupled receptor. *Science* 289: 739–745

159 Istvan ES, Deisenhofer J (2001) Structural mechanism for statin inhibition of HMG-CoA reductase. *Science* 292: 1160–1164

160 Ban N, Nissen P, Hansen J et al (2000) The complete atomic structure of the large ribosomal subunit at 2.4 A resolution. *Science* 289: 905–920

161 Berchtold H, Hilgenfeld R (1999) Binding of phenol to R6 insulin hexamers. *Biopolymers* 51: 165–172

162 Ohlenschläger O, Ramachandran R, Gührs KH et al (1998) Nuclear magnetic resonance solution structure of the plasminogen-activator protein staphylokinase. *Biochemistry* 37: 10635–10642

163 Steinmetzer K, Hillisch A, Behlke J et al (2000) Transcriptional repressor CopR: structure model-based localization of the deoxyribonucleic acid binding motif. *Proteins* 38: 393–406

Modern Methods of Drug Discovery
ed. by A. Hillisch and R. Hilgenfeld
© 2003 Birkhäuser Verlag/Switzerland

9   NMR-based screening methods for lead discovery

Martin Vogtherr and Klaus Fiebig

Institut für Organische Chemie, Johann-Wolfgang von Goethe-Universität Frankfurt/Main, Marie-Curie Str. 11, D-60431 Frankfurt (Main), Germany

9.1   Introduction

Traditionally the role of nuclear magnetic resonance (NMR) in drug discovery has been as an analytical tool to aid chemists in characterizing small molecule compounds and identifying novel natural products (see Chapter 5). Today, due to the development and massive application of x-ray crystallography and NMR to determine atomic-resolution structures of proteins, nucleic acids, and their complexes, NMR has established itself as a key method in structure based drug design [1] (see also Chapter 8). Unfortunately, structure-based drug design by NMR becomes more difficult with increasing molecular weight of the target. Thus, although rational drug design has proven to be effective for several targets, there has been a major shift to high throughput screening (HTS) of large libraries of compounds. HTS methods only require a sufficiently robust assay which, when miniaturized, is used to screen hundreds of thousands of small molecule compounds.

The recent emphasis on high throughput screening (HTS) methods has also greatly influenced the role of NMR in drug discovery. Since obtaining a robust HTS assay system for drug activity can be difficult, NMR has emerged as a powerful and versatile screening technique for ligand binding [2, 3]. Furthermore, a significant fraction of NMR based screening strategies is independent of the molecular weight of the target molecule. NMR also provides a generic binding assay, which is independent of the target's biological function. This may enable screening of proteins with unknown function, which now are identified at an ever-increasing rate through genomic sequence analysis (see Chapter 1).

NMR employs the same feature to detect binding as other spectroscopic methods, by monitoring changes in spectral properties brought about by the binding process [4–6]. Compared to UV-based screening methods NMR is rather insensitive. However, there have been significant advances in sensitivity due to instrumentation design such that NMR can be used at much reduced sample concentrations. In spite of its inferior sensitivity NMR possesses several advantages over other techniques. First, NMR is a method with extremely high analytical value. Using NMR techniques, it has been possible to elucidate virtually any molecular structure up to 3D structures of proteins and nucleic acids. Second, there is often a multitude of spectral properties, which change due to the binding of a ligand. Third, modern NMR techniques combined with isotopic labeling strategies can record changes in spectral properties of the desired compounds virtually without background.

NMR has been used in countless studies to explore molecular binding and interaction processes (reviewed in [4–6]). Many of these date back to the early days of NMR well before the era of high-resolution protein structure determination by NMR. In most of these studies a well-defined set of interacting partners was examined, and no attempt was made to find novel ligands through "screening" of large libraries of compounds. However, many of these studies have used the same features of NMR spectra and their specific changes caused by binding of a ligand as modern screening methods do. In particular, broadening or shifting of resonance lines can be monitored very easily using simple 1D NMR techniques.

Most screening experiments in pharmaceutical research investigate the interaction between a macromolecular target and numerous potential inhibitors or ligands either individually or as mixtures. For NMR based screening, both the target and the ligand display NMR signals, which may be perturbed by an interaction. Hence there arises a natural classification: methods which monitor the NMR signals of the target protein, and methods which monitor those of the ligands. As will be seen in the next two sections, this distinction is sensible both conceptually and methodologically.

9.2 Target-based methods—SAR by NMR

One of the first and perhaps best known screening method utilizing NMR is the "SAR by NMR" (structure activity relationship by NMR) approach developed by Fesik et al. at Abbott Laboratories [7–10]. This technique monitors all amide chemical shifts of a uniformly ^{15}N-labeled target protein, which can be produced quite cost effectively by overexpression in bacteria. Chemical shifts of both ^{1}H- and ^{15}N-amide atoms are recorded as peaks in a two-dimensional spectrum, the ^{15}N-^{1}H-HSQC (heteronuclear single quantum correlation) spectrum. A small region of such a ^{15}N-^{1}H-HSQC spectrum is shown in Figure 1a (black contours). In such a spectrum each amino acid residue except proline give rise to one single peak that can be assigned using standard high-resolution multidimensional NMR techniques [11].

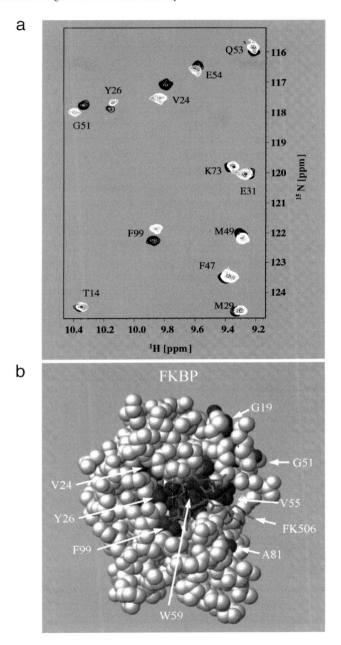

Figure 1. Monitoring binding activity by inspection of 2D heteronuclear correlation spectra. (a) Shown is a small region of ^{15}N-^{1}H-HSQC spectra of uniformly ^{15}N labeled FKBP (0.25 mM) without (black) and with (white) 2-phenylimidazole at a concentration of 1.0 mM. (b) Amide resonances, which shift upon binding of 2-phenylimidazole, are mapped onto the structure of FKBP (black). FK506 is shown binding to the active site as a reference.

If a small molecule binds to the protein, the nuclear spins of the nitrogen and hydrogen atoms in the binding site are affected and the corresponding peaks in the ^{15}N-^1H-HSQC spectrum shift (Fig. 1a, white contours). If the ligand interacts with the protein backbone, then these perturbations are especially strong. Due to the good amide chemical shift dispersion of folded proteins most resonances in the 2D ^{15}N-^1H-HSQC are well resolved such that even small perturbations in only one of the dimensions can be detected. If amide ^{15}N and ^1H chemical shifts have been assigned previously, and if the structure of the target protein is known, then the ligand binding site can be identified and mapped onto the surface of the structure (Fig. 1b).

^{15}N-^1H-HSQC spectra can be acquired in 10 to 30 min for moderately concentrated protein solutions (approximately 0.3 mM). Thus, up to 1000 compounds can be screened per day if mixtures of 10 are evaluated simultaneously (pool 10 approach). This includes time spent on mixtures with binding compounds, which need to be deconvoluted by performing experiments on each of the single compounds.

Prior to SAR by NMR there have been several published reports which have used 2D NMR spectroscopy to detect and localize ligand binding (e.g., [12, 13]). However, SAR by NMR was the first method to propose a ^{15}N-^1H-HSQC based screening technique to identify ligands from a library of compounds. The crucial and new point in the SAR by NMR concept was to use information about the localization of the binding sites of different ligands to synthesize new higher affinity ligands as shown in (Fig. 2a). First, ligand 1 and the corresponding binding site 1 are identified and optimized (i). Second, screening in the presence of an excess of ligand 1 is carried out to identify and then optimize the neighboring ligand 2 (ii). Third, experimentally derived structural information is used to guide the design of linkers between ligand 1 and ligand 2 (iii). Although the individual ligand fragments may only bind weakly due to their small size, tethering of the two fragments will result in a molecule displaying a binding affinity equal to the product of the two fragment binding affinities plus the affinity gained by the linker. This principle is well known in chemistry as the "chelate effect." In Figure 2b several other strategies are shown that may be used to optimize binding affinities of initially found leads. Fragment optimization (i) and fragment linking (ii) have already been discussed. Generation of a directed library (iii) based on an initial fragment may be a viable alternative if second site ligands cannot be identified. If several ligands can be identified which bind to the same site, then fragment-merging (iv) using classical SAR approaches may be used. Typically similar fragments are superimposed and docked into the binding site using the differential chemical shift perturbation they cause in the protein ^{15}N-^1H-HSQC. Furthermore, a strategy of fragment optimization via disassembly, shown in Figure 2b (v), can be employed to enhance the binding of larger molecules found by other screening techniques.

In the initial SAR by NMR publication [7] the Abbott group demonstrated using the FK506 binding protein FKBP-12 that several fragments with micromolar to millimolar binding affinity could be identified and linked to form a nanomo-

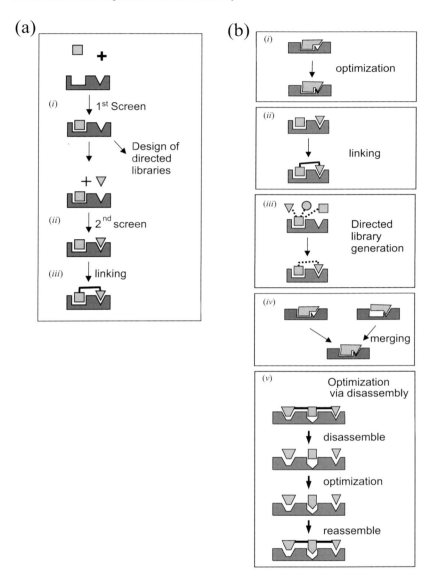

Figure 2. (a) Schematic principle of the SAR by NMR approach. (b) Several approaches to design enhanced ligands by evaluation of binding sites. In each information about the localization of the bound ligand within the protein is utilized to streamline synthesis of tight-binding ligands.

lar inhibitor of FKBP-12. Although the resulting FKBP-12 inhibitors were of little therapeutic value, demonstration of the SAR by NMR concept set the stage for subsequent studies on more interesting targets: the matrix metalloproteinase stromelysin [8], the E2 protein from the human papilloma virus (HPV E2) [9], and the Erm Methyltransferase [10]. Three important conclusions can be drawn from these studies: 1) all studies employ ^{15}N-^{1}H-HSQC based screening to identify initial weakly-bound ligands with K_D's in the millimolar range, which are then optimized using SAR. Several of these initial compounds would have not been detected using a high throughput screen. 2) Linking of optimized fragments of neighboring binding sites produced inhibitors in the low nanomolar and micromolar range for stromelysin and HPV E2, respectively. For neither of these targets high throughput screening of more than 100,000 compounds of the Abbott cooperate library produced inhibitors with superior activities. 3) Diverse lead structures can be discovered and optimized rapidly because chemical synthesis efforts can be highly directed. In the case of stromelysin a 15 nanomolar inhibitor was found in less than 6 months [8]. For HPV E2 micromolar inhibitors which bind to the DNA-binding site were discovered [9]. Remarkable is that with only little prior pharmacological knowledge complex macromolecular DNA-protein interactions could be inhibited.

A large advantage of the ^{15}N-^{1}H-HSQC screening method is that it is highly sensitive for detecting even weakly bound species. Dalvit et al. [14] were able to detect shifts in FKBP caused by weakly associated DMSO (which in most studies of this kind is used for the preparation of ligand stock solutions). Based upon this finding, the authors could not only localize the binding site, but also could design sulfoxide-based inhibitors that bind to FKBP much more tightly. This study elegantly emphasizes the power of NMR screening for low affinity ligands that are optimized in a second step for tighter binding.

Currently SAR by NMR is being applied to a multitude of low molecular weight (<30 kDa) targets and certainly many suitable targets remain to be discovered. Nevertheless protein target selection for SAR by NMR has been severely limited by two factors: 1) the size of the target protein for which ^{15}N-^{1}H-correlated spectra can be recorded and 2) the relatively large concentrations of protein and ligand required.

To overcome the first limitation, transverse relaxation optimized spectroscopy [15] (TROSY) in combination with large magnetic fields is expected to extend the molecular weight barrier past 100 kDa (see Chapter 8). Pellecchia et al. [16] have successfully used TROSY technology to map the binding interactions in the 74 kDa complex of the Pilus chaperone FimC with the adhesin FimH. In such large proteins, spectral overlap may become a severe problem since the number of peaks increases linearly with the protein size. To evaluate such spectra, a combination of selective labeling techniques and smart analysis software will be essential. Pattern recognition approaches will be needed to detect the minute changes of the amide resonances in a small fraction of the protein's amino acid residues.

The second major limitation, the solubility and protein availability limitation, can only be overcome by increasing the sensitivity of NMR instrumentation itself. New helium-cooled cryogenic NMR probes can enhance the performance of high resolution NMR up to four times by cooling the radio frequency (RF) detection coil to 25 K (−248°C). Recently Hajduk et al. have applied this technology to SAR by NMR to achieve sensitivity gains of a factor of 2.4 when compared to conventional probe technology [17]. This allows a reduction in protein and ligand concentrations to 50 µM, which shifts the stringency of the screen toward tighter binding molecules. As a consequence, the number of compounds per mixture can be increased from 10 to 100. Using cryogenically cooled NMR probes libraries of 200,000 compounds can be tested in less than a month with much reduced protein consumption. Throughput of ^{15}N-^{1}H-HSQC-based screening is thus greatly enhanced and may now be comparable to the throughput of other HTS techniques (see Chapter 4).

Up to now, the SAR by NMR method has mainly been applied to enzymes. In contrast to conventional assay based HTS methods, SAR by NMR can be easily applied to other types of proteins, to nucleic acids, or to protein-protein and protein-nucleotide complexes, since no enzymatic activity is required. The power of SAR by NMR lies in the direct assignment of binding activity to regions within the protein, which would not have been possible by other methods. One disadvantage of ^{15}N-^{1}H-HSQC based screening is that it cannot give any information about the ligand that has bound. This information is usually obtained by a deconvolution step, which consists of recording ^{15}N-^{1}H-HSQC spectra of the individual compounds. Alternatively to deconvolution, ligand-based NMR methods described in the next section may be used.

9.3   Ligand-based methods

9.3.1   Dynamic NMR—fast and slow exchange

Before ligand-based methods are discussed it is important to recapitulate some basic concepts of NMR time-scales as applied to binding equilibria. Principally, two limits are distinguished in NMR of exchanging systems. In the special case of ligand binding "exchange" refers to the ligand switching between its bound and unbound states. In the slow exchange limit two separate peaks are detectable, one for bound and one for unbound ligand. In the fast-exchange limit only one average peak reflecting the weighed properties of both forms in equilibrium can be observed (Fig. 3a). "Slow" and "fast" is connected to the difference in chemical shift between the free and the bound state. Reactions that are faster than this frequency difference (in Hz) are classified to be in the fast exchange limit.

Most ligand-based screening methods employ the fast-exchange limit. Some of the NMR parameters, for instance, the linewidth, are drastically different for free *versus* complexed ligand. Therefore, in the fast-exchange limit, an average linewidth is observed which strongly depends on the protein concentration. Large changes in linewidth allow the use of large excesses (10–100 fold) of ligand, so the protein only gives rise to a broad background or hump that does not interfere with the ligand signals.

The requirement for fast exchange also implies limitations in the binding affinity range that can be studied. Viewed from the ligand, a binding equilibrium can be seen as a reaction from the free to the bound state and *vice versa*. Binding affin-

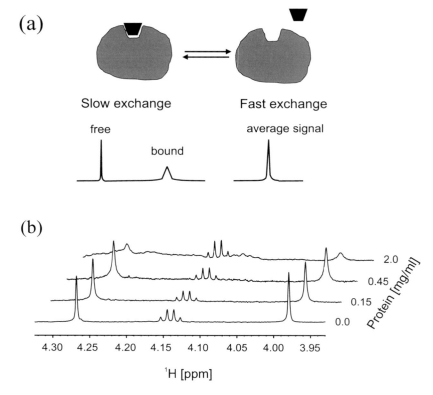

Figure 3. (a) The distinction between slow and fast exchange on the NMR timescale. If the release of the ligand is slower than the frequency difference between the signals, then two signals are observed. If the release is faster than the frequency difference, then only an averaged signal can be acquired. (b) A binding study that was conducted by measuring T_2 relaxation rates by observation of line broadening. Spectra were recorded of varying concentrations of an unknown 200,000 kDa protein from industrial research (0.0 to 2.0 mg/ml) to which a mixture of 10 ligands, each at 1 mM concentration, was added. Shown is a region of the 1D spectrum containing signals of a binding (outer lines) and of a nonbinding ligand (inner multiplet).

ity is measured by the dissociation constant K_D which is strongly linked to kinetics, since K_D is equal to the quotient between on- and the off-rate, k_{on}/k_{off}, of a ligand. The on-rate is usually fast enough to be regarded as diffusion controlled, therefore only the magnitude of the off-rate k_{off} is of importance. Very tight binding molecules, which are characterized by slow off-rates, will fall into the slow-exchange limit. Thus typical dissociation constants for ligand-based binding studies are limited to the micro- and millimolar range.

This situation is different for ^{15}N-^{1}H-HSQC based screening of protein signals (SAR by NMR). Here, weak binding molecules can be detected as well as ones that bind tightly. Generally an excess of ligand is used for ^{15}N-^{1}H-HSQC based screening so that only resonances of the fully ligand saturated protein are observed.

9.3.2 Methods based on relaxation properties

NMR relaxation is associated with the disappearance of magnetization after excitation. If magnetization relaxes fast, NMR signals are broadened. The significance of relaxation properties for screening is due to the fact that relaxation is caused by fluctuating electric fields which are a consequence of molecular rotations. These are tightly coupled to the molecular size, and thus change dramatically for a ligand bound to its large target protein as compared to a free ligand [19].

The most prominent relaxation pathway in macromolecules is the transverse or T_2 relaxation. It increases monotonously with molecular size and is directly related to the observed linewidth. For large proteins T_2 relaxation will be dominant and a severe limitation for structural studies. Figure 3b shows an example of a ligand-protein interaction that was characterized by line-broadening. In the depicted titration study 1D ^{1}H spectra of an unknown protein and 10 unknown ligands were recorded at different protein-ligand ratios. With increasing protein concentration signals of the ligand that binds to the protein become broadened, whereas linewidths of the nonbinding ligand remain unchanged. Although used since the early days of NMR this method remains popular even nowadays due to the trivial acquisition and evaluation of the spectra [4–6 and ref. cited therein, 20]. Line-broadening based methods allow the screening of mixtures of compounds provided that mixtures are carefully assembled such, that ligand signals are not heavily overlapped. This constraint typically limits the number of compounds to 10 per mixture.

Closely related to T_2 relaxation is $T_{1\rho}$ relaxation, the rotating frame relaxation in a weak radio frequency field. By application of a so-called radio frequency spinlock filter it is possible to effectively quench the signals of the protein [21] including those of bound ligands [22]. Since this technique leads to the disappearance of bound ligand signals, difference spectroscopy must be employed by subtracting these signals from a suitable reference spectrum [22]. Unfortunately, in practice, differences between different types of spectra may lead to artifacts, which may make interpretation of the resulting spectra difficult.

Another relaxation phenomenon is the nuclear Overhauser enhancement (NOE) effect (see Chapter 8). It is the result of longitudinal cross relaxation involving two spins in spatial neighborhood, typically less than 6 Å apart. Small molecules up to a few 100 Dalton have weak positive or zero NOEs. Macromolecules, in contrast, are characterized by large and negative NOEs (Fig. 4a). These large negative NOEs also develop within a ligand bound to a protein target and are then transferred into solution upon dissociation of the protein-ligand complex. This so-called transferred NOE (TrNOE) effect [23] can be measured using a simple NOESY experiment. Evaluation is straightforward, since only changes in sign of certain cross peaks in the 2D NOESY spectrum need to be observed. Application of this principle to mixtures [24, 25] directly leads to the use of TrNOE in screening. Figure 4b shows an example. Using transferred NOEs, Henrichsen et al. [25] were able to identify an E-selectin antagonist from a mixture of 11 possible ligands. In this case application of SAR by NMR would have

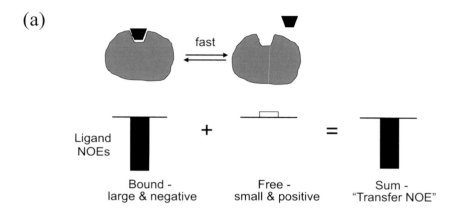

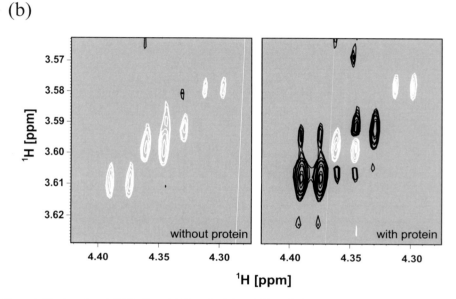

Figure 4. The transferred NOE effect. (a) The principles of this technique: In the fast exchange limit, the signals for bound and unbound ligand add up, and the sum is dominated by the much larger negative NOE effect arising from the bound ligand. (b) A TrNOE study (adapted from [37]). NOESY spectra of samples, which contain four isomeric methylated β-methylgalactosides, at concentrations of 1 mM each. When a galactose-binding lectin from elderberry is added to one of the samples (right spectrum), signals of two of the ligands change their sign, indicating that two of the four isoforms bind to the lectin protein. Positive peaks are displayed as white contours, while negative peaks are shown as black contours.

been impossible since the molecular weight of E-selectin at 220 kDa is far beyond the present limit for the SAR by NMR screening technique.

The main requirement for the application of TrNOEs is the existence of pairs of ligand protons that are both close in space and well separated in the NMR spectrum to avoid spectral overlap with diagonal peaks. To overcome this overlap problem the TrNOE effect can also be used as a filter in multidimensional spectra such as the 3D TOCSY-TrNOESY spectra [26]. Such spectra are rather time-consuming and thus are not appropriate for fast screening applications.

Besides its value in screening, TrNOE spectra can yield important information of the structure of the bound ligand [23]. This may be crucial when attempting to superimpose several flexible ligands for the construction of a pharmacophore model. A whole arsenal of other NMR methods such as isotope-filter methodology [27] exists to obtain detailed information about the binding mode of the ligand. Isotope-filters and isotope edited NOE spectra can yield direct distance constraints between ligand and protein and therefore will allow precise docking of the ligand into the binding site. These methods cannot be termed "screening" any more, but provide a direct connection to structure-based drug design [1–6], a link between structural and functional methodology, which is unique in biophysical chemistry.

9.3.3 Diffusion methods

The use of pulsed field gradients in high-resolution NMR readily allows the measurement of diffusion rates of solutes in liquids [28]. The concepts inherent in this technique are illustrated in Figure 5. Application of a pulsed field gradient causes a spatial change in the B_0 magnetic field of the NMR instrument and results in a spatially encoded dephasing of magnetization. After switching off the gradient all nuclear spins are dephased and thus would add up to zero signal. Fortunately the process of dephasing is reversible, and applying exactly the same gradient with opposite sign will rephase spins. Due to the spatial encoding of the phases the amount of magnetization that can be restored by the second gradient pulse is inversely proportional to the diffusion that has taken place in the meantime. The less a molecule has moved, the more magnetization can be recovered by this so-called gradient echo. This principle is easily applied to screening. If a small molecule is bound to a large protein, its diffusion will be slowed down due to time spent in the bound state and the much slower diffusion rate of the protein. Thus less magnetization will be recovered for non-binding molecules, which diffuse more rapidly, than for ones that bind, which diffuse more slowly.

Diffusion is measured by systematically varying diffusion time, gradient length, or gradient strength. The relationships between these parameters and the Stokes-Einstein diffusion coefficient D can then be used to obtain the values of D from several experiments. For screening purposes, this would still mean rather long measuring times. It is therefore easier to compare relative signal intensities from diffusion experiments with and without protein. In a difference spectrum of ligand signals recorded with and without protein, only the signals of bound ligands should remain visible [23, 30–33]. The diffusion filter building block can also be incorporated into 2D pulse sequences. Especially the TOCSY spectrum, where correlation peaks between all protons belonging to the same spin system

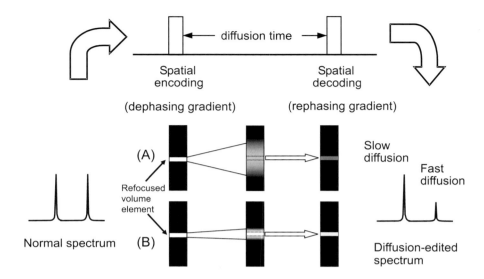

Figure 5. Principle of diffusion-based NMR spectroscopy. A dephasing gradient pulse encodes each volume element along the magnet axis. Fast diffusion as in (A) spreads this encoded magnetization over a larger volume than slow diffusion (B). By a second, rephasing gradient only the signals of molecules that did not change their position during the diffusion time are recovered. This scheme can be incorporated into 2D sequences or more complex 1D schemes such as the NOE pumping experiment discussed below (Fig. 7a).

are observed, can aid in the deconvolution of otherwise highly overlapped 1D spectra. The diffusion weighed TOCSY (or DECODES [29]) spectrum has proven valuable in this respect [30–33]. Alternatively, provided that the difference in diffusion behavior between bound and non-bound ligands is large enough, one can selectively suppress free ligands to yield 1D- or 2D-spectra of the bound species [23, 30–33]. This approach has been applied to screen peptide libraries for binding to vancomycin [33] and to monitor the binding of a dye to DNA [32].

9.3.4   "Combination" methods

In the comparatively simple methods discussed so far only one NMR parameter is exploited to detect binding. The more powerful "combination" methods discussed subsequently employ a more complex strategy. The Saturation Transfer Difference (STD) [34] technique depicted in Figure 6a is an example for such a combined method, which uses selective irradiation of the protein in combination with magnetization transfer effects due to ligand binding.

In the STD technique a long and selective irradiation pulse (several seconds) is used to saturate protein spins, resulting in the equilibration of the population of α and β spin states. Thus saturation can be viewed as the bleaching out of protein

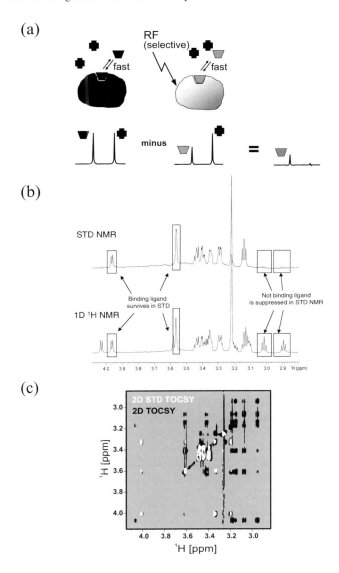

Figure 6. The saturation transfer difference (STD) NMR. (a) Illustration of the technique: The protein magnetization is saturated ("bleached out") via selective radio frequency (RF) irradiation of protein resonances. Saturation then spreads through the whole protein by spin-diffusion. Saturation only affects bound ligands, but not free ligands. This effect can be seen in a difference spectrum, in which only signals of bound ligands survive. Panel (b) and (c) show 800 MHz STD NMR spectra of 1 mM β-methylglucoside, 1 mM β-methylgalactoside and 0.02 mM of a galactose-binding lectin from elderberry. (b) 1D STD NMR spectrum (top) as compared to a reference 1D spectrum (bottom). (c) A 2D STD-TOCSY (white) is overlayed on top of a standard 2D-TOCSY (black) spectrum. In the black spectrum spinsystems of all ligands are visible, whereas in the white STD filtered TOCSY spectrum only spinsystems of bound ligands are seen. Measuring time was 12 min for the 1D-STD spectrum in (b) and 12 h for the 2D STD-TOCSY shown in (c).

resonances since the difference in the population of α and β spin states is directly proportional to the NMR signal. It is important to carefully select the frequency of the saturation pulse by centering it on a few isolated resonances of the protein such that signals of unbound ligands are not bleached out as well. Although only few resonances of the protein are hit initially, saturation spreads out through the whole protein mediated by NOEs between spatially close protons. This process is called spin diffusion and increases in efficiency with the protein's molecular size. Naturally, bound ligand will also be saturated. Upon dissociation, saturation of the bound ligand is transferred to the free ligand population, which slightly attenuates signals of ligands that bind to the target protein. The small decrease in intensity of a bound ligand can be detected by subtracting the spectrum from a reference spectrum. In the resulting difference spectrum, only signals of bound ligands survive. Moreover, this concept can be incorporated into virtually any modern NMR experiment. As can be seen in Figure 6b and c, 1D and 2D STD homonuclear correlation spectra allow the unique identification of the bound β-methylgalactoside from a mixture of β-methyl-galactoside and β-methyl-glucoside. Furthermore, STD spectroscopy can be combined with high-resolution magic angle spinning (HR-MAS) techniques, commonly used for solid state NMR purposes, to allow screening using an immobilized protein. Immobilized protein may significantly aid recovery of precious ligands. As a proof of principle, HR-MAS in combination with STD NMR was utilized to detect a lectin-sugar interaction [35] when the lectin was tethered to controlled porous glass.

Very similar to STD spectroscopy is the NOE pumping experiment [36] illustrated in Figure 7a. In this technique protein magnetization is selected not by spin diffusion, but via a diffusion filter similar to the one described above for diffusion-based NMR screening. The resulting protein magnetization is transferred onto a bound ligand by intermolecular NOEs. Subsequently ligand magnetization is shuttled to the solution upon dissociation of the ligand-protein complex. In contrast, molecules that do not bind to the protein target will not build up magnetization in solution. Therefore, only magnetization from bound ligands will be detected in a 1D experiment.

The basic idea of selecting protein signals using the protein's spectroscopic properties can also be applied to exclusively select ligand signals. This principle, which is complementary to the other two NOE based methods, has been termed "reverse NOE pumping" (RNP) [37] and functions as illustrated in Figure 7b. First, ligand signals are selected by a simple T_2 filter, which utilizes the much faster T_2 relaxation of a protein as compared to a ligand. After selection of the ligand magnetization, those molecules that bind to the protein will lose some of their magnetization to the (unmagnetized) protein scaffold. Consequently, signals of ligands that bind to the protein will be attenuated, whereas signals of non-binding small molecules will remain unchanged. In a difference spectrum, again only resonances of bound ligands remain.

The common scheme of these methods is that they consist of two steps. In the first step either ligand or protein magnetization is selected by utilizing the specific NMR properties of the ligand or protein. In the second step, magnetization is

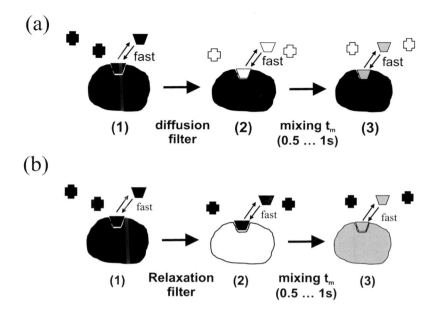

Figure 7. Simple and reverse NOE pumping experiments. (a) The NOE pumping experiment starts with initial magnetization uniformly distributed among all molecules (1). From this magnetization only protein magnetization is selected (2) via the slow diffusion properties of the protein (diffusion filter). This magnetization is then transferred to bound ligands (3). Ligand signals of bound ligands can be detected, whereas unbound ligands remain invisible. No subtraction is necessary, which reduces experiment time and avoids subtraction artifacts. (b) The reverse NOE pumping experiment (RNP) also starts from an initial uniform magnetization of protein and ligands (1). Then, only the magnetization of the ligands is selected (2) by utilizing the more favorable relaxation properties of the ligands (relaxation filter). Bound ligands lose magnetization while they are associated with the protein, which is not the case for free ligands. The lost magnetization can be detected by subtraction from a reference spectrum.

transferred to the binding partner by intermolecular NOEs. In the case of STD and RNP, subtraction from a reference spectrum follows. All three methods end up with ^{1}H magnetization solely of bound ligands, which can subsequently be used for a whole range of 2D-NMR techniques. However, multidimensional NMR spectra utilizing NOE pumping or RNP have not yet been reported in the literature.

9.4   Compound libraries for NMR-based screening

Construction of diverse and well-behaved small molecule libraries for screening remains a topic of much interest but also one of controversy (see Chapter 7 and 10). One contributing factor to this state of affairs is that there is only little published information about small molecule libraries and much of existing information is proprietary. Also, many cooperate libraries have been constructed by accumulation of compounds synthesized in-house over several decades. Therefore many of these libraries are random collections instead of systematic assemblies of substances.

The availability and ease of generating vast libraries using combinatorial chemistry (see Chapter 6) has prompted most of the discussion about diversity of current libraries. Unfortunately, combinatorial libraries of limited chemical diversity are often of little use during the initial screening process. On the other hand directed combinatorial libraries can prove highly effective once initial low affinity ligands have been identified. By using a very low diversity library of methylated sugars in combination with NMR screening it was possible to obtain information about critical sugar hydroxyl groups [38]. Two binding ligands from a mixture of 20 of these sugar derivatives with identical sum formulae could be identified.

There are several key criteria that should be taken into account when building up a library for NMR screening from scratch. Compounds should be commercially available in milligram to gram quantities and their solubility should not be below 1 mM in H_2O. The library should be a set of sufficiently chemically diverse compounds, which are chemically and isomerically pure, and non-reactive and non-toxic. Of significant importance is that compounds consist of chemical modules with functional groups or linkers that are amenable to simple high-yield chemistry. For ligand based screening a simple well resolved ^1H NMR spectrum for each compound is highly desirable.

Libraries currently used for NMR screening purposes come in two main types: large libraries of 50,000 to 200,000 compounds and small but maximally diverse libraries of approx. 200 to 500 compounds. Naturally, the size of the library will critically depend on the screening power or throughput of the NMR method employed. On the one hand, if the particular NMR method, such as ^{15}N-^1H-HSQC-based screening using cryogenically cooled NMR probes [17] allows screening of large numbers of small molecules, then emphasis will be placed on obtaining a sufficient number of such compounds. In this case not all molecules need to be mutually diverse (for a definition of diversity see Chapter 7). In fact, small clusters of compounds with reduced diversity that, for instance, only differ by a methyl group, may even be desirable to exploit potential differences in binding affinities due to small steric perturbations.

On the other hand, design of a small library for NMR-based screening methods should place significantly more emphasis on diversity. A particular philosophy for the design of small and diverse libraries has been published from a group of researches at Vertex cooperation [39–41]. This group has used data mining tools to analyze the Comprehensive Medicinal Chemistry (CMC) databank of all known drugs [39, 40]. Surprisingly, they find that when stripped of their side chain atoms half of the known drugs can be described with only a limited set of 32 chemical frameworks [39]. Furthermore, 60% of all (non-carbonyl) side chains of these molecules can be mapped to 20 distinct functional groups [40]. Using this data Fejzo et al. [41] have constructed a small but diverse library of approximately 200 compounds consisting of representatives of all 32 frameworks combined with a diverse distribution of side chains selected from the 20 dominant functional groups. Using this library they have developed the SHAPES strategy for NMR screening [41] in which weak binding compounds identified via diffusion edited spectroscopy are used to guide computational clustering of compound databases and therefore significantly bias compounds that then undergo high-throughput screening. This NMR-based prescreening strategy resulted in a fourfold increase in HTS hit rates when compared to a randomly selected set of compounds.

9.5 Summary and conclusions

Diversity and robustness of NMR based screening methods make these techniques highly attractive as tools for drug discovery. Although not all screening techniques discussed here may be applicable to any given target, there is however a good chance that at least one of the described methods will prove productive in finding several medium affinity ligands. A comparison of each of the methods is given in Table 1. For drug targets of molecular weight <30 kDa SAR by NMR appears to be the method of choice since it yields detailed information about the location of the binding site. It remains to be seen whether ^{15}N-^{1}H-TROSY based screening techniques will prove useful for larger protein targets, especially considering the added effort needed for spectral assignment and the increased complexity due to spectral overlap. Nevertheless, with the application of new cryo-cooled NMR probes, ^{15}N-^{1}H-HSQC based screening can now be considered a high throughput method.

Ligand-based NMR screening methods can be used for protein targets of virtually any size, but are restricted in the ligand's binding affinity range. Because sufficient ligand-protein dissociation rates are needed, only binding of ligands with low (milimolar) to intermediate (micromolar) affinities is detectable. It is expected that cryo-cooled NMR probe technology will also advance ligand detected NMR screening to the high throughput level. Certainly protein and ligand concentrations can be lowered drastically and experiment times can be shortened with increased sensitivity. However, spectral overlap will be of major concern when mixtures of up to 100 compounds are to be screened. For such applications only techniques for which the signals of bound ligands survive will be useful, and sophisticated software will be needed to deconvolute the spectra of multiple bound ligands. Although only ligands with medium to low affinities can be found, ligand based NMR screening has been used as an effective prescreening tool for assay based high throughput screening. Identifying a large ensemble of medium affinity ligands may not only aid in building a binding site pharmacophore model (see Chapter 11), but also may yield crucial information for overcoming tissue availability, toxicity, or even intellectual property related problems.

Although NMR based screening is only one of the more recent additions to the bag of tools used in drug discovery [1, 2], its simplicity and wide range of application (including protein-protein and protein-nucleic acid interactions) has attracted much attention. Advances in NMR instrumentation and methodology have already paved the road for NMR based screening to become a high throughput technique. In addition to this, NMR is exceptional in the amount of detailed struc-

Table 1. Comparison between the screening techniques discussed in the text. Given concentrations are those stated in the cited references. The cryo-cooled probe technology was not taken into account for the ligand-based methods, where it would also facilitate detection of binding at much lower concentrations.

Protein based	Protein size	Mixture size	Typical published concentrations and additional requirements	Characteristics *Typical measuring time*	Ref.
Conventional	< 30 kDa	< 10		Observation of protein backbone chemical shifts. Identification of binding site possible *10–30 min (2D-HSQC)*	[7–9]
TROSY	< ca. 100 kDa	< 10	1 mM ligand, 0.3 mM ^{15}N-labelled protein		[15]
Cryocooled probe		< 100	50 μM ligand and protein		[16]

Ligand based	Ligand size	Mixture size	Typical published concentrations and additional requirements	Characteristics *Typical measuring time*	Ref.
Line broadening		< 10	1 mM ligand, 50–200 μM protein ligand signals must not be overlapped	Fast and simple *<5 min (1D)*	[3–6]
T$_2$ filtering		< 10	50 μM ligand and protein	Fast, difference method *<5 min (1D)*	[21]
Transfer NOE	Small (500– 1000 Da)	< 20	1 mM ligand, 50 μM protein Spatially close protons are needed	Time-consuming Simple evaluation *1–4 h (2D-NOESY)*	[22–24]
Diffusion		< 10	100 μM–10 mM ligand and protein (equimolar)	Fast, extension to 2D possible *<10 min (1D)*	[21, 28-33]
STD		< 20	1 mM ligand, 10-20 μM protein	Fast, sensitive, extension to 2D possible *< 10 min (1D)* *12–24 h (2D)*	[33,34, 37]
NOE pumping		< 10	10 mM ligand, 100 μM protein	Fast	[35]
RNP		< 10	1 mM ligand, 20 μM protein	*<10 min (1D)*	[36]

tural information it can provide. Not only can NMR readily reveal the binding site (^{15}N-^{1}H-HSQC screening) or the conformation of the bound ligand (transfer NOE), but it can also supply information that enables precise docking of the ligand to the protein's binding pocket (isotope-filtered NOESY). NMR data can therefore provide a natural connection between experimental HTS and combinatorial chemistry techniques with computational methods such as 3D-database searching (see Chapter 10), virtual screening (docking) and structure-based ligand design (see also Chapter 8).

9.6   References

1 Roberts GCK (1999) NMR spectroscopy in structure-based drug design. *Curr Opin Biotech* 10: 42–47
2 Moore JM (1999) NMR screening in drug discovery. *Curr Opin Biotech* 10: 54–58
3 Roberts GCK (2000) Applications of NMR in drug discovery. *Drug Discovery Today* 5: 230–240
4 Feeney J, Birdsall B (1993) NMR studies of protein-ligand interactions. In: Roberts GCK (ed): *NMR of macromolecules*. Oxford University Press, Oxford, 183–215
5 Jardetzki O, Roberts GCK (1981) *NMR in molecular biology*. Academic Press, San Diego
6 Craig DJ, Higgins KA (1998) NMR studies of ligand-macromolecule interactions. *Annu Rep NMR Spectrosc* 22: 61–138
7 Shuker SB, Hajduk PJ, Meadows RP et al (1996) Discovering high-affinity ligands for proteins: SAR by NMR. *Science* 274: 1531–1534
8 Hajduk PJ, Sheppard G, Nettesheim DG et al (1997) Discovery of potent nonpeptide inhibitors of stromelysin using SAR by NMR. *J Am Chem Soc* 119: 5818–5827
9 Hajduk PJ, Dinges J, Miknis GF et al (1997) NMR-based discovery of lead inhibitors that block DNA binding of the human papillomavirus E2 protein. *J Med Chem* 40: 3144–3150
10 Hajduk PJ, Dinges J, Schkeryantz JM et al (1999) Novel inhibitors of Erm methyltransferases from NMR and parallel synthesis. *J Med Chem* 42: 3852–5859
11 Sattler M, Schleucher J, Griesinger C (1999) Heteronuclear multidimensional NMR experiments for the structure determination of proteins in solution employing pulsed field gradients. *Prog NMR Spectrosc* 34: 93–158
12 Rizo J, Liu ZP, Gierasch LM (1994) ^{1}H and ^{15}N resonance assignments and secondary structure of cellular retinoic acid-binding protein with and without bound ligand. *J Biomol NMR* 4: 741–760
13 Hensmann M, Booker GW, Panayotou G et al (1994) Phosphopeptide binding to the N-terminal SH2 domain of the p85· subunit of PI 3'-kinase: A heteronuclear NMR study. *Protein Science* 3: 1020–1030
14 Dalvit C, Floersheim P, Zurini M et al (1999) Use of organic solvents and small molecules for locating binding sites on proteins in solutions. *J Biomol NMR* 14: 23–32
15 Pervushin K, Riek R, Wider G et al (1997) Attenuated T_2 relaxation by mutual cancellation of dipole-dipole coupling and chemical shift anisotropy indicates an avenue to NMR structures of very large biological macromolecules in solution. *Proc Natl Acad Sci USA* 23: 12366–12371
16 Pellecchia M, Sebbel P, Hermanns U et al (1999) Pilus chaperone FimC-adhesin FimH interactions mapped by TROSY-NMR. *Nat Struct Biol* 4: 336–339
17 Hajduk PJ, Gerfin T, Boehlen JM et al (1999) High-throughput nuclear magnetic resonance-based screening. *J Med Chem* 42: 2315–2317
18 Lian LY, Roberts GCK (1993) Effects of chemical exchange on NMR spectra. In: GCK Roberts (ed): *NMR of macromolecules*. Oxford University Press, Oxford, 153–182
19 van de Ven FJM (1995) *Multidimensional NMR in liquids*. VCH, Weinheim
20 Limmer S, Vogtherr M, Nawrot B et al (1997) Specific recognition of a minimal model of aminoacylated tRNA by the elongation factor Tu of bacterial protein biosynthesis. *Angew Chem Int Ed Engl* 36: 2485–2489
21 Scherf T, Anglister J (1993) A T1 rho-filtered two-dimensional transferred NOE spectrum for studying antibody interactions with peptide antigens. *Biophys J* 64: 754–761
22 Hajduk PJ, Olejniczak ET, Fesik SW (1997) One-dimensional relaxation- and diffusion-edited NMR methods for screening compounds that bind to macromolecules. *J Am Chem Soc* 119: 12257–12261
23 Ni F (1994) Recent developments in transferred NOE methods. *Prog NMR Spectrosc* 26: 517–606
24 Meyer B, Weimar T, Peters T (1997) Screening mixtures for biological activity by NMR. *Eur J Biochem* 246: 705–709
25 Henrichsen D, Ernst B, Magnani JL et al (1999) Bioaffinity NMR spectroscopy: Identification of an E-selectin antagonist in a substance mixture by transfer NOE. *Angew Chem Int Ed Engl* 38: 98–102
26 Herfurth L, Weimar T, Peters T (2000) Application of 3D TOCSY-TrNOESY for the assignment of bioactive ligands from mixtures. *Angew Chem Int Ed Engl* 39; 2097–2099
27 Breeze AL (2000) Isotope-filtered NMR methods for the study of biomolecular structure and interactions. *Prog NMR Spectrosc* 36: 323: 372
28 Stejskal EO, Tanner JE (1965) Spin diffusion measurement: Spin echoes in the presence of a time-dependent field gradient. *J Chem Phys* 42: 288–292
29 Lin M, Shapiro MJ (1996) Mixture analysis in combinatorial chemistry. Application of diffusion-resolved NMR spectroscopy. *J Org Chem* 61: 7617–7619

30 Lin M, Shapiro MJ, Wareing JR (1997) Screening mixtures by affinity NMR. *J Org Chem* 62: 8930–8931

31 Lin M, Shapiro MJ, Wareing JR (1997) Diffusion-edited NMR—affinity NMR for direct observation of molecular interactions. *J Am Chem Soc* 119: 5249–5350

32 Anderson RC, Lin M, Shapiro MJ (1999) Affinity NMR: Decoding DNA binding. *J Comb Chem* 1: 69–72

33 Bleicher K, Lin M, Shapiro MJ et al (1998) Diffusion edited NMR: Screening compound mixtures by affinity NMR to detect binding ligands to vancomycin. *J Org Chem* 63: 8486–8490

34 Mayer M, Meyer B (1999) Characterization of ligand binding by saturation transfer difference NMR spectra. *Angew Chem Int Ed Engl* 35: 1784–1788

35 Klein J, Meinecke R, Mayer M et al (1999) Detecting binding affinity to immobilized receptor proteins in compound libraries by HR-MAS STD NMR. *J Am Chem Soc* 121: 5336–5337

36 Chen A, Shapiro MJ (1998) NOE pumping: A novel NMR technique for identification of compounds with binding affinity to macromolecules. *J Am Chem Soc* 120: 10258–10259

37 Shapiro MJ, Chen A (2000) NOE pumping. 2. A high-throughput method to determine compounds with binding affinity to macromolecules by NMR. *J Am Chem Soc* 122: 414–415

38 Vogtherr M, Peters T (2000) Application of NMR based binding assays to identify key hydroxy groups for intermolecular recognition. *J Am Chem Soc* 122: 6093–6099

39 Bemis GW, Murcko MA (1996) The properties of known drugs. 1. Molecular frameworks. *J Med Chem* 39: 2887–2893

40 Bemis GW, Murcko MA (1999) Properties of known drugs. 2. Side chains. *J Med Chem* 42: 5095–5099

41 Fejzo J, Lepre CA, Peng JW et al (1999) The SHAPES strategy: An NMR-based approach for lead generation in drug discovery. *Chem Biol* 6: 755–769

Modern Methods of Drug Discovery
ed. by A. Hillisch and R. Hilgenfeld
© 2003 Birkhäuser Verlag/Switzerland

10 Structure-based design of combinatorial libraries

John H. van Drie[1], Douglas C. Rohrer[2], James R. Blinn[2] and Hua Gao[2]

[1] Vertex Pharmaceuticals, 130 Waverly St, Cambridge, MA 02139, USA
[2] Pharmacia, Discovery Research, 7000 Portage Road, Kalamazoo, MI 49008, USA

10.1 Introduction

During the 1980s, modern drug discovery was dominated by the idea of "structure-based drug design." The promise was, given an x-ray structure of the protein target, one could design the ideal protein ligand in a small number of iterations. If our computational methods could predict accurately which ligands would bind a protein best, this approach would be wildly successful; however, our methods are not that accurate, and in the end many experimental iterations have been required (for a review, see [1]). It was during this era that a small number of groups began experimenting with novel approaches to discovering leads *in silico*, now generically called "3D database searching." Here, either a pharmacophore or a protein pocket was used to sift through a database of conformers of molecules, and those molecules which met the criteria were submitted to a biological assay (for a review, see [2]).

In the late 1980s and early 1990s, combinatorial chemistry emerged, with the promise of making billions of molecules (see also Chapter 6). The pendulum had swung away from the notion of designing the one perfect molecule and had moved to the antipodal position: make billions of molecules, screen them all to look for the hot molecule (see Chapter 4). Although the initial efforts in combi-chem relied on peptides [3], it was not until the first organic libraries were prepared by Bunin and Ellman [4] and DeWitt et al. at Parke-Davis [5] that this became a technology enthusiastically embraced by the pharmaceutical industry.

By the mid-1990s, the impracticality of making so many molecules became apparent, and the pendulum began to swing back again. Led by Eric Martin and his colleagues at Chiron among others [6], attention was directed at selecting an ideal, diverse subset of the billions of possibilities that combi-chem offered. This spawned many Platonic discussions about what constituted the "ideal diverse library" (compare Chapter 7).

Now, the pendulum continues to swing back even more. Chemists are now directing their attention less to the use of combi-chem in screening libraries (with library size in the range of 100,000–1,000,000 molecules), and more to the use of combi-chem in the design of directed libraries (with library size in the range 100–1000 molecules). With directed libraries, the aim is to design a library where the molecules should possess biological activity against a given target. Now, the challenge becomes to exploit either the earlier SAR of molecules against that target, or to exploit a protein structure, to drive the design of these directed libraries to maximize the number of compounds in the library which are active. This process has been termed *structure-based combi-chem* [7]. The techniques of 3D database searching developed in the 1980s have been adapted to meet this challenge. Rather than screening a database of molecules readily available for testing, we are now screening *virtual libraries*: a computational description of the molecules that a given combinatorial scheme is capable of generating. The pharmacophore or protein pocket is then used to select a subset of that virtual library, according to how well the molecule satisfies the constraints of the pharmacophore or protein pocket.

In this chapter we consider multiple methods for the structure-based design of combinatorial libraries. In cases where the only prior knowledge about the target is from the structure-activity relationship, three possibilities are considered: similarity searching, "binary QSAR," and pharmacophore discovery coupled with 3D database searching. In cases where a macromolecular structure of the target is available, three software tools are described: CombiBUILD, Omega, and SLIDE. The relative advantages and disadvantages of each method are outlined. Examples of successful applications of virtual screening and 3D database searching in the literature are reviewed.

10.2 3D database searching in lead discovery

Our own recent experiences have focused less on the application of 3D database searching in lead discovery, and more on their use in combinatorial library design. Nonetheless, it is worthwhile to examine the experiences of other groups that have been reported in the literature on 3D database searching in lead discovery. These applications were last reviewed comprehensively in 1995 [2]. Since that review, a number of new applications of 3D database searching to lead discovery have appeared, and the pace of such publications is accelerating. In the following, we do not aspire to make a comprehensive review of that literature; these reports highlighted here are a subset of the literature of pharmacophoric 3D database searching which has appeared since that 1995 review. For a review on "virtual screening," see Walters et al. [23] and Shoichet and Bussiere [8].

G. W. Milne's laboratory at the NCI has been exceptionally prolific in the application of pharmacophoric 3D database searching to lead discovery, and unusually methodical in probing the assumptions underlying the use of these methods. Subsequent to their work already cited in the 1995 review, they have reported on the discovery of novel HIV-1 protease inhibitors [9]. A pharmacophore was derived from the x-ray co-crystal structures, containing three features (two hydroxyls, one carbonyl, with defined geometric constraints). They performed a search of the open NCI database (206,000 compounds), retrieving 2400 hits, which was further filtered by novelty and the presence of an additional hydrophobic feature, to a final 50 compounds submitted to biological testing. The two best hits were shown to possess sub-µM activity against the enzyme; one showed antiviral activity in a cell-based activity at 12 µM.

Milne's laboratory also reported on the discovery of HIV-1 integrase inhibitors by 3D database searching [10]. Using known HIV-1 integrase inhibitors, a pharmacophore was developed, containing three features (chosen from 2 hydroxyls, and two carbonyl oxygens). Searching the open NCI database, 340 hits were found. This led to 10 novel, structurally-distinct HIV-1 integrase inhibitors, four of which were below 30 µM.

A second report from Milne's laboratory on the discovery of HIV-1 integrase inhibitors via pharmacophoric 3D database searching described a different tack [11]. Natural product inhibitors of this enzyme were used to construct two three-feature pharmacophores, each containing two oxygens and one hydroxyl with differing geometric constraints. These were both used to search the open NCI database, yielding 800 unique hits from the two pharmacophore searches combined. Twenty-seven hits showed inhibition better than 100 µM.

Kaminski et al. [12] discovered novel farnesyl protein transferase (FPT) inhibitors via 3D database searching. Using known FPT inhibitors, a five-feature pharmacophore was constructed in an automated fashion using the Hypothesis Generation module of Catalyst [13]. Four of these five features were "hydrophobic" features, the fifth a H-bond acceptor, exhibiting what many have argued is a flaw in the Hypothesis Generation procedure—a tendency to exaggerate the contribution of hydrophobic features in a pharmacophore. Nonetheless, despite this non-selective, "greasy" pharmacophore, Kaminski et al. identified 718 hits in the Schering-Plough corporate database, five of which showed *in vitro* FPT inhibition better than 5 µM. This led to a novel series of dihydrobenzothiophene FPT inhibitors.

Llorens et al. [14] describe the discovery of amygdalin as a CD4 receptor ligand via pharmacophoric 3D database searching. The structure-activity relationship (SAR) of analogs of peptide T led them to propose a four-feature pharmacophore for CD4-peptide T interaction (a phenyl, an amide, a carbonyl, and a hydroxyl in a specific geometric arrangement). Searching the NCI, Derwent, Maybridge, and Biobyte databases, they retrieved an unspecified number of hits, one of which was amygdalin, which was found to bind to CD4 with a sub-µM IC_{50}.

Marriott et al. [15] describe the discovery of novel muscarinic (M3) antagonists via pharmacophoric 3D database searching. Using known SAR against the M3 receptor, they developed a four-feature pharmacophore (two H-bond donors, one H-bond acceptor, and a basic, tertiary amine, in a specific geometric arrangement) using the automated procedure of DISCO [16]. With this pharmacophore, they performed a 3D search against their corporate database. They retrieved 177 hits, three of which were found to have potent M3 antagonist activity, many structurally-distinct from known M3 antagonists.

Kiyama et al. [17] report the discovery of novel angiotensin II receptor antagonists, by finding replacements for the biphenyltetrazole moiety of DuP 753. Using a three-feature pharmacophore (2 aryl groups, and the third group one of: carboxyl, ester, amide, or tetrazole, in a specific geometric arrangement), they performed a search of the MDDR database of 94.000 compounds, retrieving 139 hits, which ultimately led to a series of novel AII antagonists containing a tricyclic dibenzocycloheptene system, with affinities in the range 0.29–12 nM.

One of the most ingenious applications of pharmacophoric 3D database searching was described by Greenidge et al. [18]. Beginning with the x-ray co-crystal structure of the thyroid hormone bound to its receptor, they define a pharmacophore consisting of seven features (five hydrophobic features, 2 H-bond acceptors) augmented critically by ~100 exclusion spheres to represent the steric boundaries of the active site. Searching the Maybridge database of 47,000 compounds, this pharmacophore retrieved only one hit, a compound whose IC_{50} against the thyroid hormone receptor was 69 μM (by contrast, the pharmacophore devoid of exclusion constraints retrieved four hits). Those authors appear to express surprise that the Catalyst 3D database searching software was capable of employing numerous exclusion spheres. It should be noted that this software originally had been designed with such a capability, based on the experiences with ALADDIN [19], which similarly allowed the description of steric constraints with an unlimited number of exclusion spheres.

An impressive contribution was made by a group at the Georgetown Institute for Cognitive and Computational Science. Wang et al. [20] report on the discovery of a novel dopamine transporter inhibitor via 3D database searching. Their pharmacophore contained three features (phenyl, carbonyl oxygen, secondary or tertiary basic amine, in a specific geometric arrangement). Using this pharmacophore to search the NCI database, they retrieved 4,000 hits, which was filtered heuristically to 385 hits, one of which was used as a lead for further optimization, ultimately leading to a compound whose binding affinity was 11 nM.

To our knowledge, the *non plus ultra* of lead discovery via 3D database searching in the 1990s was the little-noticed contribution of Leysen and Kelder in the discovery of Org-8484, a exceptionally-selective 5-HT2c agonist [21]. Beginning with the known structure-activity relationship of various serotonin (5-HT) ligands, they first report the result of 2D substructure searching, next the result of application of the principles of bioisosterism, and finally the result of pharmacophoric 3D database searching. Their 3D pharmacophore was quite simple: two features (phenyl ring, basic amine), with geometric constraints based on the putative receptor site point with which the amine interacts. Using this pharmacophore to search their corporate database, one hit was especially noteworthy: Org-8484, showing 6 nM activity against 5-HT2c and 1600 nM against 5-HT2a. This selectivity for 5-HT2c over 5-HT2a is unparalleled in the serotonin literature.

10.3 3D database searching in combinatorial library design

Our own recent interest has focused on the use of 3D database searching and related techniques in combinatorial library design. The process we follow proceeds in three stages:

Stage I is the domain of the synthesis chemist. A scaffold is chosen, usually based on its prior success at leading to active molecules against the given target; this scaffold is frequently found via screening. The chemist chooses the scheme for performing combi-chem on that scaffold, which leads to the criteria by which reagents can be selected from reagent libraries. For example, the scheme selected by Kick and Ellman for their cathepsin D library ([22], see Fig. 1) led to the identification of 700 amines, and 1900 acylating agents which could be used to construct molecules.

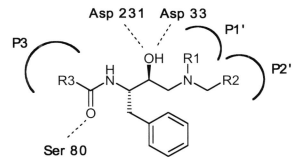

Figure 1. Combinatorial library design of Kick Roe et al. in the design of cathepsin D inhibitors.

Stage II is the construction and filtering of a "virtual library," a description of all molecules which could be made by that scaffold and those reagents selected in Stage I. When no filtering is applied, this often leads to virtual libraries of the order of 10^9 molecules, as in the Kick and Ellman library. Two groups have independently proposed methods for filtering these virtual libraries. Mark Murcko and his colleagues at Vertex [23] describe REOS (Rapid Elimination of Swill) to apply simple rules to eliminate molecules that have unlikely to be interesting even if they turn up active. Chris Lipinski and his colleagues at Pfizer [24] boldly proposed the "rule of 5," a series of simple rules that filter out molecules that are unlikely to have acceptable pharmacokinetic properties, based on well-known principles governing such properties [25] (see Chapter 12). According to Lipinski's rules," only molecules possessing the following properties should be considered:

- molecular weight <500
- calculated logP <5
- number of hydrogen-bond donors <5
- number of H-bond donors + number of H-bond acceptors <10

In stage III, a subset of this virtual library is chosen, based on the likelihood those molecules will be active at the given biological target. In our laboratories, we have been experimenting with a variety of approaches for designing this final library, based on the chemical structure, known biological activities, and possibly a macromolecular structure of the target:

- 2D similarity searching to find molecules in that virtual library most like known actives;
- "binary QSAR," in which a model is derived based on the known structure-activity relationship, with the selection of compounds from that virtual library based on their predicted activity according to that model;
- DANTE pharmacophore discovery with 3D database searching, in which a pharmacophore is derived based on the known SAR, and a 3D database is constructed for the virtual library. 3D database searching is performed using that pharmacophore against that 3D database to select molecules from the virtual library that are either most likely to be active, or which probe regions of space hitherto unexplored by the SAR.
- Using a protein structure, CombiBUILD constructs its own virtual library, and selects from that those molecules which best complement the protein active site. This relies on a model of how the scaffold binds to the active site.
- Using a protein structure, Omega analyzes an entire virtual library, and selects from that those molecules which best complement the protein active site. This relies on a model of how the scaffold binds to the active site.
- Using a protein structure, SLIDE can analyze an entire virtual library, and selects from that those molecules which best complement the protein active site. This relies on a model of how the scaffold binds to the active site. SLIDE also takes into account the potential for the protein to change its conformation, to optimally accommodate the ligand.

What follows describes our experiences with these methods. Our experience thus far is limited, and the details are still proprietary; nonetheless, we can summarize at a general level what we've learned in the application of these methods. Our confidence in these methods is based in part on the fact that they are primarily ones that have grown out of older, well-documented and -established methods. It is a testament to how quickly this field is evolving that when we last reviewed our experience with methods for structure-based combi-chem [26], only two of the six methods above were described.

Not discussed here are two noteworthy advances in the structure-based design of combinatorial libraries, as we have yet no experience with them. These are: 1) the construction of custom reagent libraries, based on fragments which often appear in sets of biologically active [27]; and 2) the identification of "drug-like" properties and their use in filtering virtual libraries [28–29].

10.4 Construction and filtering of virtual libraries

10.4.1 Enumeration

Given a scaffold and a description of the types of chemistry that can be performed combinatorically, and sets of reagents for each point of elaboration, enumeration is the computational process of constructing each individual molecule that could be made with that library. Broadly, one may characterize the methods for enumeration into two classes: "chemical" enumeration, in which the rules of synthetic chemistry are considered as the reagents are attached to the scaffold; and "plastic model" enumeration, in which molecules are put together as one puts plastic models together. Both can be directed to yield the same result. Chemical enumeration tends to be easier-to-use, especially by the synthesis chemist, while plastic-model enumeration tends to be much faster, and better suited for the generation of libraries in excess of one million compounds.

For chemical enumeration, we rely on the PC-based Afferent software [30]. One simply specifies the scaffold, the chemistry, and provides the reagent lists as .sd files, and Afferent will construct the virtual library. It typically takes about 1 h to construct an 10,000-molecule library. One feature of Afferent that stands out is its ability to handle multi-center reactions, such as a Diels-Alder reaction (though these tend not to be common in combi-chem). Afferent can write the output either to an .sd file, or to Oracle tables.

For plastic-model enumeration, we rely on MOE (Molecular Operating Environment) [31]. It can enumerate a 200,000-molecule library in 1 h. It writes the output as an .sd file. These capabilities are comparable to most molecular modelling packages, which now routinely include plastic-model enumeration capabilities. As shall be seen, MOE allows ready customization, and interfaces easily with other components of our multi-vendor software environment.

10.4.2 Filtering of virtual libraries

In our setting, we have two filtering steps for virtual libraries. The first filtering step, PUREOS-1, allows one to reject *reagents* with objectionable properties. The second filtering step, PUREOS-2, examines the whole molecule, and allows one to reject *products* with objectionable properties.

PUREOS-1 can filter SD files generated from ACD, CRCD, or other chemical reagent databases. Twenty-one filters are contained in this program including number of special atoms like Cl, Br, I, metal, number of functional groups like NH2, NO2, COCl, CONH2, SO2NH2, double bonds, triple bonds, Esters, and physical properties like molecular weight and LogP(o/w). Each filter represents a particular functional group or physical-chemical property of a chemical compound. Users can select a set of filters for a particular project and define the range for each filter. Using PUREOS-1, chemists can eliminate reagents containing unwanted functional groups or with unfavorable physical-chemical properties before actual library enumeration. PUREOS-1 has been found to be a very useful tool in the design of a library of inhibitors against a kinase of therapeutic interest. In that case, using PUREOS-1, 7000 out of 9000 amines from the ACD were eliminated.

PUREOS-2 is a program developed to screen virtual combinatorial library according to Lipinski's "rule of five" and similar rules. It filters a virtual library to eliminate virtual compounds with undesirable predicted physico-chemical properties. The current filters for PUREOS-2 include molecular weight (Weight), hydrophobicity (log P(o/w)), number of rotatable bonds (n_rotatable), number of hydrogen bond donors (HB_don), and number of hydrogen bond acceptors (HB_acc) and molecular polar surface area. In the design of several combinatorial libraries, it has been found that nearly 50% of the virtual library compounds can be eliminated through PUREOS-2.

Both PUREOS-1 and PUREOS-2 were implemented under the Molecular Operating Environment, MOE [31]. This implementation is quite flexible, making it easy to expand the types of constraints one wishes to apply at each step.

Both PUREOS-1 and PUREOS-2 are fast, typically processing 100,000 molecules per hour.

10.5 Design of directed libraries

10.5.1 SAR-driven design

In cases where no x-ray structure is available, one can still develop empirical models based on the structure-activity-relationship (SAR), and use these models to direct the library design. One may term this process SAR-driven design. We describe here three methods for this type of design: 1) using 2D molecular similarity, 2) using "binary QSAR," and 3) using pharmacophore discovery and pharmacophoric 3D database searching.

By 2D similarity

The simplest method for SAR-driven design is borrowed from our experiences in processing the hits from high-throughput screening. There, one commonly performs similarity searches on each hit and submits those similar molecules to biological testing. Analogously, in a combi-chem setting, one may construct a large virtual library, synthesize and test a small subset, and submit those for biological testing. Similarity searches can be performed on the virtual library against the actives, and new combinatorial libraries can be built focused on these presumed "islands of activity" (see Fig. 2). This is the quickest and easiest method for SAR-driven design. Such similarity searches on 100,000-compound virtual libraries typically take less than 1 min in Pharmacia's Cousin database system (which runs on a multiprocessor P6 computer). Cousin's similarity metric is a modified Tanimoto coefficient (see references in [26]). The disadvantage of this method is that the design is based solely on 2D structural similarity, and focuses the libraries narrowly on a small number of templates and a small number of reagents.

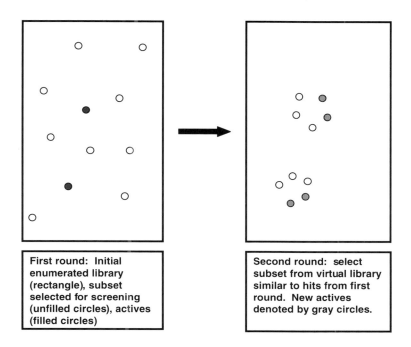

| First round: Initial enumerated library (rectangle), subset selected for screening (unfilled circles), actives (filled circles) | Second round: select subset from virtual library similar to hits from first round. New actives denoted by gray circles. |

Figure 2. Schematic representation of the simplest "structure-based" method for library design, in which molecules are selected based on the 2D similarity of their structure to known actives.

Using B-QSAR

Binary QSAR is QSAR-like method developed by Paul Labute which correlates structures, using molecular descriptors, with a binary expression of biological activity (i.e., 1 for active and 0 for inactive) and calculates a probability distribution for active and inactive compounds in a training set [32]. This binary QSAR model then can be used to predict the probability for a new compound to be active. It is distinct from the typical Hansch-style QSAR, in that a regression line is *not* computed; B-QSAR relies strictly on Bayesian inference.

Binary QSAR estimates the probability density $Pr(Y = 1|X = x)$ from a training set with biological activity Y (0 for inactive and 1 for active) and molecular descriptors X. A principal components analysis (PCA) is conducted on the training set to calculate an n by p linear transform, Q, and an n-vector, u, such that the random p-vector $Z = Q(X-u)$ has mean and variance equal to the p by p identity matrix, where p is called the number of principal components.

The original molecular descriptors, X, are transformed by Q and u to get a decorrelated and normalized set of descriptors. The probability density is then estimated by Bayes' theorem and assuming that the transformed descriptors are independent:

$$\Pr(Y = 1 | X = x) \approx \left[1 + \frac{\Pr(Y = 0)}{\Pr(Y = 1)} \prod_{i=1}^{p} \frac{\Pr(Z_i = z_i | Y = 0)}{\Pr(Z_i = z_i | Y = 1)} \right]^{-1}$$

$$Z = Q(X - u) = (Z1, \ldots, Zp)$$

Each probability density is approximated by constructing a histogram. Once all of the $2p + 2$ probability densities have been obtained from the training set, the desired density $\Pr(Y = 1 | X = x)$ is calculated according to the above formula.

Y is a Bernoulli random variable (take on value 0 or 1) representing "active" or "inactive," and X is a n-vector of real numbers (a collection of molecular descriptors). One of us (HG) has already reported on the use of this binary QSAR method to screen a combinatorial library for carbonic anhydrase II inhibitors, in which were found six compounds with IC_{50} values in the sub-μM range [33]; a group at Pharmacopaeia [34] also reported the use of B-QSAR-based design of combinatorial libraries. Most recently, we have used this technique in the combinatorial library design of inhibitors against a kinase of therapeutic interest.

The advantages of using B-QSAR in SAR-driven library design is that it is fast, and that it makes no prior assumptions about the nature of how the biological activity is elicited. The key disadvantage, like the similarity-based method, is that it tends to reinforce the structural patterns already apparent in the SAR.

Typically, MOE can enumerate 100,000 compounds per hour, and can screen 20,000 compounds per hour depending on the complexity of the B-QSAR model.

Using DANTE pharmacophore discovery and 3D database searching
Pharmacophore discovery is the process of taking a set of molecules and their biological activities, performing exhaustive conformational analysis on all of these molecules, and inferring a pharmacophore: a set of 3D structural characteristics common to all the actives, and possibly which discriminate those actives from the inactives. Pharmacophoric 3D database searching is the companion technique which, given such a 3D pharmacophore, searches a database of existing compounds and reports all of those which can adopt an energetically-reasonable conformation consistent with the pharmacophore.

The field of pharmacophore discovery is relatively new. In contrast, successes in discovering novel active compounds using 3D database searching were first reported in the 1980s, as described earlier. One of us (JVD) has developed a novel method for pharmacophore discovery, DANTE, which is especially suited to its use in conjunction with 3D database searching and in its application to SAR-driven combinatorial library design [7, 35–37].

A DANTE pharmacophore discovery analysis begins by performing exhaustive conformational analysis. This step is performed outside of DANTE, and a number of tools are available for this: Catalyst (MSI, [38]), Omega (details to be given in a later section), and CONFORT (unpublished, available from R. S. Pearlman, U. of Texas). The first step in DANTE is to identify all possible chemical features, based on a standard set: H-bond acceptor, H-bond donor, hydrophobe, aromatic ring, positive- and negative-charge. Following the method of Mayer et al. [39], DANTE identifies candidate pharmacophores as those sets of geometric arrangements of features common to most or all of the actives. Unique to DANTE, these candidate pharmacophores are ranked by their selectivity, a mathematical expression reflecting the probability that such a pattern could emerge at random. Finally, also unique to DANTE, all molecules are superimposed using the most selective pharmacophore, and the "shrink-wrap" procedure is applied to infer the steric boundaries of the binding site.

Figure 3 depicts a typical DANTE pharmacophore, one for benzodiazepine CCK antagonists based on the data of Evans et al. [40]. Four chemical features are used to superimpose all conformers shown, constraining their conformation according to the geometric constraints attached to each feature. In addition, unique to DANTE is the presence of a full description of the steric constraints associated with that SAR: the core of the pharmacophore is surrounded in all directions by opaque patches of surface symbolizing a putative steric boundary on the receptor.

Figure 4 shows how a benzodiazepine combinatorial library could be designed with such a pharmacophore. Once a 3D database has been constructed from the virtual library, that database can be searched using that pharmacophore, with those reagents leading to steric clashes being rejected during the search. The final, designed library consists of those molecules which were hits during this 3D database search.

Even more importantly than avoiding steric clashes is the notion of exploring *terra incognita*. A common misperception in the modelling literature is that the

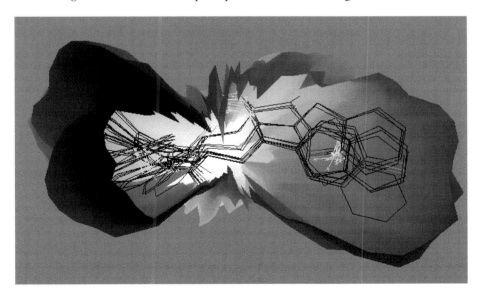

Figure 3. "Shrink-wrap" representation of inferred binding site for CCK antagonists based on the data of Evans et al., constructed using the DANTE software.

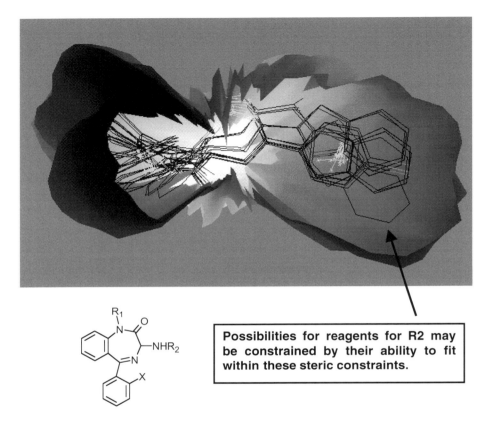

Possibilities for reagents for R2 may be constrained by their ability to fit within these steric constraints.

Figure 4. How the shrink-wrap surface may be used in the design of a benzodiazepine combinatorial library.

complement of sterically-forbidden regions in an SAR is sterically-allowed regions. In fact, as was first pointed out by [7], the complement of sterically-forbidden regions is *terra incognita*, regions of space the SAR has not explored. (This name is used in explicit recollection of those 17th century maps, which show the known world, the New World, and in which bottom is labeled *terra incognita*, where Australia and Antarctica are now known to be). One can invert the database search, and explicitly look for molecules which protrude into *terra incognita*, designing libraries with the explicit intent to probe hitherto-unexplored regions.

The pharmacophore produced by DANTE can be converted into a 3D database query suitable for input into Catalyst's 3D database search. Catalyst's conformational analysis and database construction tools are used to construct a 3D version of the virtual library.

Speeds for the 3D database construction are in the range of 4000 molecules/h, and speeds for the search vary from 1 min up to 30 min, depending on the nature of the pharmacophore and size of the database. Construction of the pharmacophore is quick (under 30 min), compared to the slow step of performing high-quality conformational analysis of the molecules used to construct the pharmacophore, which usually ranges from 2 to 24 h.

10.5.2 Protein-structure-driven design

In those cases where the directed combi-chem library is against a target for which we have a macromolecular structure, we have the greatest potential for designing libraries which have a high proportion of actives. Initially, we began by using 3D database searching, in a manner similar to what was described in the previous section. Those experiences were not satisfying, initiating a quest for the optimal protein-structure-driven combinatorial library design tool. This may be a Holy Grail. We describe here our experiences thus far with three software tools: CombiBUILD, Omega, and SLIDE.

CombiBUILD
CombiBUILD is software developed by Diana Roe in the Kuntz lab at UCSF [41], an outgrowth of earlier work on a *de novo* design program BUILDER [42]. CombiBUILD is the software used in the most remarkable published success to date in protein-structure-based combi-chem, the design of cathepsin D inhibitors [43]. In this completely *prospective* study, synthesis chemist Ellen Kick in Jon Ellman's lab at UC Berkeley devised the synthetic combinatorial scheme shown in Figure 1. Facing the prospect of more than 3×10^9 molecules, she approached the Kuntz group for assistance in designing a smaller library. In response, Diana Roe in the Kuntz group developed CombiBUILD. They chose 1000 molecules for synthesis that CombiBUILD had deemed to optimally complement the active site of cathepsin D, whose x-ray structure had at that time recently been published [44]. Of those 1000 molecules designed by CombiBUILD, 67 had sub-µM activity, 23 had <330 nM activity, and seven had <100 nM activity. This stimulated another round of combi-chem, after which four molecules emerged with <20 nM activity.

CombiBUILD functions by performing its own enumeration and performing nearly-exhaustive, recursive conformational exploration in the active site, with the initial placement of the molecule determined by overlaying the scaffold to user-defined points. (By contrast, CombiDOCK, also from the Kuntz group [45], docks each molecule to determine the initial binding geometry). Each torsional angle is rotated through a set of predefined ideal values based on the nature of the bond, and rigid body minimization is performed in the active site for each conformation. Special bump grids are employed to prune the conformational space. The intermolecular energy, as determined by a simplified AMBER force-field, is used to compute a score, by which each member of the library is ranked.

We have yet to duplicate the stunning success of CombiBUILD against a target of interest to us, though it has generated libraries superior to those constructed randomly or with diversity-based methods. Our own experience in retrospective studies leads us to question whether scoring function used in CombiBUILD is optimized for all applications. In our setting, with separate enumeration software and filtering programs like PUREOS to trim virtual libraries, the fact that CombiBUILD cannot use as input a virtual library is a disadvantage (it performs its own enumeration given a scaffold and sets of reagents as input). And, finally, it is not as fast as we would like, typically processing 15,000 molecules in 24 h.

Another drawback associated with CombiBUILD's enumeration on the fly approach is that only *one* reaction type can be considered at each reaction site. Enumeration prior to the virtual screening step means that any type of reaction product can be included in the screening set.

Omega

Omega (Optimized Molecular Ensemble Generation Application) is software recently made available by Matt Stahl at OpenEye Software [46] for performing exhaustive, unconstrained conformational search, and for performing active-site-constrained conformational search. Though the approach is unpublished, Omega has an extensive lineage, as both an updated version of the WIZARD conformational analysis software [47, 48] and as an updated version of the MONGOOSE protein-structure-based combinatorial library design software [23]. Verbal accounts of MONGOOSE and its successor Skizmo indicate that it has been used with success in protein-structure-based combinatorial library design at Vertex.

Omega has the ability to do gas-phase and restricted active-site searches. The gas-phase ensemble generation uses a deterministic breadth-first, best-first torsion search of conformational space, using an energy cutoff to exclude high-energy branches of the search tree. Discrete torsion values (defined by user-modifiable rules) are used; there is no variation of bond lengths, angles, or ring torsions. The restricted active site searching (restricted docking) uses an initial guess (provided by the user) about the orientation of a common ligand substructure in the active site to orient the gas phase search results, which are then evaluated using a modified version of the scoring function of Eldridge et al. [49], with the position of the top-scoring conformations being optimized.

Gas-phase conformational searching generates structures of lower quality than would be generated with a typical all-degree-of-freedom energy minization procedure, but with that tradeoff is able to perform full conformational searches very quickly. Full searches on drug-like molecules typically take a few tenths of a second; highly flexible molecules may take up to a few seconds. Molecules with extremely high flexibility (e.g., >17 rotatable bonds) are typically excluded, as the results would likely be meaningless in any case, but Omega prunes the search and limits the total number of conformations being considered (to typically 200,000 at any given time). The quality of the generated structures is dependent on the accuracy of the rules used to generate rotamers, and also depends on the quality of the initial 3D structure provided as input (which defines bond lengths, angles, and ring geometries). The torsion rules are contained in an customizable file, with substructures defined as SMARTS strings followed by the torsion values to be sampled. The ability to generate full conformational models for thousands of compounds per hour allows small virtual libraries to be processed interactively, and large libraries overnight. Also, since Omega is starting with fully-enumerated libraries, and all structures are evaluated independently, the search can be readily split over multiple processors.

The active-site search constitutes a restricted flexible docking method, the restriction being the requirement that the user have a hypothesis about the binding mode of the compounds in question, and that they share some common substructure which can be used to establish an initial approximate location within the protein active site. The active-site search is also very fast, nearly as fast as the gas-phase search alone. In contrast to CombiBUILD, Omega does not enumerate the virtual library in the course of doing a search through conformational/chemical space. Virtual libraries are enumerated externally and an initial 3D structure is generated as is the case for a gas-phase search. Each compound to be docked first undergoes a gas-phase conformational search; the resultant low-energy structures are then oriented into the active site and scored following a rigid-body geometrical optimization against the intermolecular scoring function. The best n structures are then written out. If the scoring function is to be used directly, a single conformation/compound suffices, but if further processing (e.g., energy minimization, scoring by another method, etc.) is envisioned then more structures can/should be written. Each structure written out has both an intermolecular pseudoenergy from the scoring function and an intramolecular strain energy from the gas-phase search. No attempt is made to combine the two energies, as it would be unclear how to properly scale them. The lack of a true minimum-energy docked conformer is ameliorated by the rigid-body optimization and rather soft potentials of the scoring function, with the result that good-scoring conformations are generally quite reasonable. Searches on ligands from crystal structures will typically find a docked conformation which is very similar to the crystallographic result, but it will not necessarily be the best-scoring conformation. Due to the limitations on both the conformational generation (discrete, unminimized conformations) and docking (rigid body optimization, scoring function limitations), as well as the requisite assumption about docking mode, differences of a few kcal/mole are not likely to be distinguished.

Overall, our impression of Omega is incomplete, but at this point it appears to be fast (4000 structures/h) and uses as input a virtual library, which is convenient in our setting. Our experiences lead us to be concerned about scoring functions here as well.

SLIDE

Mark Twain, speaking about the weather, once complained "Everybody talks about it, but nobody *does* anything about it." The same has been true about the issue of the flexibility of the protein in performing protein-structure-based ligand design. We have seen the importance of this on numerous occasions in a variety of protein-structure-based design projects. In one example, the chemists had synthesized a novel, potent compound, which we could not determine how it fit into the protein based on the x-ray structure we had. Once this compound appeared in a co-crystal structure, we saw that the protein had flexed to accommodate this ligand, and with this x-ray structure it was easy for any computational docking procedure to place the ligand in the proper binding orientation. DeGrado and colleagues have reported similar observations [50]. There are many tools to deal with the conformational flexibility of the *ligand* (e.g., CombiBUILD, Omega, FlexX [51]), while none deal with the more difficult problem of dealing with the conformational flexibility of the *protein*.

Until recently. SLIDE (Screening for Ligands by Induced-Fit Docking) is a program developed by Volker Schnecke and Leslie Kuhn at Michigan State University that is designed for searching large libraries to discover ligands best complementing a protein active site [52–54]. While we are still preparing to apply this software to the design of combinatorial libraries, it is already apparent from their published data that it should have the speed and other attributes that we need as a tool for protein-structure-based combinatorial library design.

The binding site of the protein is represented by surface residues, water molecules, and a set of favorable interactions points above the protein surface. SLIDE docks ligands into the site using geometric hashing techniques to evaluate shape and chemical complementarity. Induced flexibility of the protein side chains and ligand are modeled with equal status, applying directed minimal rotations to the rotatable single bonds in the interface. Lowest-cost conformational changes in *both* ligand and protein that generate a shape-complementary surface guide the search. The *backbone* of the protein is held fixed; only side-chain induced-fit is considered.

Astonishingly, they report execution times on libraries in the range of 50,000–100,000 molecules ranging from under 1 h to 24 h. As we are just arranging to put this software to the test in our own environment, we do not yet have any independent experience to report on the use of SLIDE applied to combinatorial library design. Nonetheless, their reported experience, and the published algorithmic details, leave us sanguine about its potential.

10.6 Conclusions

As combinatorial chemistry shifts from a technology for the production of screening libraries, to one for the production of directed libraries aimed at a specific target, the need for design strategies that exploit our knowledge of the target is rapidly growing. When no macromolecular structure is available, one can distill from the known SAR information useful in guiding this combinatorial library design, e.g., via pharmacophore discovery (see Chapters 7 and 11). Even with a protein structure in hand, the uncertainties in our understanding of protein-ligand interactions makes the process of protein-structure-based design an inherently inaccurate process. But marrying protein-structure-based-design to combi-chem, resulting in the ability to readily make hundreds of molecules which have been designed for their complementarity to an active site, we anticipate that the development of potent, specific ligands will be greatly accelerated.

Overall, our experiences thus far are limited, and indicate that the software tools for performing structure-based combi-chem are still crude, but rapidly evolving. Success stories will be slow to appear in the scientific literature, however, as the majority of this work is occurring in pharmaceutical and biotech companies, which traditionally publish their results many years after the initial discovery.

10.7 Acknowledgements

We thank our colleagues in the combinatorial chemistry group at Pharmacia Kalamazoo for numerous discussions and continued enthusiasm, especially R. A. Nugent and S.D. Larsen. The support and enthusiasm of K. L. Leach has also been appreciated. Thanks also to D. C. Roe for her continued assistance in the use of CombiBUILD, and to M. T. Stahl for his on-going help in the use and development of Omega. The provision of their preprints from V. Schnecke and L. A. Kuhn is also gratefully acknowledged.

Unless otherwise noted, all timings were performed on a single SGI R10K processor. As all the operations described here are highly-parallelizable, and readily spread across multiple processors, the actual elapsed time is much smaller.

References

1 Erickson JW, Fesik SW (1992) Macromolecular x-ray crystallography and NMR as tools for structure-based drug design. *Annu Rep Med Chem* 27: 271–289
2 van Drie JH (1995) 3D database searching in drug discovery: http://www.netsci.org/Science/Cheminform/feature06.html
3 Furka A, Sebestyen F, Asgedom M, Dibo G (1991) General method for rapid synthesis of multicomponent peptide mixtures. *Int J Peptide Protein Res* 37: 487–493
4 Bunin BA, Ellman JA (1992) A general and expedient method for the solid-phase synthesis of 1,4-Benzodiazepine derivatives. *J Amer Chem Soc* 114: 10997–10998
5 DeWitt SH, Kiely JS, Stankovic CJ et al (1993) "Diversomers": an approach to nonpeptide, nonoligomeric chemical diversity. *Proc Natl Acad Sci U S A* 90: 6909–6913
6 Martin EJ, Blaney JM, Siani MA et al (1995) Measuring diversity: experimental design of combinatorial libraries for drug discovery. *J Med Chem* 38: 1431–1436
7 Van Drie JH, Nugent RA (1997) Addressing the challenges posed by combinatorial chemistry: 3D databases, pharmacophore recognition and beyond. *Env Res* 9: 1–21
8 Shoichet BK, Bussiere DE (2000) The role of macromolecular crystallography and structure for drug discovery: Advances and caveats. *Curr Opin Drug Disc Dev* 3: 408–422
9 Wang S, Milne GW, Yan X et al (1996) Discovery of novel, non-peptide HIV-1 protease inhibitors by pharmacophore searching. *J Med Chem* 39: 2047–2054
10 Hong H, Neamati N, Wang S et al (1997) Discovery of HIV-1 integrase inhibitors by pharmacophore searching *J Med Chem* 40: 930–936
11 Neamati N, Hong H, Mazumder A et al (1997) Depsides and depsidones as inhibitors of HIV-1 integrase: discovery of novel inhibitors through 3D database searching *J Med Chem* 40: 942–951
12 Kaminski JJ, Rane DF, Snow ME et al (1997) Identification of novel farnesyl protein transferase inhibitors using three-dimensional database searching methods *J Med Chem* 40: 4103–4112

13 Anonymous (1993) Hypotheses in Catalyst. This is the sole documentation on the theoretical and mathematical basis of Hypothesis Generation, privately distributed by BioCAD, unpublished

14 Llorens O, Filizola M, Spisani S et al (1998) Amygdalin binds to the CD4 receptor as suggested from molecular modeling studies. *Bioorg Med Chem Lett* 8: 781–786

15 Marriott DP, Dougall IG, Meghani P et al (1999) Lead generation using pharmacophore mapping and three-dimensional database searching: application to muscarinic M(3) receptor antagonists. *J Med Chem* 42: 3210–3216

16 Martin YC, Bures MG, Danaher EA et al (1993) A fast new approach to pharmacophore mapping and its application to dopaminergic and benzodiazepine agonists. *J Comp-Aided Mol Des* 7: 83–102

17 Kiyama R, Honma T, Hayashi K et al (1995) Novel angiotensin II receptor antagonists. Design, synthesis, and *in vitro* evaluation of dibenzo[a,d]cycloheptene and dibenzo[b,f]oxepin derivatives. Searching for bioisosteres of biphenylyltetrazole using a three-dimensional search technique. *J Med Chem* 38: 2728–2741

18 Greenidge PA, Carlsson B, Bladh LG et al (1998) Pharmacophores incorporating numerous excluded volumes defined by X-ray crystallographic structure in three-dimensional database searching: application to the thyroid hormone receptor. *J Med Chem* 41: 2503–2512

19 Van Drie JH, Weininger D, Martin YC (1989) ALADDIN: an integrated tool for computer-assisted molecular design and pharmacophore recognition from geometric, steric, and substructure searching of three-dimensional molecular structures. *J Comput Aided Mol Des* 3: 230–255

20 Wang S, Sakamuri S, Enyedy IJ et al (2000) Discovery of a novel dopamine transporter inhibitor, 4-hydroxy-1-methyl-4-(4-methylphenyl)-3-piperidyl 4-methylphenyl ketone, as a potential cocaine antagonist through 3D-database pharmacophore searching. Molecular modeling, structure-activity relationships, and behavioral pharmacological studies. *J Med Chem* 43: 351–360

21 Leysen D, Kelder J (1998) "Ligands for the 5-HT2c receptor as potential antidepressants and anxiolytics". In: v.d. Goot H (ed.) Trends in Drug Research II. Elsevier, 49–60

22 Kick EK, Ellman JA (1995) Expedient method for the solid-phase synthesis of aspartic acid protease inhibitors directed toward the generation of libraries. *J Med Chem* 38: 1427–1430

23 Walters PW, Stahl MT, Murcko MA (1998) Virtual Screening: An Overview *Drug Discovery Today* 3: 160–178

24 Lipinski CA, Lombardo F, Dominy BW et al (1997) Experimental and computational approaches to estimate solubility and permeability in drug discovery and development settings. *Adv Drug Deliv Rev* 23: 3–25

25 Hansch C, Leo A (1995) Exploring QSAR: Fundamentals, Applications in Chemistry, Biology ACS Publications Washington DC Hansch C, Bjorkroth JP, Leo A (1987) Hydrophobicity and central nervous system agents: On the principle of minimal hydrophobicity in drug design. *J Pharm Sci* 76: 663–687

26 Van Drie JH, Lajiness MS (1998) Approaches to virtual library design, *Drug Discovery Today* 3: 274–283

27 Lewell XQ, Judd DB, Watson SP et al (1998) RECAP—retrosynthetic combinatorial analysis procedure: a powerful new technique for identifying privileged molecular fragments with useful applications in combinatorial chemistry. *J Chem Inf Comp Sci* 38: 511–522

28 Ajay A, Walters WP, Murcko MA (1998) Can we learn to distinguish between "drug-like" and "nondrug-like" molecules? *J Med Chem* 41: 3314–3324

29 Sadowski J, Kubinyi H (1998) A scoring scheme for discriminating between drugs and nondrugs *J Med Chem* 41: 3325–3329

30 Afferent Systems, see: http://www.afferent.com

31 Chemical Computing Group, see: http://www.chemcomp.com

32 Labute P (1999) Binary QSAR: A new method for the determination of quantitative structure-activity relationships. In: Altman R.B, Dunker AK, Hunter L et al (eds): *Pacific Symposium on Biocomputing '99*. World Scientific, New Jersey, 444–455

33 Gao H, Bajorath B (1999) Comparison of binary and 2D QSAR analyses using inhibitors of human carbonic anhydrase II as a test case. *Mol Diver* 4: 115–130

34 Lauri G, Lynch D, Brown, RD "Application of B-QSAR to Pharmacopaeia HTS data"; *unpublished work*

35 Van Drie JH (1996) An inequality for 3D database searching and its use in evaluating the treatment of conformational flexibility. *J Comput Aided Mol Des* 10: 623–630

36 Van Drie JH (1997) Strategies for the determination of pharmacophoric 3D database queries. *J Comput Aided Mol Des* 11: 39–52

37 Van Drie JH (1997) "Shrink-wrap" surfaces: a new method for incorporating shape into pharma-cophoric 3D database searching. *J Chem Inf Comp Sci* 37: 38–42

38 Catalyst is software originally developed at BioCAD (1990–1994), and is now marketed by MSI, see: http://www.msi.com

39 Mayer D, Naylor CB, Motoc I et al (1987) A unique geometry of the active site of ACE consistent with structure-activity studies. *J Comput Aided Mol Des* 1: 3–16

40 Evans BE, Rittle KE, Bock MG et al (1998) Methods for drug discovery: development of potent, selective, orally effective cholecystokinin antagonists. *J Med Chem* 31: 2235–2246

41 Roe DC, Kuntz ID (1995) BUILDER v.2: improving the chemistry of a *de novo* design strategy. *J Comp-Aided Mol Des* 9: 269–282. (The CombiBUILD software is available from UCSF, after purchase of the DOCK 4.0 suite)

42 Lewis RA, Roe DC, Huang C et al (1992) Automated site-directed drug design using molecular lattices. *J Mol Graph* 10: 66–78

43 Kick EK, Roe DC, Skillman AG et al (1997) Structure-based design and combinatorial chemistry yield low nanomolar inhibitors of cathepsin D. *Chem Biol* 4: 297–307

44 Baldwin ET, Bhat TN, Gulnik S et al (1993) Crystal structures of native and inhibited forms of human cathepsin D: implications for lysosomal targeting and drug design. *Proc Natl Acad Sci USA* 90: 6796–6800

45 Sun Y, Ewing TJ, Skillman AG et al (1998) CombiDOCK: structure-based combinatorial docking and library design. *J Comp-Aided Mol Des* 12: 597–604

46 OpenEye, see: http://www.eyesopen.com

47 Dolata DP, Leach AR, Prout K (1987) WIZARD: AI in conformational analysis. *J Comput Aided Mol Des* 1: 73–85

48 Dolata DP, Walters WP (1993) MOUSE: a teachable program for learning in conformational analysis. *J Mol Graph* 11: 106–111

49 Eldridge MD, Murray CW, Auton TR et al (1997) Empirical scoring functions 1: The development of a fast empirical scoring function to estimate the binding affinity of ligands in receptor-ligand complexes. *J Comput Aided Mol Des* 11: 425–445

50 Rockwell A, Melden M, Copeland RA et al (1996) Complementarity of combinatorial chemistry and structure-based ligand design: Application to the discovery of novel inhibitors of matrix metalloproteinases. *J Amer Chem Soc* 118: 10337–10338

51 Rarey M, Kramer B, Lengauer T et al (1996) A fast flexible docking method using an incremental construction algorithm. *J Mol Biol* 261: 470–489

52 Schnecke V, Swanson CA, Getzoff ED et al (1998) Screening a Peptidyl Database for Potential Ligands to Proteins with Side-chain Flexibility. *Proteins* 33: 74–87

53 Schnecke V, Kuhn LA (1999) Database Screening for HIV Protease Ligands: The Influence of Binding-Site Conformation and Representation on Ligand Selectivity. In: Lengauer T, Schneider R, Bork P et al (eds.): *Proceedings ISMB 99, 7th International Conference on Intelligent Systems for Molecular Biology.* AAAI Press, Menlo Park, CA

54 Schnecke V, Kuhn LA (1999) Flexibly Screening for Molecules Interacting with Proteins. In: Thorpe MF, Duxbury PM (eds): *Applications in Rigidity Theory.* Plenum Publishing, New York, 385–400: 242–251

11 3D QSAR in modern drug design

Glen E. Kellogg[1] and Simon F. Semus[2]

[1] *Virginia Commonwealth University, Department of Medicinal Chemistry, School of Pharmacy, Richmond, VA 23298-0540, USA*
[2] *Wyeth-Ayerst Research, Department of Biological Chemistry, CN8000, Princeton, NJ 08543, USA*

11.1 Introduction

The belief that there is a direct relationship between chemical structure and biological activity of therapeutic agents is fundamental to the field of medicinal chemistry. Indeed, the efforts of early medicinal chemists focused on well-defined structural modifications to active lead compounds as molecules were developed into drugs. The relationship between chemical properties like solvent partitioning and biological activity was recognized over a century ago [1]. Almost 40 years ago, Hansch, Fujita and co-workers invented the field of Quantitative Structure-Activity Relationships (QSAR) [2]. In this approach whole molecule parameters such as $LogP_{o/w}$ (the partition coefficient for 1-octanol/water partitioning),[3] molar refractivity, shape and topology indices [4], etc. for groups of related compounds are statistically correlated with measures of biological activity to obtain a QSAR equation. This equation relates easy to measure (or predict/calculate) values for molecules to the more difficult to measure biological activities. Once a QSAR is obtained, verified and found to be "predictive," the biological activity for new chemical entities that may or may not exist can easily be predicted. Numerous success stories from QSAR over the past four decades validate the fundamental relationship between structure and activity.

One frustrating aspect of traditional (2D) QSAR is the difficulty of "inverting" QSARs into better drugs. That is, given an equation relating several parameters to activity, what *chemistry* is needed to optimize the activity. It is frequently not enough to know, for example, that increasing $\text{LogP}_{o/w}$ will increase activity. Most molecules have several locations where chemical modification can be made and not all of them will tolerate the addition of hydrophobic bulk and maintain or enhance activity. Clearly, more specific structural information about the molecules and their modes of association or binding with their interacting receptor, enzyme or macromolecule is needed to *design* new compounds with QSAR.

Most calculated parameters for QSAR are based on adaptations of graph theory. "Flat" representations of molecules are, of course, of limited use in understanding structure and biomolecular interactions. Advancement of the QSAR paradigm required the more recent availability of high performance calculation and visualization computer hardware, rapid and reliable methods for calculating three-dimensional structures for molecules from two-dimensional graphs, and lastly, a very clever insight tying QSAR and molecular modeling together. Richard Cramer of Tripos, Inc. discovered a new way to represent molecules for QSAR. His observation was that the 3D field properties of molecules, whether from electrostatic, steric or other atomic/molecular properties, can be used as unique parameter sets for QSAR. This was called Comparative Molecular Field Analysis (CoMFA) [5] and was the first commonly accepted and widely used method for 3D QSAR.

11.2 Computational technologies

11.2.1 True 3D methods

Most of the truly three-dimensional QSAR methods use a similar computational technology as illustrated in (Fig. 1). Simply, the molecules in the data set are, one-by-one, inserted into a 3D cage of test atoms that are arrayed in a Cartesian coordinate system. Each test atom probes each atom of the molecule and the results are summed and stored as a map value for that test atom, or more generically for that set of coordinates. The nature of the probe is dependent on the property being mapped. Equations (1), (2) and (3) are typical mathematical functions for steric, electrostatic and lipophilic potentials, respectively as used by the CoMFA (steric, electrostatic) [5, 6] and HINT/CoMFA [7] (lipophilic) methods.[1]

$$S_t = \sum \varepsilon_{it} \,[1.0/p_{it}^{12} - 2.0/p_{it}^{6}], \; p_{it} = r/(R_i + R_t) \tag{1}$$

$$E_t = 332.17 \sum Q_i \, Q_t/D_{it} \, r \tag{2}$$

$$L_t = \sum a_i \, S_i \, e^{-r} \tag{3}$$

---

[1] Definitions for symbols: ε_{it} is the van der Waals constant, r is the distance between the test atom (t) and the molecular atom (i), R_i and R_t are van der Waals radii, 332.17 is a unit conversion constant, Q_i and Q_t are partial atomic charges, D_{it} is the value of the dielectric function for i and t, a_i is the hydrophobic atom constant and S_i is the solvent accessible surface area for i.

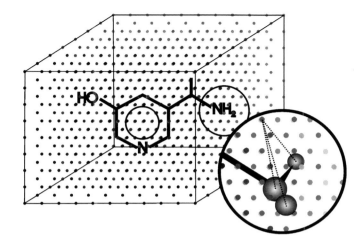

Figure 1. A molecule in a cage of test atoms at predefined grid points. The map value at each test atom is the sum of all measured "interactions" between that test atom and all atoms in the molecule. The interactions are measured or scored by the type of physical property being obtained, utilizing a functional relationship between atomic properties of the test and molecular atoms as varied by distance.

For CoMFA and related technologies, the "field" around the molecule is of more importance than the actual structure of the molecule. In fact, grid points that are located within the van der Waals surface of the molecule are truncated at predefined values for two significant reasons. First, this eliminates the computationally unpleasant possibility of division by zero if a grid point coincidentally lands on an atom (c.f., Eqs. (1) and (2)). Second, and more important is that these maps provide 3D profiles of their associated properties, in a manner that may correspond to receptor recognition. Clearly macromolecule receptors are more keyed to molecular shapes (steric field), electrostatic and lipophilic potentials than they are to whole molecule descriptors or specific atoms or functional groups. Other field types describing hydrogen bonding, electronic (molecular orbital) properties, polarizability, topology etc. are possible [8–10].

The HASL method of Doweyko [11] is similar. In it, each grid point within the van der Waals surface of the molecule represents a triad of counters noting hydrophobes, hydrogen bond donors and hydrogen bond acceptors.

In CoMSIA (Comparative Molecular Similarity Indices Analysis) [12] the implementation of the fields differs from CoMFA, although the underlying methodology is consistent. In CoMSIA five different similarity fields are calculated: steric, electrostatic, hydrophobic, hydrogen bond donor and hydrogen bond acceptor. These fields were selected to cover the commonly accepted major contributions to ligand binding. Similarity indices are calculated at regularly spaced grid points for the pre-aligned molecules. A comparison of the relative shapes of CoMSIA and CoMFA fields is shown below (Fig. 2) [6, 12]. For the distance dependence between the probe atom and the molecule atoms a Gaussian function

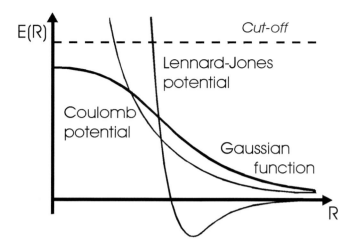

Figure 2. Graph showing distance-dependent behavior of energy functions: Coulomb potential, Lennard-Jones potential and Gaussian (Adapted from CoMSIA documentation, Tripos, Inc.).

is used. Because of the different shape of the Gaussian function, the similarity indices can be calculated at all grid points, both inside and outside the molecular surface. One consequence of this alternative approach is the ability to display the resultant fields directly onto the molecules being studied rather than (only) projecting them into the enveloping space.

At this point, we generally are dealing with biological activities as functions of 3D descriptor arrays. The statistical techniques to solve this problem treat this data as *linear* arrays of descriptors that can then be solved with multilinear regression, partial least squares (PLS) [5, 13], genetic algorithms (GA) [14], etc. In effect the 3D information is maintained external to the statistics engine. This is a significant point that reveals why the superimposition of the molecules in the data set is critical (*vide infra*). In order for there to be correspondence between 3D descriptors of different molecules in the data set, the structures of the molecules must be aligned in space and sampled in the same cage of grid points. We focus on the "alignment problem" in the next section. The end result is a single equation relating biological activity to the calculated values at grid points. Thus, the biological activity of new molecules that can be adequately superimposed on molecules of the learning data set can, in principle, be predicted. Statistical validation techniques such as bootstrapping or cross-validation or the use of test sets of known compounds are used to verify the predictiveness of 3D-QSAR models.

11.2.2 Pseudo-3D methods

Alternative methods of 3D QSAR have been developed that are independent of alignment. These programs, such as EVA [15] and HQSAR [16], use statistical models that correlate either two or three dimensional dependent descriptors with biological activity. In fact, the former method employs hashed fingerprints that do not require information pertaining to the three-dimensional nature of the ligands. The latter method, like any other that utilizes descriptors that are dependent upon the three-dimensional structure, is highly influenced by the conformation of the molecules. EVA is a vector descriptor based on EigenVAlues corresponding to individual molecular vibrational modes. This approach makes use of values derived from semi-empirical quantum chemical calculations that are related to spectroscopic data. Normal coordinate frequencies are calculated and the resulting eigenvalues are projected onto a vibrational profile. A Gaussian smoothing function is applied to each vibration and the resultant profile is sampled at fixed intervals to provide a set of values that describes each molecule. This approach averts the use of normal coordinate frequencies where the number of normal modes varies with the number of atoms in the molecule and thus ensures the use of a constant number of descriptors across the data set.

The premise of HQSAR [16] is that since the structure of a molecule is encoded within its 2D fingerprint and that structure is the key determinant of all molecular properties (including biological activity), then it should be possible to predict the activity of a molecule from its fingerprint. It is claimed that this assumption is valid because of the high degree of success of 2D similarity searching of chemical databases in which the similarity measure is the Tanimoto coefficient between fingerprints [17] (for a detailed discussion on 2D descriptors see Chapter 7). HQSAR uses an extended form of fingerprint, known as a Molecular Hologram, which encodes more information, for example, branched and cyclic fragments as well as stereochemistry, than the traditional 2D fingerprint. The key difference, however, is that a Molecular Hologram contains all possible molecular fragments within a molecule, including overlapping fragments, and maintains a count of the number of times each unique fragment occurs. The authors maintain that this process of incorporating information about each fragment, and each of its constituent sub-fragments, implicitly encodes 3D structural information. While this approach may provide information as to the relative arrangements of fragments within the molecule it clearly does not provide any spatial parameters. There are a number of descriptors available that do encode such information and have been successfully applied in QSAR studies. These may depend upon the internal coordinates of the molecule alone or on the absolute orientation. An example of a conformation independent descriptor would be the magnitude of the dipole moment, whereas the x, y or z component of the dipole moment would be influenced by the orientation. Examples of 3D descriptors include principal moment of inertia, radius of gyration, volume, surface area and polar solvent accessible surface area.

Finally, in an attempt to capture 3D information pertaining to size, shape, symmetry and atomic distribution, Weighted Holistic Invariant Molecular (WHIM) descriptors [18] have been developed. The method consists of performing a principal component analysis (PCA) on the centered molecular coordinates using a weighted covariance matrix that is obtained by applying different weighting schemes to the constituent atoms. The atoms may be weighted by their number, mass, volume, electronegativity, polarizability or electrotopological index. Directional or non-directional WHIM descriptors are then calculated dependent on whether one includes information related to the principal axes. The use of WHIM descriptors has proven of value in the development of 3D QSAR models; however, their principal disadvantage is the lack of feedback information that can be applied to further drug design. In a similar fashion to the topological indices from which they may be derived it is essentially impossible to employ the descriptor information in the design of future analogs or to readily interpret the variation of structure with descriptor value.

11.3 The 3D QSAR alignment problem and pharmacophore hypotheses

Since the ability to satisfactorily superimpose, structurally, the interesting molecules in a study is so important, a number of alternative alignment methods have been investigated, including the use of structures from x-ray crystallography, by the generation of an automated pharmacophore alignment [19], with a field similarity fit method [20], or the use of a manual alignment based upon the investigators' chemical intuition.

In some cases there may be a very simple alignment of a common core where the only variants in three-dimensional space are the substituent moieties. Application of 3D QSAR to these cases is often unnecessary because in this situation the model may be no better or informative than a conventional two-dimensional QSAR model since the essential three-dimensional nature of the molecules is discarded by use of a common structure. The variants in the model are simply the stereo or electrostatic properties of the substituents themselves. In a collection of molecules where one varies the substitution on an aromatic ring and one then determines the biological activity it may be found that, for example, a 4-chlorine substitution is beneficial and independently a fluorine in position 3 similarly enhances activity. One should assume based on simple intuition that the combination of a 4-chloro, 3-fluoro- substitution pattern into a single molecule would at the very least equal the two parents. In practical terms, the model constructed in such an alignment and the subsequent evaluation by CoMFA may simply reinforce that assumption. In fact, the data set will be limited because the model is only knowledgeable about the single substitution effects of the 4-chlorine and 3-fluorine positions, so that the only conclusion the model is able to make is that the 3-fluoro, 4-chloro disubstituted analogue should be potent. In reality this may not be case since there are interacting effects of the two moieties that are not accounted for in the model. So what has one accomplished in this example by the use a complex computational approach over the act of simple chemical intuition and analysis? The answer is probably very little other than reinforcement of a previously conceived notion. The literature is replete with examples where the sole benefit of CoMFA was the confirmation of a previously determined model. Can we look beyond this limited information and reveal more significant insights in our model?

By the very nature of the technique, the most crucial step in this 3D approach is the relative orientation of the test molecules in space. That is, the chosen alignment of the compounds in the training set is going to have the most profound impact on the predictive ability of the model. We have already noted the plethora of methods available to us for structural superimposition. From a purely drug design perspective, in those instances where one has no knowledge of the three-dimensional shape of the receptor, a set of alignment rules based upon a pharmacophore hypothesis will probably be the most valuable. Indeed, perhaps the principal value of this methodology is the *evaluation of such alignments* based upon the predictive power of the derived model. One may conclude that the greater the predictive power of the model the more the alignment reflects the bioactive conformation of the molecules.

Thus, the real value of this approach is probably not in the evaluation over a single congeneric series, but more where one is trying to align molecules that do not possess a common core structure. In such a case, as in the development of pharmacophore hypotheses, one is often trying to identify features from the molecules of separate series that may present common interaction points with the receptor. For example, the presence of H-bond donors or acceptors, the placement of hydrophobic moieties, or the presence or absence of heteroatoms, may form a basis of such a pharmacophore alignment. In addition, for mathematical reasons in any QSAR method, each molecule needs to have a similarly developed set of descriptors; that is, each molecule must be described with the same number of independent variables. This is a simple matter in grid-based 3D QSAR—each molecule is analyzed by fields of the same dimensions and resolution. The researcher will develop a number of alternate pharmacophore hypotheses, by the manual manipulation of the data set or by automated means, such that the evaluation of each of the models may be achieved by the use of such grid-based methods.

Can a limited model be extended to a non-congeneric series of molecules? A large number of research groups have clearly shown that it can. A number of years ago, we demonstrated that CoMFA could be used to predict the biological activity of a series of classical and non-classical cannabinoids [21]. Although their structures are not dramatically different and the derivation of the non-classical

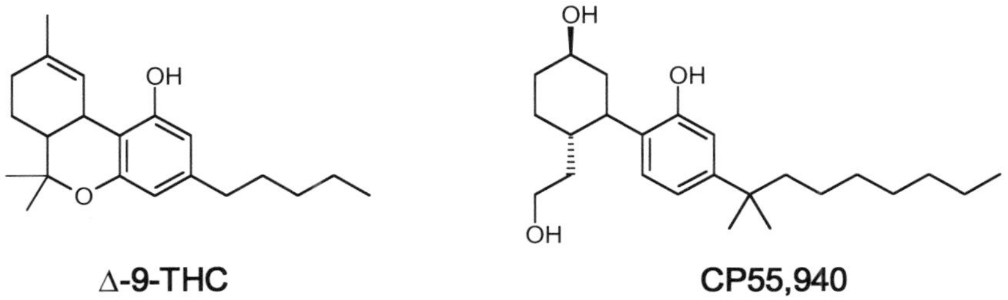

Δ-9-THC   CP55,940

Scheme 1.

series is perhaps now obvious, a common pharmacophore hypothesis and an embracing QSAR had not been previously demonstrated.

The principal psychoactive constituent of cannabis, Δ-9-tetrahydrocannabinol (Δ-9-THC), is representative of the classical cannabinoids. CP55,940, a product of the Pfizer research group, is the prototypical non-classical cannabinoid. The compounds were first described in the mid-1980s and were being developed as analgesics with the hope of avoiding the detrimental side-effects of the opiates. Cannabinoids are currently being employed as anti-emetics in chemotherapy and as appetite stimulants for AIDS patients [22]. Recent interest has focused on their potential as neuroprotective agents and in the treatment of multiple sclerosis [23]. However, despite the beneficial medical effects of these compounds, their use has been impaired by the inability to separate out the hallucinogenic properties of these molecules. We were able to develop a CoMFA model that not only encompassed a wide variety of structural variants, but also demonstrated a strong relationship between structure, binding and intrinsic activity in four *in vivo* animal assays. This experiment was a very clear demonstration of the utility of the technique in not only confirming a pharmacophore hypothesis, but also correlating an *in vitro* binding assay with whole animal behavioral models.

The dependence of Grid-based methods such as CoMFA or CoMSIA upon molecular alignment is often regarded as problematic. However, it is our notion that the alignment "problem" is an *advantage* of the methods, in that quantitative information about pharmacophore hypotheses can be revealed. In fact, it is arguable that multi-molecule alignment or pharmacophore hypotheses are merely alternative conformational representations. Thus, if the value in these predictive models is the ranking of such alignments, then the 3D QSAR approach may be regarded as a scoring function. To that end, we have recently described HIFA [24]—a grid based method of predicting binding affinities of ligand-protein complexes that utilizes the calculation of hydropathic *interactions*. Of course, it would be a boon to medicinal chemistry if 3D QSAR methods easily found the best and most chemically relevant alignment (and thus pharmacophore model) without relying on the intuition of the chemist! However, that is not currently the case and is unlikely to materialize soon.

11.4 Applications of 3D QSAR

Based on the above discussion, perhaps the most crucial step in designing and evaluating three-dimensional quantitative structure-activity relationships for drug discovery is the designation of a model for superimposing the molecules of the study. Sophisticated methods are often necessary to define the superposition. The only experimental approach is to rely on x-ray crystallographic determinations of ligand-receptor complexes (see Chapter 8), but often only one or just a few complex structures are known. In these cases, the coordinate data from the ligand of one complex is used as a template upon which other congeneric ligands in the data set can be superimposed with a fairly high level of confidence. There have also been 3D QSAR studies where the crystal structures of multiple complexes were known. The most complete of these (1993–1994) by Garland Marshall et al. of Washington University in St. Louis examined data and structures for HIV-1 protease inhibitors from several laboratories [25–27].

In this section we wish to review the Marshall 3D QSAR study, and also present for comparison some recent results we have obtained using the entire range of 3D QSAR techniques on another series of HIV-1 inhibitors published more recently by Holloway et al. of Merck Laboratories [28]. The models for these complexes were obtained by analogy to crystal structures of a handful of actual HIV-1/inhibitor complexes. The drug design and development efforts based on these two landmark modeling studies have contributed significantly to the development of the currently marketed class of HIV-1 protease inhibitors (for a list of HIV-1 protease inhibitors which have reached the market see Chapter 8). Our plan in the latter case is to didactically describe the use of these techniques as we present the results. This will put us in a position to comment on the relative merits of the methods. The last portion of this section will review a small number of recent success stories that feature 3D QSAR.

11.4.1 CoMFA of experimental HIV-1 protease complexes

In one of the earliest published and most influential applications of CoMFA, Garland Marshall and members of his group, Tudor Oprea, Christopher Waller and Alessandro Giolitti, performed a very detailed CoMFA analysis on the structures of the (then) available HIV-1 protease inhibitors [25–27]. One unique aspect of this three-part study on 59 compounds is that the bound conformations of several of the ligands were known from x-ray crystallography. This is not often the case—usually 3D QSAR is performed in the absence of structural data for the target biomacromolecule. The backbone atoms of the HIV-1 enzyme for seven enzyme/inhibitor crystal structures were superimposed via root mean square (RMS) fit, and the ligands were extracted and used without further structural modification. Additional ligands were added to the model by field-fitting their structure over the closest analog for which crystallographic data existed. Several other alignment models were prepared and examined in this study but, not surprisingly, the crystallographically derived one was the most internally consistent (highest q^2 or cross-validated r^2) [25]. That alignment also was most accurate at predicting the activities for the test set molecules [26].

Several interesting observations were made in this study that have general relevance to 3D QSAR. First, the crystallographic data provided critical insight into the alignment rule. The HIV-1 inhibitors examined were extremely flexible molecules—it would be very easy to align one or more of these molecules in an inverted orientation or with a completely inappropriate conformation. Second, field fit minimization of structures to a known template molecule to expand the alignment rule is preferable to molecular mechanics minimization within the active site. This is because subtle effects of solvent, etc. that constrained the original (crystallographically-determined) ligand will be mimicked in the field fit, but absent in simple site-constrained minimizations. Third, while the steric and electrostatic fields of CoMFA can represent much of the free energy of binding, there are other contributions, e.g., from solvation and hydrophobicity, that are being ignored. The addition of this type of data to the CoMFA model was one of the foci of the third part [27] of the study. Fourth, creation of a satisfactory *internally* self-consistent QSAR for the learning set is not sufficient to validate a 3D QSAR model. It must be tested on a *external* test set of compounds that were not part of the learning set [26]. Selection of good test sets involves a) consideration of quality and compatibility of biological data, b) the range of activities should be comparable to but not exceed that of the learning set, and c) the test set should have a balance of active and inactive compounds.

In fact, responses to many of these observations were incorporated into the more contemporary 3D QSAR methods. The three critical factors are: alignment rule, mathematical modeling and rational selection of physically and chemically meaningful "field" properties, and selection of an intelligent and statistically valid data set. Below, we look at another data set of HIV-1 protease inhibitors with several modern 3D QSAR methods.

11.4.2 Arsenal of 3D QSAR methods

We have made a comparison of the grid-based methods, such as CoMFA, CoMSIA and the recently described HIFA, with the pseudo-3D method HQSAR. CoMFA calculates steric fields using a Lennard-Jones potential and calculates electrostatic fields using a Coulombic potential (see Eqs. 1 and 2). While this approach has been widely accepted and exceptionally valuable, it is not without problems. In particular, both potential functions are very steep near the van der Waals surface of the molecule, causing rapid changes in surface descriptions, and requiring the use of cut-off values so calculations are not done inside the molecular surface. In addition, a scaling factor is applied to the steric field, so both fields can be used in the same PLS (partial least squares) analysis.

The data set employed in this study has been utilized to correlate binding affinity with intermolecular molecular mechanics calculated energy [28]. The subsequent addition of HINT calculated hydropathic binding descriptors to that model and by the incorporation of solvent effects made only small improvements in the QSAR [29]. This data set was chosen for the basis of our current study because the inhibition of HIV protease by small molecules is well understood, and the Holloway et al. [28] study is very well documented as to structures and activity data.

Holloway et al. reported a simple correlation for the compounds (Tab. 1) of;

$$pIC_{50} = -0.170 \, E_{inter} - 15.846, \tag{4}$$

where E_{inter} is the sum of intermolecular van der Waals and electrostatic interac-

Table 1. Structures and HIV protease binding affinities of molecules employed in the study

Compound	R1	R2	R3	IC_{50} (nM)
I000	CH_2Ph	H	-	0.25
1PAM	CH_2Ph	CH_3	-	7.70
1PC2	$CH_2CH_2CH_2Ph$	H (3-OH)	-	0.19
1PCF	CH_2-4-CF_3Ph	H	-	0.26
1PCI	(E)-$CH_2CH=CHPh$	H	-	0.23
1PF5	$CH_2C_6F_5$	H	-	0.60
1PME	CH_2-4-CH_3Ph	H	-	0.29
1PNH	CH_2-4-NH_2Ph	H	-	0.31
1PNO	CH_2-4-NO_2Ph	H	-	0.27
1PNS	H	H	-	2934.00
1POH	CH_2-4-OHPh	H	-	0.16
1PPE	$CH_2CH=CH_2$	H	-	27.50
1PPI	CH_2-4-IPh	H	-	0.72
1PPO	$CH_2C(O)Ph$	H	-	5.42
1PPY	CH_2-4-pyridyl	H	-	0.53
1PSP	CH_2SPh	H	-	0.25
1PTB	CH_2-4-t-butylPh	H	-	0.17
2PBA	-	-	CH_2Ph	114.00
2PCP	-	-	5-(3-hydroxycyclopent-3-enyl)	9.53
2PIN	-	-	1-indanyl	34.25
2PPA	-	-	$CH(CH_2OH)CH_2Ph$	690.00
2PVO	-	-	$CH(CH_2OH)CH(CH_3)_2$	161.00
2PIE	-	-	1-(2-carboxymethylindanyl)	66.30
2PCO	-	-	2-hydroxycyclohexyl	121.80
2PTE	-	-	1-(2-hydroxytetralinyl)	0.70
2PCM	-	-	$CH_2C_6H_{11}$	30000.00
2PPE	-	-	$CH(CH_3)C_6H_5$	146.00
2PIO	-	-	1-(2,3-dihydroxyindanyl)	0.10
2PPO	-	-	$CH(CH_2OH)Ph$	38.60
2PTO	-	-	(S,S)-4-(3-hydroxychromanyl)	0.18
2PT2	-	-	(R,R)-4-(3-hydroxychromanyl)	40.50

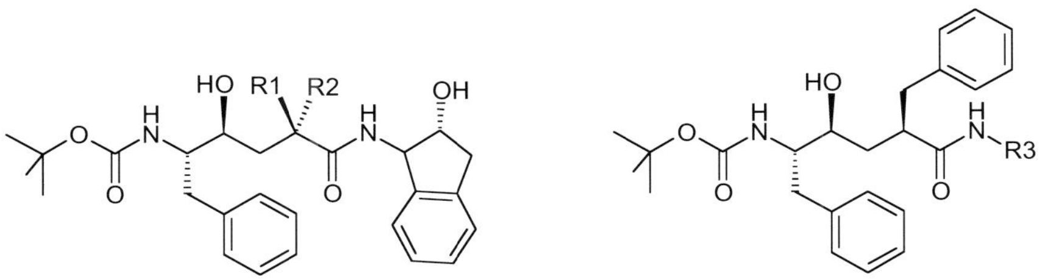

Scheme 2.

tions between the ligand and enzyme. The model gave $r^2 = 0.797$ and a standard error of 0.675. This is a surprisingly simple and yet robust model considering the complexity of the molecules. In fact this model is superior to that obtained with a standard CoMFA model (see Tab. 2): Thirty-one compounds bound to HIV protease were analyzed in a standard CoMFA model.[2, 3]

This model yielded a cross-validated correlation $(q^2) = 0.297$, with a standard error of prediction of 1.249 with two components. A fitted PLS regression employing the same number of components yielded a higher r^2 of 0.623, with a standard error of estimate of 0.919.

Comparison of CoMFA to CoMSIA on the same data set reveals similar results, in line with previously published observations.[4] There is a nominal improvement in the steric/electrostatic model by incorporation of the hydrophobic and hydrogen bond field descriptors, although it is interesting to note that the hydrophobic field alone is of greater predictive power.

The HINT intermolecular field analysis (HIFA) [24] model of the ligands *in* the receptor complex yielded a cross-validated correlation $(q^2) = 0.696$, with a standard error of prediction of 0.958 with five components.[5] A fitted PLS regression employing the same number of components yielded a higher r^2 of 0.912, with a standard error of estimate of 0.477. The superiority of this model is a consequence of the additional structural information that is derived from measurement of the *interaction* between the ligands and their receptor, unlike CoMFA and CoMSIA that are performed via measurement of only the properties of the ligand molecules.

The hologram QSAR method (HQSAR) does provide a better analysis than the usual grid based 3D approaches.[4] The researcher is able to regulate the type of fingerprint that may be generated and subsequently employed in the analysis.

---

[2] Computational studies were initiated on the x-ray coordinates of HIV-1 protease complexed with the Merck inhibitor, L-689,502,[28] utilizing the Sybyl suite of molecular modeling programs [6]. All crystallographic waters, with the exception of the tightly bound water (residue 407), were removed. The oxygen of this water, together with the protein heavy atoms were kept rigid during geometry optimizations. Hydrogen atoms were added to the protein-water-inhibitor complexes and the structure were subsequently optimized with the Merck Molecular Force Field (MMFF) [30]. All binding affinities were expressed as pIC_{50}. A 3D region defining the boundaries of the grid-based calculations was constructed that surrounds the ligand-binding domain, with spacing between sampling points of 2Å.

[3] For the CoMFA study, the optimized ligands were extracted from the protein binding site and assigned charges by the Gasteiger-Hückel method. The optimum model was determined by the standard PLS algorithm with a leave-one-out cross-validation, using up to ten components. A CoMFA was performed on the 31 ligands using an sp^3 C probe bearing a unit positive charge, employing the described 3D region.

[4] As for CoMFA, the optimized ligands were extracted from the protein and assigned Gasteiger-Hückel charges for the CoMSIA and HQSAR studies. CoMSIA was performed in a similar manner to CoMFA, using the standard steric-electrostatic, donor-acceptor and hydrophobic fields. HQSAR was performed using the default hologram lengths [53, 59, 61, 71, 83, 97, 151, 199] and selecting the "best" model that gave the highest cross-validated r^2.

[5] In the HIFA study the HIV protease-water complex was partitioned in HINT employing the protein dictionary method of determination, using all hydrogens. The ligand was partitioned in HINT by the explicit calculation method, also using all hydrogens. Hydropathic interaction maps were calculated using the HINT intermolecular protocol, specifying all interactions constrained by the described 3D region. The resultant maps were imported into Sybyl using the same region definition. A QSAR data table was constructed containing the molecules that are reported in Table 1 and a column generated for the map files as a CoMFA field-type, using the externally created fields.

Table 2. 3D QSAR results for the HIV-1 protease inhibitor data set

Method	"Field" Information	q^2	r^2	Standard error of estimate
CoMFA	steric, electrostatic	0.297	0.623	0.919
HIFA	hydropathic interaction	0.696	0.912	0.477
CoMSIA	steric, electrostatic	0.279	0.470	1.090
	donor, acceptor	0.292	0.462	1.098
	hydrophobic	0.367	0.741	0.789
	all fields	0.389	0.812	0.660
HQSAR	a,c,d	0.528	0.611	1.049
	a,b,c,d	0.566	0.637	0.902
	a,b,c,h,d	0.584	0.666	0.866
	a,b,c,h,ch,d	0.600	0.682	0.845

Features that may be considered in unique fingerprint generation are atom type (a), bond type (b), atomic connection (c), hydrogen bond donor or acceptor (d), chirality (ch) and inclusion of hydrogen atoms (h). Thus, in the case where all options are suppressed, an identical fingerprint would be generated for benzene and pyridine. However, unique fingerprints would be obtained if one employs either the atom type specifier where the difference in the C and N atom type would be observed, or by inclusion of the hydrogen atoms where one would now count the six hydrogen atoms in benzene compared to the 5 in pyridine.

From the Table 2, it can be seen that a modest improvement in the HQSAR analysis can be obtained by addition of the hydrogen bond donor/acceptor and hydrogen atom count. The chirality flag appears to have little effect, which is entirely predictable from the data set (Tab. 1), since there is only one diastereomeric pair and there is little separation in potency. An important role of a QSAR model, besides predicting the activities of untested molecules, is to provide hints about what molecular fragments may be important contributors to activity. In HQSAR, following the calculation of atom contributions to activity, the molecule display is color coded to reflect the individual atomic contributions. In this example, the colors at the red end of the spectrum (red, red orange, and orange) reflect favorable (or positive) contributions, while colors at the green end (yellow, green blue, and green) reflect unfavorable (negative) contributions. Atoms with intermediate contributions are colored white. In the case of the inactive analog 2PCM shown below (Fig. 3), the phenyl ring projects into a region that the corresponding CoMSIA derived hydrophobic field associates with poor activity. In fact, there is generally good agreement in the graphical output of these two methods. The highly active compound 2PIO occupies steric regions that are associated with good binding affinity and matches well to the electrostatic pattern (not shown) determined by CoMSIA.

It is particularly interesting to note that HQSAR, an inherently 2D method, provides a more predictive model than its standard 3D grid based counterparts. This begs the question as to why this should be? In part, this would suggest that at a superficial level the variation in biological activity is entirely encoded within the

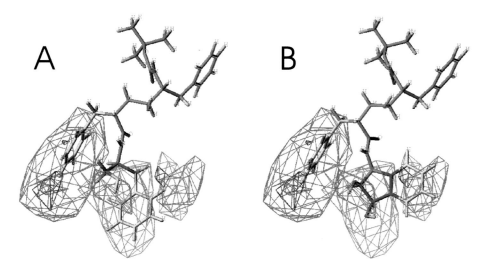

Figure 3. Projection of steric (green/red) and hydrophobic (orange/blue) CoMSIA maps onto the HQSAR color coded models of A) 2PCM and B) 2PIO.

structure of the ligands and thus is conformationally independent. However, it is probably also a consequence of the respective methodologies themselves. In the hashed fingerprint method of HQSAR one correlates the positive effect, for example, of a hydroxyl group on the indane ring with binding affinity. Whereas, in the grid based methods this moiety is reduced to a combination of steric, electrostatic and/or hydrophobic field effects. In so doing, the measurement becomes less precise as the substituent effect is dispersed in sampling space, with a sampling grid perhaps as coarse as 2 Å. (Using less coarse grids can help resolve this issue, but has a concomitant cost of added "noise" from more, mostly useless, independent variables in the model.)

These results highlight the limitation of employing 3D grid based methods on congeneric series and provide clues as to why similar force field based scoring algorithms in general correlate poorly with measured binding. The fact that the HIFA approach provides the best analysis indicates the value of incorporating information regarding the *interaction* of these ligands with their receptor, when it is available. Perhaps this key point is obvious: the more detailed structural data one has about the ligands and receptors, the more accurate the QSAR models can be.

11.4.3   Recent success stories of 3D QSAR

There are, of course, innumerable other applications of 3D QSAR. As of this writing, the original Cramer, Patterson and Bunce paper [5] describing the CoMFA procedure has been cited nearly 700 times in the ISI Citation data base. It should be pointed out, if it is not already obvious, that it is, at least, several years after the completion of a pharmaceutical drug discovery project before the strategies, etc. behind the project are disclosed by publication. We are thus certain that published studies of 3D QSAR are only the tip of the iceberg with respect to the actual usage of this paradigm for drug discovery and development in industry. We have chosen, for brief review here, a few recent examples that we find particularly interesting for a variety of reasons. In particular, we highlight studies that effectively use a combination of 2D and/or 3D techniques.

In a study where there is little information about the receptor, a 3D QSAR was created and used in a drug design project at Astra involving muscarinic ligands [31]. To initiate the study, a potent, conformationally rigid muscarinic agonist, dihydro-syn-4'-methylspiro[1-azabicyclo[2.2.1]heptane-3,5'(4'H)-furan]-3'-one [31], was examined with calculated electrostatic and GRID [32] maps. This data afforded information as to the optimal position of potential interaction sites between the ligand and its receptor. Consequently, the requisite groups were orientated in space to achieve maximal interaction with the compound and thus provided a pseudo-receptor structure. For example, as suggested by GRID, a carboxylate group was placed adjacent to the protonated ring nitrogen of the rigid compound. Similarly, aliphatic hydroxyl groups were constructed to maximally interact with both the ring ether and carbonyl oxygen atoms. The complex structure was then geometry optimized. A diverse subset of muscarinic agonists were "docked" into the pseudo receptor site model defined by these studies and the resultant alignment was employed in the 3D QSAR. A CoMFA was performed on 80 ligands modeled in this pseudo receptor site, including an extensive sample of acyl hydrazones, hydrazides and oximes. The model obtained had a cross-validated r^2 of 0.556, employing six components, with a final r^2 of 0.900, standard error of 0.278 [31]. This model was further refined with the use of alternative probe atoms of differing charge, the use of smaller grid spacing and the incorporation of hydropathic fields calculated by HINT [7]. The final model was employed in the design and affinity prediction of a number of potential muscarinic receptor ligands [31].

In another study, the inhibitory activity values (IC_{50}) of 54 phenyltropanes at the dopamine (DA), serotonin (5-HT) and noradrenaline (NA) transporters were quantitatively examined by use of different QSAR-techniques and PLS [33]. The use of 36 physicochemical and quantum mechanical parameters resulted in informative 2D models that were simple to interpret. For the purpose of comparison, additional CoMFA models using the standard fields were constructed from a molecular alignment maximizing the similarity of shape and electrostatic potential. Highly significant models with good fitting and predictive abilities were developed for each transporter. The major structural requirements revealed from the CoMFA analyses were in good agreement with the findings of the individual 2D analyses. The models obtained suggest a close similarity between the examined transporters in terms of binding interaction. A comparison of the CoMFA contour maps provided a rational basis for the design of transporter-selective ligands.

Pajeva and Wiese have recently reported molecular modeling studies on multidrug resistance (MDR) modifiers based on phenothiazines and related compounds [34, 35]. MDR is a major impediment to the treatment of metastatic cancers. It is a broad-spectrum invulnerability to chemotherapy, involving a number of mechanisms. A number of drugs have been identified, called MDR reversing agents or modifiers, that are not cytotoxic by themselves but can reverse MDR. The authors performed QSAR and CoMFA studies on phenothizines, thioxanthenes and related compounds to identify and quantitatively assess structural features important for anti-MDR activity. The key factor of membrane penetrations and interactions, expected to be related to hydrophobicity, was demonstrated in CoMFA models with cross-validated statistics (q^2) as high as 0.93 when the HINT field was combined with the steric and electrostatic fields [35]. In effect, hydrophobicity is a molecular property that significantly influences MDR reversals. Furthermore, describing hydrophobicity as a space-directed molecular property, i.e., a field, is preferable to the use of scalar ($LogP_{o/w}$) representations of hydrophobicity.

Predictive correlations have been obtained for angiotensin-converting enzyme inhibitors in a number of different protocols [36–38], including CoMFA and Catalyst [39] over the past several years. However, certain inadequacies of the CoMFA technique have been noted, primarily that the standard steric and electrostatic fields alone do not fully characterize the zinc-ligand interaction. This has been partially rectified by the inclusion of indicator variables into the QSAR table to designate the class of zinc-binding ligand. Also, using molecular orbital (MO) fields derived from semi-empirical calculations as additional descriptors in the QSAR table produced predictive correlations based on CoMFA and MO fields alone [40].

11.5 Conclusions

We have attempted to demonstrate the variety of 3D QSAR technology currently available for modern drug discovery. We have not mentioned every method in use, but feel that we have sampled a broad spectrum. It should be obvious from the discussion above that it is prudent to apply more than one technique to the problem at hand. Just as one synthetic method will not synthesize every compound, one 3D QSAR method will likely not work with every data set. In addition, one should not limit the application of analytical methods to those based solely on three-dimensional information. The value of combining traditional 2D and "pseudo-3D" with the grid based 3D methods has been amply demonstrated. Indeed, the choice of methods that the researcher employs should be based upon the nature of the dataset under consideration. It would appear that grid based 3D methods offer little, if no advantage, over traditional 2D methods when applied to data sets comprised of a single congeneric series. However, 3D QSAR is unrivaled in its ability to evaluate alternate structural alignments of diverse data sets. It should also be emphasized that the point of undertaking a QSAR study should be to learn about a set of existing compounds so that new ones with better properties can be found. In that context, it is important to design QSAR studies so that *chemically* meaningful information can be extracted—the kind of information that can be directly applied to drug design. Since the development of pharmacophore hypotheses is fundamental to the researcher in the absence of structural information, the ability to assess the relevance and predictability of such models is crucial. Grid based 3D approaches, in particular, not only provide valuable scoring data, but also deliver graphical information correlating structural modification with biological activity. Finally, to reiterate the conclusions of Oprea, Waller and Marshall [25], there is a difference between *internally*-consistent QSAR models and *externally*-predictive QSAR models. In many ways this represents the errors that occur when extrapolating from a model rather than interpolating within a model.

11.6 Acknowledgements

The authors acknowledge the support of their organizations: VCU and Wyeth-Ayerst. We thank Tripos, Inc. for their help over the past decade of our research in developing and using 3D QSAR applications. Finally, we are grateful to Derek J. Cashman (VCU) for help with manuscript graphics.

References

1 Meyer H (1899) Zur Theorie der Alkoholnarkose; erste Mittheilung: Welche Eigenschaft der Anaesthetica bedingt ihre narkotische Wirkung? *Arch Exp Pathol Pharmakol* 42: 109–118
2 Hansch C, Fujita T (1964) ρ-—π analysis. A method for the correlation of biological activity and chemical structure. *J Am Chem Soc* 86: 1616–1626
3 Rekker RF (ed) (1977) *The hydrophobic fragmental constant. Its derivation and applications. A means of characterizing membrane systems.* Elsevier, New York
4 Hall LH, Kier LB (1991) The molecular connectivity chi indexes and kappa shape indexes in structure-property relations. In: Boyd D, Lipkowitz K (eds): *Reviews of computational chemistry.* VCH Publishers, Weinheim, Germany, 367–422
5 Cramer RD, III, Patterson DE, Bunce JD (1988) Comparative molecular field analysis, 1. Effect of shape on binding of steroids to carrier proteins. *J Am Chem Soc* 110: 5959–5967
6 Sybyl molecular modeling suite, Tripos Inc., 1699 S. Hanley Road, St. Louis MO 63144, USA
7 Kellogg GE, Semus SF, Abraham DJ (1991) HINT – a new method of empirical hydrophobic field calculation for CoMFA. *J Comput Aided Mol Design* 5: 545–552
8 Vaz R (1997) Use of electron densities in comparative molecular field analysis (CoMFA): A quantitative structure activity relationship (QSAR) for electronic effects of groups. *Quant-Struct Act Relat* 16: 303–308
9 Kellogg GE, Kier LB, Gaillard P et al (1996) The E-State fields. Application to 3D QSAR. *J Comput Aided Mol Des* 10: 513–520
10 Kellogg GE (1997) Finding optimum field models for 3D QSAR. *Med Chem Res* 7: 417–427
11 Woolfrey JR, Avery MA, Doweyko AM (1998) Comparison of 3D quantitative structure-activity relationship methods: Analysis of the *in vitro* antimalarial activity of 154 artemisinin analogues by hypothetical active-site lattice and comparative molecular field analysis. *J Comput Aided Mol Des* 12: 165–181
12 Klebe G, Abraham U, Mietzner T (1994) Molecular similarity indexes in a comparative-analysis (CoMSIA) of drug molecules to correlate and predict their biological-activity. *J Med Chem* 37: 4130–4146

13 Wold S, Johansson E, Cocchi M (1993) PLS – Partial Least Squares projections to latent structures. In: Kubinyi H (ed): *3D QSAR in drug design. Theory methods and applications.* Escom, Leiden, 523–550

14 Rogers D, Hopfinger AJ (1994) Application of genetic function approximation to quantitative structure-activity-relationships and quantitative structure-property relationships. *J Chem Info Comput Sci* 34: 854–866

15 Ferguson AM, Heritage T, Jonathon P et al (1997) EVA: A new theoretically based molecular descriptor for use in QSAR/QSPR analysis. *J Comput Aided Mol Des* 11: 143–152

16 Tong W, Lowis DR, Perkins R et al (1998) Evaluation of quantitative structure-activity relationship methods for large-scale prediction of chemicals binding to the estrogen receptor. *J Chem Info Comput Sci* 38: 669–677

17 Willet P, Winterman V, Bawden D (1986) Implementation of nearest-neighbor searching in an online chemical structure search system. *J Chem Info Comput Sci* 26: 36–41

18 Todeschini R, Gramatica P (1997) 3D-modelling and prediction by WHIM descriptors. Part 5. Theory development and chemical meaning of WHIM descriptors. *Quant Struct-Act Relat* 16: 113–119

19 Martin YC, Bures MG, Danaher EA et al (1993) A fast new approach to pharmacophore mapping and its application to dopaminergic and benzodiazepine agonists. *J Comput Aided Mol Des* 7: 83–102

20 Klebe G, Abraham U (1993) On the prediction of binding properties of drug molecules by comparative field analysis. *J Med Chem* 36: 70–80

21 Thomas BF, Compton DR, Martin BR et al (1991) Modeling the cannabinoid receptor: A three-dimensional quantitative structure activity analysis. *Mol Pharmacol* 40: 656–665

22 Pertwee R (1997) Cannabis and cannabinoids: Pharmacology and rationale for clinical use. *Pharm Sci* 3: 539–545

23 Baker D, Pryce G, Croxford JL et al (2000) Cannabinoids control spasticity and tremor in a multiple sclerosis model. *Nature* 404: 84–87

24 Semus SF (1999) A novel hydropathic intermolecular field analysis (HIFA) for the prediction of ligand-receptor binding affinities. *Med Chem Res* 9: 535–550

25 Waller CL, Oprea TI, Giolitti A et al (1993) 3-dimensional QSAR of human-immunodeficiency-virus-(I) protease inhibitors. 1. A CoMFA study employing experimentally-determined alignment rules. *J Med Chem* 36: 4152–4160

26 Oprea TI, Waller CL, Marshall GR (1994) 3-dimensional quantitative structure-activity relationship of human-immunodeficiency-virus-(I) protease inhibitors. 2. Predictive power using limited exploration of alternate binding modes. *J Med Chem* 37: 2206–2215

27 Oprea TI, Waller CL, Marshall GR (1994) 3D-QSAR of human immunodeficiency virus (I) protease inhibitors. III Interpretation of CoMFA results. *Drug Design and Discovery* 12: 29–51

28 Holloway MK, Wai JM, Halgren TA et al (1995) A priori prediction of activity for HIV-1 protease inhibitors employing energy minimization in the active-site. *J Med Chem* 38: 305–317

29 Wei DT, Meadows JC, Kellogg GE (1997) Effects of entropy on QSAR equations for HIV-1 protease: 1. Using hydropathic binding descriptors. 2. Unrestrained complex structure optimizations. *Med Chem Res* 7: 259–270

30 Halgren TA (1996) Merck molecular force field. V. Extension of MMFF94 using experimental data, additional computational data and empirical rules. *J Comput Chem* 17: 616–641

31 Wu ESC, Kover A, Semus SF (1998) Synthesis and modeling studies of a potent conformationally rigid muscarinic agonist: 1-azabicyclo[2.2.1]heptanespirofuranone. *J Med Chem* 41: 4181–4185

32 Goodford PJ (1985) A computational procedure for determining energetically favorable binding sites on biologically important macromolecules. *J Med Chem* 28: 857–864

33 Muszynski IC, Scapozza L, Kovar K-A et al (1999) Quantitative structure-activity relationships of phenyltropanes as inhibitors of three monoamine transporters: classical and CoMFA studies. *Quant Struct-Act Relat* 18: 342–353

34 Pajeva IK, Wiese M (1997) QSAR and molecular modeling of cataphiphilic drugs able to modulate multidrug resistance in tumors. *Quant Struct-Act Relat* 16: 1–10

35 Pajeva IK, Wiese M (1998) Molecular modeling of phenothiazines and related drugs as multidrug resistance modifiers: A comparative molecular field analysis study. *J Med Chem* 41: 1815–1826

36 DePriest SA, Mayer D, Naylor CB et al (1993) 3D-QSAR of angiotensin-converting enzyme and thermolysin inhibitors: A comparison of CoMFA models based on deduced and experimentally determined active site geometries. *J Am Chem Soc* 115: 5372–5384

37 DePriest SA, Shands EFB, Dammkoehler RA et al (1991) 3D-QSAR: Further studies on inhibitors of angiotensin-converting enzyme. *Pharmacochem Libr* 16: 405–416

38 Sjoestroem M, Eriksson L, Hellberg S et al (1989) Peptide QSARS: PLS modeling and design in principal properties. *Prog Clin Biol Res* 291: 313–317
39 Sprague PW (1995) Automated chemical hypothesis generation and database searching with Catalyst. *Persp Drug Discovery Design* 3: 1–20
40 Waller CL, Marshall GR (1993) 3D-QSAR three dimensional quantitative structure activity relationship of angiotension-converting enzyme and thermolysin inhibitors. II. A comparison of CoMFA models incorporating molecular orbital fields and desolvation free energies based on active-analog and complementary-receptor-field alignment rules. *J Med Chem* 36: 2390–2403

12 Physicochemical concepts in drug design

Han van de Waterbeemd

Pfizer Global Research and Development, PDM, Sandwich, Kent CT13 9NJ, UK

12.1 Introduction

A successful drug candidate has the right attributes to reach and bind its molecular target and has the desired duration of action. Binding to the target can be optimised by designing the proper three-dimensional arrangement of functional groups. Each chemical entity also has, through its structure, physicochemical and biopharmaceutical properties. These are generally related to processes such as dissolution, oral absorption, uptake into the brain, plasma protein binding, distribution, and metabolism. Therefore fine-tuning of the physicochemical properties has an important place in lead optimization.

In this chapter recent developments around some of the key physicochemical properties and how they are used in various phases of the discovery process are addressed.

12.2 Experimental approaches

12.2.1 Solubility

The first step in the process to enter the body from a formulation for most drugs is dissolution. Traditional dissolution and solubility measurements are quite tedious. In today's discovery environment higher throughput demands are extended to physicochemical measurements, such as assessment of solubility. Turbidometric or kinetic solubility evaluation has been reported as a reasonable first approximation to detect low solubility compounds [1, 2]. An automated potentiometric titration method for solubility measurement, and solubility-pH profiles has been introduced [3]. In addition to these new experimental approaches various efforts have been reported to estimate solubility from molecular structure (see section 12.3). However, knowledge of aqueous solubility and dissolution rate still may be insufficient to predict *in vivo* performance, since food effects may have considerable and variable impact [4].

12.2.2 Lipophilicity

One of the key properties relevant to the biopharmaceutical and pharmacokinetic profile of a compound is lipophilicity. Very often lipophilicity is taken as equivalent to octanol/water partitioning or distribution, or seen as equivalent to hydrophobicity. Therefore it is important to first repeat the recent IUPAC definitions here [5]:

> *Hydrophobicity* is the association of non-polar groups or molecules in an aqueous environment which arises from the tendency of water to exclude non-polar molecules
>
> *Lipophilicity* represents the affinity of a molecule or a moiety for a lipophilic environment. It is commonly measured by its distribution behavior in a biphasic system, either liquid-liquid (e.g., partition coefficient in 1-octanol/water) or solid-liquid (retention on reversed-phase high-performance liquid chromatography (*RP-HPLC*) or thin-layer chromatography (*TLC*) system).

Partition coefficients (log P) refer to the compound in its neutral state, while distribution coefficients (log D) are measured at a selected pH, often 7.4. n-Octanol is the most widely used model of a biomembrane. Large compilations of octanol/water distribution data are available, e.g., in the MedChem database. n-Octanol is an H-bond donor and acceptor solvent and may not mimic the characteristics of very lipophilic membranes such as the blood-brain barrier (BBB). Therefore various other solvent systems have been used in the past to estimate membrane transport (see Fig. 1). More recently a number of chromatographic systems have been suggested to measure lipophilicity. Among these immobilized artificial membranes (IAM) have been well-studied [6].

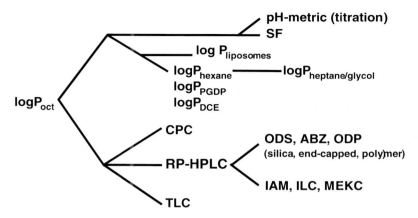

Figure 1. Experimental methods to measure lipophilicity (*modified after* [7]). Key of abbreviations—logP$_{oct}$:1-octanol/water partition coefficient, log P$_{liposomes}$: partition coefficient between liposomes and buffer, logP$_{hexane}$: 1-hexane/water partition coefficient, logP$_{PGDP}$: propyleneglycol dipelargonate/water partition coefficient, logP$_{heptane/glycol}$: a non-aqueous partitioning system, SF: shake-flask, pH-metric: logP determination based on potentiometric titration in water and octanol/water, CPC: centrifugal partition chromatography, RP-HPLC: reversed-phase high performance liquid chromatography, TLC: thin-layer chromatography, ODS: octadecylsilane, ABZ: end-capped silica RP-18 column, ODP: octadecylpolyvinyl packing, IAM: immobilized artificial membrane, ILC: immobilized liposome chromatography, MEKC: micellar electrokinetic capillary chromatography.

Along the same lines several groups have suggested that studying partitioning into liposomes may produce relevant information related to membrane uptake and absorption [8–15].

Another development using distribution or partition coefficients was to look at the difference between two solvent scales, particularly octanol/water and alkane/water. The thus obtained Δlog P values appear to encode for the H-bonding capacity of a solute. H-bonding itself was demonstrated to be an important property for membrane crossing. Further research showed that H-bonding can be more conveniently assessed by computational methods (see below).

12.2.3 Absorption models

Making an estimate of the absorption potential of a new chemical has become an important point in the discovery process. Combinatorial chemistry has put an accent on higher throughput methods to study absorption. The following absorption models are available:

- In clinic (human oral absorption studies)
- *In vivo* (rat, dog, other species)
- *In situ* (perfused rat intestine)
- *Ex vivo* (Ussing chamber)

- *In vitro* (Caco-2, TC7, MDCK and other cell lines)
- Physicochemical methods
- *In silico/in computro* (absorption prediction)

Going from the top to the bottom of this list, these methods are increasingly appropriate for higher throughput. With increasing databases and learning the confidence in rapid screening methods is also increasing. Currently most companies, for absorption estimation, compromise between throughput and reliability in the prediction to man.

A well-established model is the Caco-2 monolayer technique. This consists of measuring the flux through a monolayer of cells grown on a filter using an immortalized human colon adenocarcinoma cell line [16]. Since these cell lines take between 15–21 days to grow, other faster growing cell lines like MDCK cells [17] may offer an advantage. The TC7 cell line, a Caco-2 clone, appears to express less P-gp [18]. Both Caco-2 and TC7 also express CYP3A4, although mRNA for CYP3A4 was only detectable in TC7 cells [19]. This would make this cell line appropriate to study the interplay between CYP3A4 and P-gp as barriers to intestinal absorption.

To resemble the blood-brain barrier (BBB) cell cultures should mimic the tight junctions in the BBB. This appears to be a major difficulty. Cerebrovascular endothelial cells have been isolated and cultured as BBB model. Recently it was suggested that monolayer uptake kinetics may be an alternative to transmonolayer flux measurements as a more reliable model of the BBB [20].

Percutaneous absorption and other routes can also be modeled with *in vitro* systems. For an excellent overview, see [21].

12.3 Computational approaches

A wide range of different structural and physicochemical properties can be computed. Many programs are available and offer variations of lipophilicity, size, electronic and H-bonding descriptors. These properties can be used to predict other biopharmaceutical of pharmacokinetic properties, or can be used to describe combinatorial libraries and are used to prioritize screening hits. Less useful for this purpose are the many topological indices, apart from those that correlate strongly to size and shape properties (see Chapter 7).

Sufficient solubility is one of the first prerequisites to be absorbed from the GI-tract. Using the general solvation equation methodology, Abraham and co-workers developed following equation to predict solubility in aqueous environment from molecular structure [22]

$$\log S_w = 0.52 - 1.00\ R_2 + 0.77\ \pi_2^H + 2.17\ \Sigma\alpha_2^H + 4.24\ \Sigma\beta_2^H - 3.36\ \Sigma\alpha_2^H\Sigma\beta_2^H - 3.99\ V_x$$

$$n = 659;\ r^2 = 0.920;\ s = 0.56;\ F = 1256$$

where R_2 is the excess molar refraction, π_2^H is the solute dipolarity/polarizability,

$\Sigma\alpha_2^H$ is the overall or summation hydrogen-bond acidity, $\Sigma\beta_2^H$ is the overall or summation hydrogen-bond basicity, $\Sigma\alpha_2^H\Sigma\beta_2^H$ is a mixed term dealing with hydrogen-bond interactions between acid and basic sites in the solute, Vx is McGowan's characteristic volume. These properties can be calculated using the program ABSOLV [22]. No information is given on the potential intercorrelation of the terms used in this equation. Future research should be directed to predict solubility in buffered media simulating the intestinal fluid. Another approach in solubility prediction is based on molecular topology and neural network modeling [23]. The overall correlation between predicted and observed solubilities was $r^2 = 0.86$, which looks reasonable. However, the individual estimates may be 1 log unit or more wrong, which in the case of poorly soluble compounds is unacceptable. Accurate measured values are still required to overcome shortcomings of current solubility predictors.

Probably the best known calculated lipophilicity property is CLOGP, the calculated log P value of a compound in its neutral state [24]. Although in many cases useful, more relevant are log D values accounting for ionization at a selected pH, often 7.4. Log D values can be estimated by combining estimates of log P and pKa and using the appropriate Henderson-Hasselbach equations. Recent reviews of log P computation can be found in, for example, [25, 26]. A number of programs are available to calculate pKa values. The quality of such data depend on the complexity of the structures. More innovative structures are more likely to be difficult cases.

H-bonding capability of a drug has been found to correlate well with human intestinal absorption [27, 28] and uptake in the brain [29, 30]. All these approaches give a similar picture, i.e., H-bonding is important for membrane crossing. This property can be simply calculated by summing up the polar surface area of all nitrogen and oxygen atoms in a molecule. Thus conformational flexibility is taken into account. There is debate whether the minimum energy conformation is sufficient [28–30] or whether a wide range of conformations should be considered and factored into a dynamic polar surface area [27]. More simple even is to count total numbers of H-bond acceptors and donors. H-bond factors for donors and acceptors can by obtained from the program HYBOT [31].

Prediction of human intestinal absorption direct from molecular structure has been studied by several groups. The above-mentioned correlation with polar surface area is one example. Others have used combinations of calculated properties [32], sometimes mixed with experimental physicochemical data such as log D values [33]. Using almost identical approaches, the permeation through Caco-2 cells [34] or uptake into the brain can be estimated [35]. All of these methods currently do not take into account the role of the biochemical barrier of a membrane. In the future such models need to be complemented by terms for P-gp efflux and contributions of other transporters, and for gut wall metabolism.

12.4 Design of combinatorial libraries

In order to make combinatorial lead optimization (see Chapter 6) an efficient process the library members should have suitable drug-like properties. Several constraints have to be considered [36, 37]. A compromise must be made between synthetic and economic feasibility, property space and diversity, and last but not least bioavailability constraints. Filters to eliminate "obviously bad" compounds from a chemical library have been suggested [37]. These include an evaluation of molecular weight, lipophilicity, reactive or toxic functional groups, floppiness and drug-likeness. Drug-likeness of the library members can be judged, for example, by looking if the new compound falls within the most popular scaffolds usually found in known drugs [37] (see also Chapter 7). Thus virtual or *in silico* screening avoids making poor compounds already at the design stage (see Chapters 8 and 10). The concept of molecular diversity has been used both in design and prioritization of combinatorial series [38] (see also Chapter 7).

12.5 Selection of subsets for HTS and hit validation

Selection of promising subsets from an existing in-house database is an important first step to avoid hits which are not drug-like or have other undesirable properties. Computer-based screening of compound databases for the identification of novel leads [39] may be a screening tool as such, or may be used to select appropriate subsets. Calculated and experimental molecular properties can both be used as a good guide in HTS hits evaluation [40] (see also Chapter 4).

12.6 Lead optimization and candidate selection

Once a lead has been discovered in terms of sufficient binding to the molecular target and having appropriate chemical attractiveness, a number of properties will need to be fine-tuned. Generally this includes further improvements in binding and selectivity, as well as optimization of biopharmaceutical and pharmacokinetic properties. It is in this stage that physicochemical profiling and absorption estimation may guide a project towards a small set of suitable candidates.

The optimal candidate meets a number of pre-set criteria. These may vary according to the desired application form. For oral absorption it is obvious to have a good estimate of predicted human absorption and bioavailability. For once or twice a day dosing reliable clearance predictions to man need to be available. The role of physicochemistry and structure in metabolism and pharmacokinetics is reasonably well understood [41, 42]. An important task is the metabolic evaluation of possible candidates (see Chapter 13). There are now well-established *in vitro* techniques for assessing the role of specific P450 isoenzymes in the metabolism of drugs and to evaluate the potential for drug-drug interactions.

12.7 Examples of the use of physicochemical properties in drug discovery

12.7.1 HIV inhibitors

Marketed inhibitors of the human immunodeficiency virus (HIV) are all relatively large in molecular size, e.g., saquinavir (MW 671) and indinavir (MW 614). Increasing solubility, lowering H-bond potential were recognized as favorable to increase bioavailability of this class of compounds, while the role of molecular size indeed remains unclear [43]. The incorporation of an ionizable center, such as an amine or similar function, into a template can bring a number of benefits including water solubility. A key step [44] in the discovery of indinavir was the incorporation of a basic amine (and a pyridine) into the backbone of hydroxyethylene transition state mimic compounds (Fig. 2a) to enhance solubility and potency (for structure-based design efforts on indinavir see Chapter 8).

In another series of potential HIV inhibitors it was tried to introduce specific functional groups to improve solubility and bioavailability [45]. It was concluded that structural features regarded as beneficial for oral uptake in one series of compounds may not enhance bioavailability in other types of inhibitors. Rather the overall structural features of a compound govern the oral bioavailability.

12.7.2 GABA uptake inhibitors

GABA (γ-aminobutyric acid) is a major neurotransmitter in mammals and is involved in various CNS disorders. In the design of a series of GABA uptake inhibitors a large difference in *in vivo* activity between two compounds with identical IC_{50} values was observed, one compound being devoid of activity [46]. The compounds have also nearly identical pKa and log D_{oct} values (see Tab. 1) and differ only in their distribution coefficient in cyclohexane/water (log D_{chex}). This results in a ΔlogD of 2.71 for the *in vivo* inactive compounds, which is believed to be too large for CNS uptake. The active compound has a Δlog D of 1.42, well below the critical limit of *ca* 2. Besides this physicochemical explanation further evaluation of metabolic differences should complete this picture. It should be noted that the concept of using the differences between solvent systems was originally developed for compounds in their neutral state (Δlog P values, see 13.2.2). In this case two zwitterions are being compared, which are considered at pH 7.4 to have a net zero charge, and thus the Δlog P concept seems applicable.

12.7.3 Antifungal fluconazole

Fluconazole (Fig. 2b) is an example where a knowledge of the relationship between physicochemical properties and drug disposition has allowed optimization of the drug's performance [41]. The project's goal was a superior compound

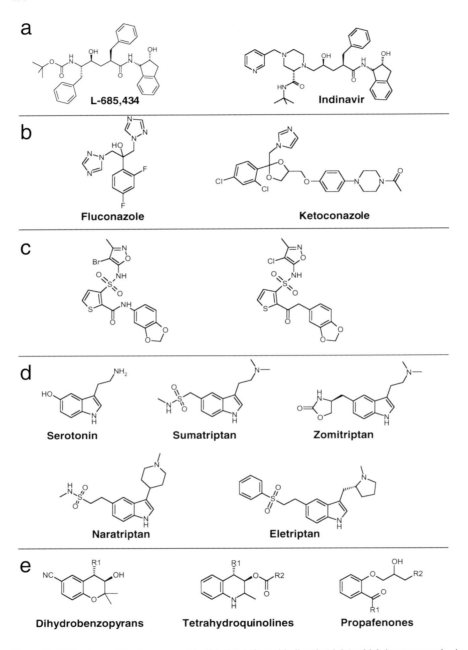

Figure 2. a) Structures of lead compound L-685,434 (left) and indinavir (right) which incorporates basic functions aiding water solubility. b) Stuctures of the antifungal agents fluconazole and ketoconazole. c) Replacement of amide with acetyl in a series of amidothiophenesulfonamide endothelin-A antagonists to improve oral bioavailability. d) Structure of serotonin (5-hydroxytryptamine) and some 5-HT$_{1B/1D}$ agonists. e) Examples of proposed MDR modulators.

Table 1. Properties of GABA-uptake inhibitors [46]

Structure	IC$_{50}$ [µM]	in vivo activity	pKa	log D$_{oct}$	log D$_{chex}$	Δlog D
	0.11	active	3.57/ 9.23	0.99	-0.43	1.42
	0.1	inactive	3.39/ 9.25	0.71	-2.00	2.71

to ketoconazole, the first orally-active azole antifungal. Ketoconazole (Fig. 2b) is cleared primarily by hepatic metabolism and shows irregular bioavailability due partly to this and also its poor aqueous solubility. Synthesis was directed towards metabolic stability and this was found in the bis-triazole series of compounds. Metabolic stability is achieved by the relativestability of the triazole moiety to oxidative attack, the presence of halogen functions on the phenyl grouping, another site of possible oxidative attack, and steric hindrance of the hydroxy function, a site for possible conjugation.

Fine-tuning of the lipophilicity of this series produced a compound with fluorine substitution (fluconazole), with a log P or D$_{7.4}$ value of 0.5. This gives high tubular reabsorption (80%), but the clearance is predominantly renal due to the high metabolic stability. The low rate of renal clearance gives a compound with a 30-h half-life, suitable for once a day administration. The moderate lipophilicity is optimum for absorption since it encompasses good water solubility and complete bioavailability.

12.7.4 Endothelin antagonists

Figure 3 shows a synthetic strategy aimed at removing H-bond donors from a series of endothelin antagonists and a resultant increase in apparent bioavailability as determined by intradueodenal AUC [47]. Noticeable CLOGP values vary only marginally with the changes in structure, values being 4.8, 5.0, 4.8 and 5.5 for compounds A, B, C and D respectively. In contrast the number of H-bond donors is reduced by 3 and the Raevsky score calculated with HYBOT95 from 28.9 (A) to 21.4 (D).

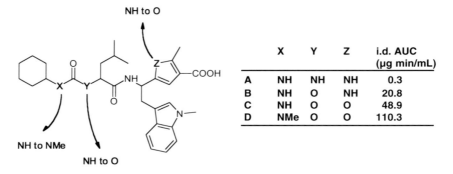

Figure 3. Removal of H-bonds as a synthetic strategy for a series of azole containing endothelin antagonists aimed at improving bioavailability by lowering H-bonding potential [47].

	X	Y	Z	i.d. AUC (µg min/mL)
A	NH	NH	NH	0.3
B	NH	O	NH	20.8
C	NH	O	O	48.9
D	NMe	O	O	110.3

A similar example, also from endothelin antagonists, is the replacement of the amide group (Fig. 2c) in a series of amidothiophenesulfonamides with acetyl [48]. This move retained *in vitro* potency, but markedly improved oral bioavailability.

12.7.5 H_1-antihistaminics

Lipophilicity and H-bonding and important parameters for uptake of drugs in the brain [49]. Their role has, for example, been studied in a series of structurally diverse sedating and non-sedating histamine H_1-receptor antagonists [50]. From these studies a decision tree guideline for the development of non-sedative antihistamines was designed (see Fig. 4).

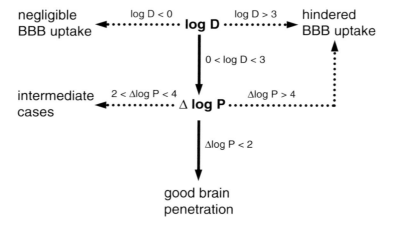

Figure 4. Decision tree for the design of non-sedative H_1-antihistaminics. Log D is measured at pH 7.4, while Δlog P refers to compounds in their neutral state.

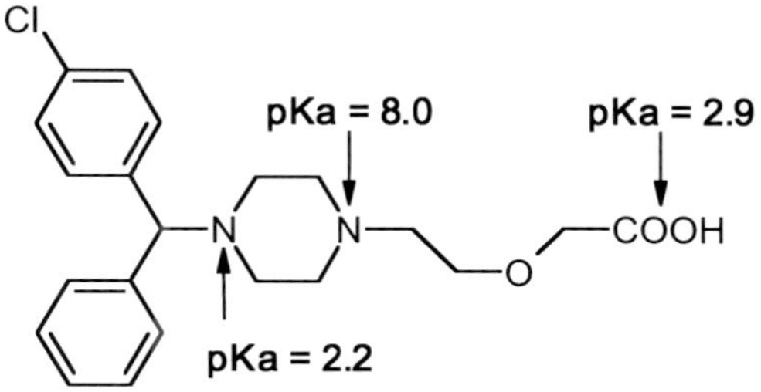

Figure 5. Molecular structure and pKa values of cetirizine.

The lipophilic properties and their dependence on conformation in relation to membrane penetration have been studied in detail for cetirizine (Fig. 5) [51]. Conformational analyses showed that zwitterionic cetirizine can exist in a folded and unfolded form, which have different overall polarity. In the folded form the two charges cancel each other, called partial intramolecular neutralization, more than in the more unfolded form. This may have interesting consequences for, for example, permeation through membranes. It was suggested that cetirizine has the ability to penetrate cell membranes sufficiently to give a biological effect without a tendency to accumulate in tissues. Through a combination of low volume of distribution and poor brain penetration cetirizine can be seen as a third generation antihistamine [51].

12.7.6 5-HT$_{1B/1D}$ receptor agonists as anti-migraine agents

A number of serotonin receptor 5-HT$_{1B/1D}$ agonists, such as sumatriptan, are marketed as anti-migraine drugs (Fig. 2d). SAR of 5-substituted tryptamines made clear that acceptable levels of oral absorption can only be obtained when the compounds are small and not too hydrophilic [52]. It is believed that such molecules can be absorbed through aqueous pores using the para-cellular pathway. First in class compound sumatriptan has a log D of −1.2 [7] and only 14% bioavailability, while zomitriptan has an estimated log D = −1.0 and is 40% bioavailable [52]. More recent compounds such as naratriptan (CLOGP = 1.696) and eletriptan (CLOGP = 3.08) have higher CLOGP values than sumatriptan (CLOGP = 0.863), and are designed to improve oral absorption.

12.7.7 Renin inhibitors

Many companies have tried to develop peptidic renin inhibitors. Unfortunately, these are rather large molecules and not unexpectedly poor absorption was often observed. The role of physicochemical properties has been discussed for this class of compounds [53–55]. One of the conclusions was that compounds with higher

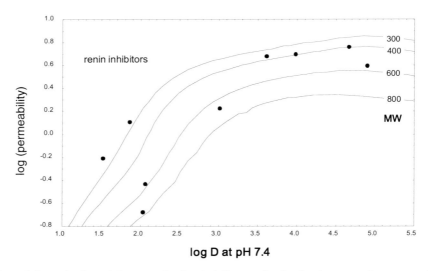

Figure 6. Iso-molecular weight curves showing the influence of molecular size on membrane permeability with increasing lipophilicity [58].

lipophilicity were better absorbed from the intestine [54]. Absorption and bile elimination rate both are MW-dependent. Lower MW gives better absorption and less bile excretion. The combined influence of molecular size and lipophilicity on absorption of a series of renin inhibitors can be seen in Figure 6. The observed iso-size curves are believed to be part of a general sigmoidal relationship between permeability and lipophilicity [56, 57].

12.7.8 Modulation of multidrug resistance (MDR)

A major problem in the treatment of cancer is the resistance of tumor cells to a wide range of cytotoxic agents, which is known as multidrug resistance (MDR). This is mainly due to an upregulation of the multidrug transporter P-glycoprotein (P-gp). Several classes of drugs such as ion channel blockers and cyclosporin analogues have been identified as potent modulators of MDR by inhibiting P-gp and possibly other efflux transporters. Co-administration of MDR modulators may be an elegant way to improve the effectiveness of anti-cancer drugs. Thus, e.g., dihydrobenzopyrans, tetrahydoquinolines, and propafenones have been studied as potential modulators (see Fig. 2e) [59, 60].

Log($1/EC_{50}$) values for inhibition of P-gp (P-glycoprotein) mediated efflux of daunomycin appears to be highly correlated with lipophilicity descriptors such as calculated log P and log kw measured by RP-HPLC [59], while studies with propafenones demonstrated good correlation to molar refractivity, a size descriptor. In conclusion, larger and more lipophilic members of certain classes of compounds appear to be better P-gp inhibitors than the smaller and more hydrophilic compounds in that series.

References

1 Lipinski CA, Lombardo F, Dominy BW et al (1997) Experimental and computational approaches to estimate solubility and permeability in drug discovery and development settings. *Adv Drug Deliv Revs* 23: 3–25

2 Kansy M, Kratzat K, Parrilla I et al (2000) Physicochemical high throughput screening (pC-HTS): Determination of membrane permeability, partitioning and solubility. In: K Gundertofte, FS Jørgensen (eds) *Molecular Modeling and Prediction of Bioactivity*, Kluwer Academic/Plenum Publishers, 237–243

3 Avdeef A (1998) pH-metric solubility. 1. Solubility-pH profiles from Bjerrum plots. Gibbs buffer and pKa in the solid state. *Pharm Pharmacol Commun* 4: 165–178

4 Stella VL, Martodhardjo S, Rao VM (1999) Aqueous solubility and dissolution rate does not adequately predict *in vivo* performance: a probe utilizing some N-acyloxymethyl phenytoin prodrugs. *J Pharm Sci* 88: 775–779

5 Van de Waterbeemd H, Carter RE, Grassy G et al (1997) Glossary of terms used in computational drug design (IUPAC Recommendations 1997)*Pure & Appl Chem* 68: 1137–1152

6 Stewart BH, Chan OH (1998) Use of immobilized artificial membrane chromatography for drug transport applications. *J Pharm Sci* 87: 1471–1478

7 Van de Waterbeemd H (2000) Intestinal Permeability: Prediction from Theory. In: J. B. Dressman, H. Lenner (eds): *Oral Drug Absorption: Prediction and Assessment*. Marcel Dekker, New York, 31–49

8 Ottiger C, Wunderli-Allenspach H (1997) Partition behaviour of acids and bases in a phosphatidyl-choline liposome-buffer equilibrium dialysis system. *Eur J Pharm Sci* 5: 223–231

9 Ottiger C, Wunderli-Allenspach H (1999) Immobilized artificial membrane (IAM)-HPLC for partition studies of neutral and ionized acids and bases in comparison with the liposomal partition system. *Pharm Res* 16: 643–650

10 Austin RP, Barton P, Davis AM et al (1998) The effect of ionic strength on liposome-buffer and 1-octanol-buffer distribution coefficients. *J Pharm Sci* 87: 599–607

11 Austin RP, Davis AM, Manners CN (1995) Partitioning of ionizing molecules between aqueous buffers and phospholipid veshicles. *J Pharm Sci* 84: 1180–1183

12 Barton P, Davis AM, McCarthy DJ et al (1997) Drug-phospholipid interactions. 2. Predicting the sites of drug distribution using n-octanol/water and membrane/water distribution coefficients. *J Pharm Sci* 86: 1034–1039

13 Fruttero R, Caron G, Fornatto E et al (1998) Mechanisms of liposomes/water partitioning of (p-methylbenzyl)alkylamines. *Pharm Res* 15: 1407–1413

14 Pauletti GM, Wunderli-Allenspach H (1994) Partition coefficients *in vitro*: artificial membranes as a standardized distribution model. *Eur J Pharm Sci* 1: 273–282

15 Balon K, Riebesehl BU, Müller BW (1999) Determination of liposome partitioning of ionizable drugs by titration. *J Pharm Sci* 88: 802–806

16 Artursson P, Palm K, Luthman K (1996) Caco-2 monolayers in experimental and theoretical predictions of drug transport. *Adv Drug Del Rev* 22: 67–84

17 Irvine JD, Takahashi L, Lockhart K et al (1999) MDCK (Madin-Darby canine kidney) cells: a tool for membrane permeability screening. *J Pharm Sci* 88: 28–33

18 Grès MC, Julian B, Bourrié M et al (1998) Correlation between oral drug absorption in humans, and apparent drug permeability in TC-7 cells, a human epithelial intestinal cell line: comparison with the parental Caco-2 cell line. *Pharm Res* 15: 726–733

19 Raessi SD, Hidalgo IJ, Segura-Aguilar J et al (1999) Interplay between CYP3A-mediated metabolism and polarized efflux of terfenadine and its metabolites in intestinal epithelial Caco-2 (TC7) cell monolayers. *Pharm Res* 16: 625–632

20 Johnson MD, Anderson BD (1999) *In vitro* models of the blood-brain barrier to polar permeants: Comparison of transmonolayer flux measurements and cell uptake kinetics using cultured cerebral capillary endothelial cells. *J Pharm Sci* 88: 620–625

21 Borchardt RT, Smith PL, Wilson G (eds) (1996) *Models for assessing drug absorption and metabolism*. Plenum Press, New York

22 Abraham MH, Le J (1999) The correlation and prediction of the solubility of compounds in water using an amended solvation energy relationship. *J Pharm Sci* 88: 868–880

23 Huuskonen J, Salo M, Taskinen J (1998) Aqueous solubility prediction of drugs based on molecular topology and neural network modeling. *J Chem Inf Comput Sci* 38: 450–456

24 Leo AJ (1993) Calculating log Poct from structure. *Chem Rev* 93: 1281–1306

25 Buchwald P, Bodor N (1998) Octanol-water partitioning: Searching for predictive models. *Curr Med Chem* 5: 353–380

26 Mannhold R, Van de Waterbeemd H (2001) Substructure and whole molecule approaches for calculating log P. *J Comput Aided Mol Des* 15: 337–354

27 Palm K, Luthman K, Ungell AL et al (1998) Evaluation of dynamic polar molecular surface area as predictor of drug absorption: Comparison with other computational and experimental predictors. *J Med Chem* 41: 5382–5392

28 Clark DE (1999) Rapid calculation of polar molecular surface area and its application to the prediction of transport phenomena. 1. Prediction of intestinal absorption. *J Pharm Sci* 88: 807–814

29 Van de Waterbeemd H, Kansy M (1992) Hydrogen bonding capacity and brain penetration. *Chimia* 46: 299–303

30 Clark DE (1999) Rapid calculation of polar molecular surface area and its application to the prediction of transport phenomena. 2. Prediction of blood-brain barrier penetration. *J Pharm Sci* 88: 815–821

31 Raevsky OA, Schaper KJ (1998) Quantitative estimation of hydrogen bond contribution to permeability and absorption processes of some chemicals and drugs. *Eur J Med Chem* 33: 799–807

32 Norinder U, Oesterberg T, Artursson P (1999) Theoretical calculation and prediction of intestinal absorption of drugs in humans using MolSurf parametrizaton and PLS statistics. *Eur J Pharm Sci* 8: 49–56

33 Winiwarter S, Bonham NM, Ax F et al (1998) Correlation of human jejunal permeability (*in vivo*) of drugs with experimentally and theoretically derived parameters. A multivariate data analysis approach. *J Med Chem* 41: 4939–4949

34 Camenisch G, Alsenz J, Van de Waterbeemd H et al (1998) Estimation of permeability by passive diffusion through Caco-2 cell monolayers using the drugs' lipophilicity and molecular weight. *Eur J Pharm Sci* 6: 313–319

35 Norinder U, Sjoeberg P, Oesterberg T (1998) Theoretical calculation and prediction of brain-blood partitioning of organic solutes using MolSurf parametrization and PLS statistics. *J Pharm Sci* 87: 952–959

36 Van de Waterbeemd H (1996) Design of bioactive compounds. In: Van de Waterbeemd, H. (ed): *Structure-Property Correlations in Drug Research.* Academic Press & R.G. Landes Comp., Austin, 1–9

37 Walters WP, Stahl MT, Murcko MA (1998) Virtual screening – an overview. *Drug Disc Today* 3: 160–178

38 Agrafiotis DK, Myslik JC, Salemme FR (1999) Advances in diversity profiling and combinatorial series design. *Mol Diversity* 4: 1–22

39 Finn PW (1996) Computer-based screening of compound databases for the identification of novel leads. *Drug Disc Today* 1: 363–370

40 Keighley WW, Nabney IT, Van de Waterbeemd H et al (2003) Data-handling for high throughput screening. In: E Murray (ed): *The Principle and Practice of High Throughput Screening.* Blackwell Science, London

41 Smith DA, Jones BC, Walker DK (1996) Design of drugs involving the concepts and theories of drug metabolism and pharmacokinetics. *Med Res Revs* 16: 243–266

42 Van de Waterbeemd H, Smith DA, Jones BC (2001) Lipophilicity in PK design: methyl, ethyl, futile. *J Comput Aided Mol Des* 15: 273–286

43 Kempf DJ (1994) Progress in the discovery of orally bioavailable inhibitors of HIV protease. *Perspect Drug Disc Des* 2: 427–436

44 Vacca JP, Dorsey BD, Schleif WA et al (1994) L-735,524: an orally bioavailable human immunodeficiency virus type 1 protease inhibitor. *Proc Nat Acad Sci* 91: 4096–4100

45 Lehr P, Billich A, Charpiot B et al (1996) Inhibitors of human immunodeficiency virus type I protease containing 2-aminobenzyl-substituted 4-amino-3-hydroxy-5-phenylpentanoic acid: synthesis, activity, and oral bioavailability. *J Med Chem* 39: 2060–2067

46 N'Goka V, Schlewer G, Linget JM et al (1991) GABA-uptake inhibitors: Construction of a general pharmacophore model and successful prediction of a new representative. *J Med Chem* 34: 2547–2557

47 Von Geldern, TW, Hoffman DJ, Kester, JA et al (1996) Azole endothelin antagonists. 3. Using D log P as a tool to improve absorption. *J Med Chem* 39: 982–991

48 Wu, C, Chan MF, Stavros F et al (1997) Discovery of TBC11251, a potent, long acting, orally active endothelin receptor-A selective antagonist. *J Med Chem* 40: 1690–1697

49 Van de Waterbeemd H, Camenisch G, Folkers G et al (1998) Estimation of blood-brain barrier crossing of drugs using molecular size and shape, and H-bonding descriptors. *J Drug Target* 6: 151–165

50 Ter Laak AM, Tsai RS, Donné-Op den Kelder GM et al (1994) Lipophilicity and hydrogen-bonding capacity of H_1-antihistaminic agents in relation to their central sedative side-effects. *Eur J Pharm Sci* 2: 373–384

51 Testa B, Pagliara A, Carrupt PA (1997) The molecular behaviour of cetirizine. *Clin Exp Allerg* 27: S13–S18

52 Glen RC, Martin GR, Hill AP et al (1995) Computer-aided design and synthesis of 5-substituted tryptamines and their pharmacology at the 5-HT_{1D} receptor: discovery of compounds with potential antimigraine properties. *J Med Chem* 38: 3566–3580

53 Boyd SA, Fung AKL, Baker WR et al (1994) Nonpeptide renin inhibitors with good intraduodenal bioavailability and efficacy in dog. *J Med Chem* 37: 2991–3007

54 Hamilton HW, Steinbaugh BA, Stewart BH et al (1995) Evaluation of physicochemical parameters important to the oral bioavailability of peptide-like compounds: implications for the synthesis of renin inhibitors. *J Med Chem* 38: 1446–1455

55 Chan OH, Stewart BH (1996) Physicochemical and drug-delivery considerations for oral drug bioavailability. *Drug Disc Today* 1: 461–473

56 Camenisch G, Folkers G, Van de Waterbeemd H (1996) Review of theoretical passive drug absorption models: Historical background, recent developments and limitations. *Pharm Acta Helv* 71: 309–327

57 Camenisch G, Folkers G, Van de Waterbeemd H (1998) Shapes of membrane permeability-lipophilicity curves: Extension of theoretical models with an aqueous pore pathway. *Eur J Pharm Sci* 6: 321–329

58 Van de Waterbeemd H (1997) Application of physicochemical methods to oral drug absorption estimation. *Eur J Pharm Sci 5 Suppl 2*: S26-S27

59 Hiessböck R, Wolf C, Richter E et al (1999) Synthesis and *in vitro* multidrug resistance modulating activity of a series of dihydrobenzopyrans and tetrahydrohydroquinolines. *J Med Chem* 42: 1921–1926

60 Tmej C, Chiba P, Huber M et al (1998) A combined Hansch/Free-Wilson approach as predictive tool in QSAR studies on propafenone-type modulators of multidrug resistance. *Arch Pharm Pharm Med Chem* 331: 233–240

13 Computer-aided prediction of drug toxicity and metabolism

Mark T.D. Cronin

Liverpool John Moores University, School of Pharmacy and Chemistry, Byrom Street, Liverpool, L3 3AF, UK

13.1 Introduction

All medicines must be safe for human use. This simple statement belies the complexity of regulations, both national and international, as well as time and financial resource required to ensure drug safety. To establish drug safety "hazard" is ascertained following which risk, or conversely drug safety, may be confirmed. Toxicity tests, normally using surrogate animal species, are required to ascertain drug hazard, or otherwise. If a drug is determined to be too hazardous, i.e., too toxic for its intended use, then it will not be developed further. Considering that the drug development process may take up to 8–10 years and cost hundreds of millions dollars (see Chapter 1), from discovery to registration, it is essential that drug toxicity be identified as early as possible. The aim of this chapter is to review briefly the more common toxicological assays, and then to assess the progress made by computer-aided toxicity and metabolism prediction to identify toxic drug entities.

13.2 Toxicity tests

In order to ensure drug safety a variety of toxicological evaluations are performed, to the requirement of national regulatory authorities such as the Food and Drug Administration (FDA) in the United States and the European Medicines Evaluation Agency (EMEA) in the European Union. The tests that may be performed are:
- acute toxicity (24 h)
- prolonged and chronic toxicity (30–90 days)
- mutagenicity
- carcinogenicity
- teratogenicity
- reproductive effects
- skin sensitization (including photosensitization)
- skin irritation
- eye irritation
- immunological assessment

It is not the purpose of, nor is it possible for, this chapter to review in full methodological details of the toxicological testing, but merely to overview the methods. The reader is referred elsewhere for such information [1–4].

13.2.1 Acute toxicity

Acute toxicity tests are normally performed in rats and mice and occasionally in the rabbit and guinea pig. The purpose of the assay is to determine ultimately the LD_{50}, or the dose of drug that will be lethal to 50% of a population. The time period of the test is short (normally 24 h and occasionally up to 48 h) with observation following dosing of up to 14 days. Dosing the animals may be via a number of routes including intravenous, intraperitoneal, dermal, oral and inhalation. Organization for Economic Co-operation and Development (OECD) guidelines provide experimental details for oral (OECD Guideline 401) and dermal dosing (OECD Guideline 402) as well as the inhalation route (OECD Guideline 403; details of the OECD guidelines can be obtained from their internet site listed in Table 1). Whilst the acute toxicity test is relatively simple to perform, it can provide the experienced toxicologist with a wealth of information far beyond the basic LD_{50}, including information on observational and physiological effects. The LD_{50} test is now not acceptable in many countries and has been replaced by the fixed dose, up-and-down, and acute toxic class procedures.

13.2.2 Repeated dose toxicity assays

Toxicological assessment for longer time periods is required as a drug progresses through the development process. This provides toxicological information regarding exposure to drugs, normally at sub-lethal concentrations, over a more realistic timeframe. Short-term repeated dose studies (OECD Guideline 407) last between 14 and 28 days. Dosing is graded in 3–4 concentrations with the highest dose designed to cause some toxicity, but not lethality. Normally between 5–10 rats of each sex (though mice and dogs are also utilized) are tested per group. At the completion of the test a whole host of clinical and histological evaluations are recorded, including experimental observations and whole body and individual organ analysis. Such information will clearly enhance that gathered from acute toxicity studies. Other subchronic toxicity studies are maintained for up to 90 days (OECD Guideline 408). Again animals are exposed to the drug continuously and potentially via a number of different routes. This provides much information regarding organ toxicity, but the advantages of the 90 day test over the 28 day test remain debatable.

13.2.3 Testing for carcinogenicity

Carcinogenicity assays may be considered as an extension of the chronic toxicity test. To test a pharmaceutical substance for carcinogenicity is a lengthy and expensive process. Typically, a substance is tested in two species (rats and mice) and both sexes with continuous exposure for up to 24 months. Exposure is typically via the oral route. In addition to a negative control group, each compound is tested at three concentrations. The highest dose is the maximum tolerated dose (MTD), which

Table 1. Internet addresses of software vendors and other useful sites. (Please note whilst the author has taken every measure to ensure the accuracy of the information and addresses, he is not responsible for changes of addresses, or the information that may be placed on these sites)

- **Organization for economic co-operation and development.** This site provides information on (and links to) testing guidelines as well as an on-line bookshop: http://www.oecd.org/ehs/
- **The Registry Of Toxic Effects Chemical Substances (RTECS).** This site provides background information on the RTECS database: http://www.cdc.gov/niosh/rtecs.html
- **MicroQSAR.** This site provides brief information and purchase details: http://www.tds-tds.com/ind_num2.htm#MICROQSAR
- **ECOSAR.** This site provides basic background information and the opportunity to download the software: http://www.epa.gov/opptintr/newchms/21ecosar.htm
- **TOPKAT** (and other Accelrys). This site provides basic background information and samples of its output: http://www.accelrys.com/products/topkat/
- **CASE, MultiCASE, META** etc. This site provides basic background information and examples of their use: http://www.multicase.com
- **LHASA (DEREK, StAR, METEOR).** This site provides basic background information and examples of their use and a description of the collaborative agreement: http://www.chem.leeds.ac.uk/luk/
- **HazardExpert and MetabolExpert.** This site gives basic descriptions of these programs and all others marketed by CompuDrug: http://www.compudrug.com/
- **OncoLogic.** This site provides background information: http://www.logichem.com/
- **SciVision.** This site provides background information on TOXsys, ONCOis, and all their other products: http://www.scivision.com/

attempts to invoke negligible target organ toxicity and will not decrease the life-span of the animals (other than by the production of cancers). Large numbers of animals are used, commonly at least 50 animals per sex at each dose. During and after the testing, animals are subjected to a plethora of clinical, observational, and histological assessments. Of the last, at least 30 tissues may be examined. An assessment of the capability of a drug to act specifically as a genotoxic (as opposed to non-genotoxic) carcinogen is also made. Such assessment is normally made *in vitro*. In 1997, the International Conference on Harmonization (ICH) of Technical Requirements for Registration of Pharmaceuticals for Human Use commended the use of a minimum battery of *in vitro* genotoxicity assays [5]. The battery approach attempts to ensure that all potential mechanisms of genotoxicity can be observed. The ICH recommendations include an *in vitro* assay for gene mutation and the use of both an *in vitro* and *in vivo* assay for chromosomal aberration.

The gene mutation assay is performed most commonly with *Salmonella typhimurium* (the so-called Ames test), and to a lesser extent with *Escherichia coli*. The Ames test is normally performed according to OECD Guideline 471. It is reliant upon a mutant strain of *S. typhimurium* that requires histidine for growth. The test is performed on agar that is histidine deficient (and thus will not support growth of the mutant bacterium). Exposure to a genotoxic chemical may cause mutation in the bacterium causing a reversion to the wild type, which is then observed to grow on the histidine deficient agar. A number of modifications can be made to the Ames test using different strains of bacteria, and the inclusion of metabolic enzymes.

In vitro systems for chromosomal damage include the mammalian cytogenic test, performed to OECD guideline 473. In this assay cell cultures from man, or from standard cell lines, are exposed to the drug compound. Human lymphocytes are commonly used in this assay. After the exposure period, the metaphase cells are examined microscopically for evidence of chromosome aberration. *In vivo* tests for chromosomal damage include the micronucleus test (performed to OECD guideline 474). This assay assesses the damage in immature erythrocytes of rat bone marrow cells after exposure to the drug.

13.2.4 Tests for reproductive toxicity and teratogenicity

The determination of the effects of chemicals on the reproductive ability of males and females, as well as issues such as teratogenicity, is a broad and complex area. It is complicated in terms of the multitude of effects that may arise, as well as the number and type of tests that may be used to determine these effects. It is gener-ally accepted that teratogenicity is only a part of potential reproductive toxicity. Testing for teratogenicity (abnormal foetal development) requires the *in utero* exposure of the foetus to a drug. Tests are normally performed in the rat, although the rabbit may occasionally be used. Typically three dose levels are applied, the highest being that which will induce some limited maternal toxicity; the lowest dose will cause no maternal toxicity. For rat tests 20 pregnant females will be used at each dose level. The drug is administered during the development of the major organs in the foetus (e.g., 6–15 days after conception). The animals are dosed via their drinking water or by gavage. The test is terminated on the day prior to normal delivery and foetal development, in terms of both the number of live foetuses, and of any malformations present, is assessed.

For the more accurate assessment of effects of a drug on species fecundity and between generations, a multigeneration toxicity test is required. A good example is the two-generation reproduction toxicity test (OECD Guideline 416), although these tests may also be maintained for three or more generations. Such assays are performed usually with rats at three dose levels. The highest dose level is typically one-tenth of the LD_{50}; the lowest should cause no sub-chronic toxicity. Males and females are treated with the drug for 60 days, after which they are allowed to mate. Subsequent generations are assessed for a wide variety of endpoints including: number of live births; abnormalities at birth; gender and weight at birth; histological examinations, etc.

13.2.5   Alternatives to toxicity testing

It can be concluded from this brief review of toxicological methods that the experimental assessment of drug toxicity is a time-consuming, expensive, yet essential, part of the pharmaceutical research and development process. The consequence of a drug being found to be toxic at a late stage of development could be immensely costly to a company. With this in mind, methods are constantly being sought to determine drug toxicity as early and as cheaply as possible. Much effort has been placed into the development of *in vitro* assays, cell culture techniques and most recently DNA arrays as replacements for toxicity testing (the latter is likely to see an enormous growth in use in the future, see Rockett and Rix [6] for more information). In addition, a number of computer-aided toxicity prediction methods are available. These are based on the fundamental premise that the toxicological activity of a drug will be a function of the physico-chemical and/or structural properties of the substance. Once such a relationship has been established, further chemicals with similar properties can be predicted to be toxic. The remainder of this chapter will review computer-aided methods to predict toxicity and metabolism.

13.3 Computer-aided prediction of toxicity and metabolism

The development of computer-aided toxicity and metabolism prediction techniques can be broadly classed into three areas (although it should be noted that there is considerable overlap between these three areas):
• Quantitative structure-activity relationships (QSARs)
• Expert systems based on QSARs
• Expert systems based on existing knowledge.

13.3.1 Quantitative structure-activity relationships

QSARs attempt to relate statistically the biological activity of a series of chemicals to their physico-chemical and structural properties (for 3D-QSAR methods see Chapter 11). They have been used successfully in the lead optimization of drug and pesticide compounds for over three decades. They have also been applied to the prediction of toxicity. There is insufficient space to review QSAR methodology in detail in this chapter, the reader is referred elsewhere for methodological details [7, 8] (and Chapter 11) and for reviews of their use in toxicity prediction [9–14].

All predictive methods should ideally be developed with a putative mechanism of toxic action in mind (although it is conceded that the mechanism of action may not be identifiable for many compounds). The most straightforward QSARs have been developed for acute toxicity, with relatively restricted groups of compounds, about which something of the mechanism of action is known. For instance, Cronin et al. [15] have modeled the acute toxicity of seven classes of simple aromatic compounds in the 15-min Microtox assay (this uses the photoluminescent marine bacterium *Vibrio fischeri*):

$$pT_{15} = 0.76 \; logP - 0.63 \; E_{LUMO} - 0.47$$

$$n = 63 \qquad r^2 = 0.85 \qquad s = 0.46 \qquad F = 171,$$

where: pT_{15} is the inverse of the millimolar concentration of toxicant
causing a 50% reduction in light output from *V. fischeri*
logP is the logarithm of octanol-water partition coefficient
E_{LUMO} is the energy of the lowest unoccupied molecular
orbital
n is the number of observations
r^2 is the square of the correlation coefficient adjusted for
degrees of freedom
s is the standard error of the estimate
F is Fisher's statistic

In this approach the toxicity of the compounds is described as a function of their ability to penetrate to the site of action, or accumulate in cell membranes (a hydrophobic phenomenon parameterized by log P), and their ability to react covalently with macromolecules (an electrophilic phenomenon parameterized by E_{LUMO}). A number of caveats to this model are immediately obvious. Firstly, whilst this model clearly fits the data well, it does so only for a relatively small number of structurally similar molecules. Its application to predict drug toxicity is likely to be extremely limited. It is envisaged that for the efficient prediction of acute toxicity a tiered approach combining both structural and physico-chemical rules with such QSARs will be required. Such rules may direct the prediction to be made from, for example, separate QSARs for aliphatic and aromatic molecules, or account for effects such as ionization or steric hindrance of reactive centers. It should be noted that such a tiered approach still requires more effort in the measurement of toxicological activity and modeling.

The second caveat relates to the nature of the biological activity itself. It should be no surprise that the data have been obtained from an *in vitro* toxicological assay. Such data are quickly and cheaply determined (and can be obtained in one laboratory, as in this case). Undoubtedly for drug toxicity a model based on mammalian LD_{50} data would be preferable. However, very few such QSARs are available. The reason for this is believed to be that there are not sufficient "quality" toxicity data on which to develop these QSARs. Whilst a large number of toxicity data may be available on databases such as the Registry of Toxic Effects of Chemical Substances (RTECS; see internet site listed in Table 1), there is no consistency in the data, and many attempts to model such data are clearly hindered by the excessive error and inter-laboratory variation that may exist. Johnson and Jurs [16] report a predictive model for the acute oral mammalian toxicity for 115 anilines using data retrieved from the RTECS database. The model utilizes a neural network based on five physico-chemical descriptors. It is reported to predict the toxicity of an external validation set well. Further unpublished analysis of the toxicity data suggests that there may, however, be considerable variation in the data, following a comparison of the toxicity data for positional isomers in the data set. Other workers have investigated the acute toxicity data from RTECS to rats and mice (based on an arbitrary categoric scale) to develop a model based on a decision tree approach for aliphatic alcohols [17].

QSARs have been developed for a whole range of other toxicity endpoints, especially those that provide a quantitative determination of toxicity. An example is the modeling of mutagenicity data from the Ames test. Mutagenicity is a toxicity reliant on the capability of a molecule to react covalently with DNA. Often, such reactivity can be described and modeled by the use of calculated molecular orbital properties. Hatch and Colvin [18], for instance, demonstrated the use of whole molecule parameters, as well as those based on the nitrenium ion, to model the mutagenic potency of aromatic and heterocyclic amines.

Furthermore, Hansch et al. [19] have reviewed the use of QSARs to predict a number of teratological endpoints. A number of QSARs are presented, some of which have been re-evaluated in a mechanistic light. For instance, the embryo toxicity of 4-substituted phenols (ET) was found to be well correlated to the Hammett constant for the substituent (σ):

$$\log 1/ET = -1.05 \, \sigma + 3.94$$

$$n = 10 \qquad r^2 = 0.80 \qquad s = 0.21 \qquad F \text{ not given}$$

It should be remembered that, in principle, data from any toxicological endpoint can be subjected to QSAR analysis. For instance, Bartlett et al. [20] investigated the percentage incidence of cutaneous rash due to oral penicillins using data taken from clinical trials (v % ROP). A correlation was sought with a number of para-

meters, the best being found with a measure of the shape similiarity of the penicillin molecules to benzylpenicillin (Sim.BP). The following relationship suggests
that a receptor binding phenonemon is important for this endpoint:

$$v\%ROP = 3.82 \ Sim.BP - 1.75$$

$$n = 14 \qquad r^2 = 0.82 \qquad s = 0.18 \qquad F = 55.7$$

QSARs can also be used to model the absorption, distribution, metabolism and
excretion (ADME) properties of drugs. Cronin et al. [21] demonstrated that the
permeability coefficients (Kp) of 107 heterogeneous compounds through the skin
in vitro could be well modeled by hydrophobicity and molecular size (as parameterized by molecular weight (MW)). It should be noted in this approach a passive
diffusion process is assumed; active and facilitated transport are little addressed
by QSAR modeling. Thus the combination of the following (and similar for other
membranes) equation with the Lipinski "rule of five" [22] (see Chapters 10 and
12) may provide a powerful tool to determine drug uptake:

$$\log Kp = 0.77 \ \log P - 0.010 \ MW - 2.33$$

$$n = 107 \qquad r^2 = 0.86 \qquad s = 0.39 \qquad F = 317$$

QSAR models for metabolism are also possible and may take a number of forms.
Hansch and Zhang [23] have reviewed the capability of substrates and inducers to
bind to the cytochrome P-450 enzymes and have developed mechanistically based
QSARs for these data. Typically the data sets analyzed are small (fewer than 20
compounds) which illustrates the paucity of data for modeling in this area. One of
the larger data sets analyzed is for the induction by miscellaneous compounds of
CYP1A2 (*I*), which reveals the following significant QSAR with hydrophobicity:

$$\log I = 0.28 \ \log P - 1.80$$

$$n = 27 \qquad r^2 = 0.76 \qquad s = 0.30 \qquad F \text{ is not given}$$

Other approaches have attempted to describe quantitatively specific metabolic
processes. A good example of this type of analysis, and of mechanistical development of QSARs, is given by Buchwald and Bodor [24]. For the human *in vitro*
hydrolysis of 67 ester-containing drugs (from seven non-congeric classes) the following relationship was obtained with half-life ($t_{1/2}$):

$$\log t_{1/2} = 0.17\Omega - 10.1 \ Q_c + 0.11 \ \log P - 3.81$$

$$n = 67 \qquad r^2 = 0.81 \qquad s = 0.36 \qquad F = 88.1,$$

where Ω is the inaccessible solid angle around the carbonyl sp^2 oxygen (a
measure of steric hindrance).
Q_c is the calculated charge on the carbonyl carbon.

This result exemplifies the problems associated with using compilations of human
data, namely that a number of half-life values did not fit the model. Fortunately
the authors took a pragmatic approach and developed a model on parameters that
agree with the mechanism of hydrolysis by carboxylesterases.

A further approach to the prediction of metabolism is the qualitative identification of molecular features that dictate whether or not a molecule may enter a particular pathway. A good example was the use of non-linear mapping to determine
the properties of benzoic acids and congeners which were necessary for glucoro-
nic acid or glycine conjugation [25].

13.3.2 Expert systems for toxicity prediction

For the terms of reference in this chapter "expert system" is taken to mean a computer-assisted approach to predict toxicity. Again the reader is referred elsewhere
for further details on expert systems [9–11, 26, 27].

Expert systems based on QSARs
The logical extension of the QSAR approach to make large scale predictions of
toxicity is to computerize it. A simple DOS-based QSAR program is MicroQSAR.
Following the entry of a SMILES string a wide variety of endpoints are predicted. Whilst most of these are environmental in nature, a number of human health
effects are also predicted. Since its inception, the program has not however been
developed to achieve its full potential. Further details of MicroQSAR can be
obtained from the internet site listed in Table 1. Another environmentally-based
prediction program is ECOSAR. This operates by assigning a molecule to a particular class to make a prediction (normally based on log P), and has, as such, been
criticized for the arbitrary manner in which classes are identified [28]. It should
be noted however that this software is freely available over the internet (see
Tab. 1). As such, it gives the user the opportunity to see the method (and view the
result) of the prediction that will be made by the United States Environmental
Protection Agency (U.S. E.P.A.).

TOPKAT, and the related program Q-TOX, are probably the best known and
commercially successful QSAR based expert system prediction programs. TOP-
KAT was developed by Health Designs Inc., a wholly owned subsidiary of
Accelrys. The link up with Accelrys means that the TOPKAT system is now integrated with a number of other tools, such as the TSAR molecular spreadsheet. The
power of this is clearly the ability to make predictions from the spreadsheet. TOP-
KAT makes predictions for a number of toxicity endpoints, including: carcinogenicity; mutagenicity; developmental toxicity; maximum tolerated dose; various
acute toxicities and others (full details and contact information are given on the

website). An appreciation of the ethos behind the toxicity prediction systems is important to understand their capabilities. TOPKAT models are developed from large heterogeneous databases of compounds, normally obtained from sources such as the RTECS database, and the open literature. The toxicity of these compounds may thus be measured by a variety of different methods in a number of different laboratories. Relationships are sought between the toxicity and any of 1000s of different physico-chemical and structural indices. These indices are normally based on topological and electrotopological properties of the whole molecules and individual atoms within them [29]. It is thus often difficult to assign any mechanistic meaning and thus confidence to these models. As an example, the Accelrys web pages (see Tab. 1) include details of a number of sample calculations, a typical one being for the prediction of the rat oral LD_{50} of 2-(methylamino)pyridine. The estimate of toxicity is derived from six electrotopological indices and two whole molecule topological indices. A particular strength of the TOPKAT system is that it provides an estimate of confidence that can be attached to the prediction. The so-called "Optimum Prediction Space" will indicate whether the prediction has been made from information (in terms of the training set of molecules) similar to that of the predicted molecule.

Another well recognized prediction methodology is the computer automated structure evaluation (CASE) technique. This was developed by Klopman and co-workers [30, 31] and the CASE technique drives a number of systems including CASETOX, CASE, MULTICASE and TOXALERT (further details are available from the internet site listed in Table 1). Predictive models have been developed for a number of toxicological endpoints including carcinogenicity; mutagenicity; teratogenicity; acute toxicities as well as physico-chemical properties. The CASE models are derived from large and heterogeneous data sets. Compounds are split into fragments ranging from two to n atoms (though fragments greater than eight atoms are likely to be unwieldy). The fragments are then assessed statistically to determine if they may promote the biological activity (biophores) or decrease it (biophobes). Once fragments have been identified, they can be used either as "descriptors" in a regression-type approach to predict toxicity, or occasionally as structural alerts for rule-based systems. As with TOPKAT, this approach requires large sets of toxicological data, which will inevitably include compilations from the open literature. Both techniques do, however, provide predictive models from such data in a short period of time. The CASE approach lacks a mechanistic approach to identify the fragments (it is simply a statistical analysis), although it is suggested that mechanistic interpretation of the fragments can be applied *a posteriori*. Recently the United States Food and Drug Administration have come to a Cooperative Research and Development Agreement (CRADA) with Multicase Inc. to develop the carcinogenicity model, with the inclusion of proprietary regulatory data [32].

Expert systems based on existing knowledge
There are a number of expert systems that make predictions of toxicity from a "rule-based" approach. These are expert systems in their purest form, which cap-

ture the knowledge of an expert for utilization by a non-expert. The power and utility of these systems is reliant upon two items: firstly adequate software is required to comprehend and interpret chemical structures; and secondly knowledge is needed to form the rule-base of the expert systems. The former is well developed and a number of software packages are commercially available as detailed herein; the latter, for many toxicological endpoints, is still at a rudimentary level.

The software packages developed by LHASA Ltd (see internet site listed in Tab. 1) provide a good illustration of the systems available to predict toxicological and metabolic endpoints. LHASA Ltd. itself is a unique company amongst the expert system providers. LHASA Ltd. has charitable status and is the coordinator for a collaborative group of "customers" who purchase its products (in particular DEREK (Deductive Estimation of Risk from Existing Knowledge) for Windows™). The collaborative group includes members from the pharmaceutical, agrochemical, and personal product industries, as well as regulatory agencies, from Europe and North America. At the time of writing there are over 20 members in the collaborative group. Members of the group are given the opportunity to contribute their own knowledge to the development of new rules [33].

Probably the most developed product from LHASA is the DEREK for Windows™ software, which provides qualitative predictions of toxicity from its rulebase. As with all such systems, the concept is simple: namely that if a particular molecular fragment is known to cause toxicity in one compound, if it is found in another compound the same toxicity will be observed. The system is driven by the LHASA software originally written in the CHMTRN language for the prediction of chemical synthesis and reactions. Examples of the rules (i.e., toxicity associated with a particular molecular fragment) contained within the knowledge-base are listed in Table 2. The knowledge base contains rules for a number of endpoints including: skin sensitization, irritancy, mutagenicity, carcinogenicity and many

Table 2. Examples of rules from the DEREK toxicity knowledge base

Chemical structure (fragment)	Toxicity	Reference
α-halo ketone	irritancy, lachrymation	[34]
organic isocyanate	respiratory sensitization	[34]
trialkyl tin compound	uncoupler, cerebral oedema, eye irritant	[34]
dinitrophenol	uncoupler	[34]
allyl halide	mutagenicity	[34]
resemblance to tetrachloro-dibenzodioxin, -dibenzofuran, -azobenzene, -azoxybenzene	mutagenicity, carcinogenicity, high acute toxicity	[34]
phenyl esters	skin sensitization	[35]
β-lactams	skin sensitization	[35]
catechols and o-alkyl precursors	skin sensitization	[35]
activated N-heterocycle halides	skin sensitization	[35]

others. It should be noted that there appears to be considerable variety in the number and quality of rules for different endpoints; skin sensitization for instance has a well developed set of rules, whereas other endpoints such as anticholinesterase activity are less developed, and may comprise merely a handful of rules. The rules themselves again provide the power and accuracy of the predictions. The DEREK rulebase has not recently been published and thus is not open for inspection. Examples of the rules are provided, however, by Sanderson and Earnshaw [34] who list 49 rules for a wide variety of toxicities and Barratt et al. [35] who list 40 rules specifically for skin sensitization. Other toxicological prediction software developed by LHASA includes StAR (Standardised Argument Report), which is designed to provide the user with more information following a "Logic of Argumentation" process.

The METEOR program from LHASA is a rule-based system for the prediction of metabolism [33]. It consists of a knowledge base of biotransformations that describe the metabolic reactions catalyzed by specific enzymes, or that are related to specific substrates. Again, the knowledge is utilized in a form such that if a particular biotransformation is associated with a molecular fragment, and that molecular fragment is found in another compound, then that biotransformation will be assumed to occur. There is currently little open information regarding the number or quality of the rules in the METEOR knowledge base.

HazardExpert has been developed and marketed by CompuDrug Chemistry Ltd. (see Tab. 1 for internet site). At its heart is a rule-based expert system that predicts toxicity. Predictions may be made for a variety of toxicity endpoints including mutagenicity, carcinogenicity, irritancy and several others. The knowledge has been developed from a number of sources, including the open scientific literature and the reports of the U.S. E.P.A. The software is able to identify molecular fragments within chemical structures. Additionally it takes account of other factors such as species and exposure duration, dose and route. Furthermore, it automatically calculates molecular weight, pK_a, log P and log D (the distribution coefficient). Combining physico-chemical information with the rule-base allows it to make predictions for risk, taking account of the properties and type of exposure.

HazardExpert is linked to a second CompuDrug product MetabolExpert. The latter program attempts to make qualitative predictions of metabolites of compounds. These predictions are again founded upon a molecular fragment-led rule base approach. Linking the two packages allows the user to perform also predictive toxicological risk assessment on the potential metabolites of a drug substance, as well as on the parent molecule.

Another rule-based expert system worthy of note is OncoLogic. This system provides predictions of the carcinogenicity of chemicals, and is marketed by LogiChem Inc. (see Tab. 1 for the internet address). The development of this system is again unique, in that it has been developed by the U.S. E.P.A. Office of Pollution Prevention and Toxics, i.e., a regulatory agency [36]. It is thus provided, on a commercial basis, to allow chemical producers and industries access to the knowledge on which the U.S. E.P.A. will make their decisions. As such, the OncoLogic software contains knowledge based on data for more than 10,000 chemicals, which is formed into a matrix of over 40,000 discrete rules. The clear advantage of this system is its depth of knowledge, and that it allows industrial users to follow and observe the predictions that will be made by regulatory agencies.

A further approach to the prediction of toxicity and metabolism is provided by the Computer-Optimised Molecular Parametric Analysis of Chemical Toxicity (COMPACT) technique of Lewis and co-workers [37]. Rather than being a methodology, COMPACT may be thought of as a philosophy. The aim of the philosophy is to be able to predict, from structural considerations, whether a molecule has the potential to act as a substrate for one or more of the cytochromes P450 (CYPs), or the ability to promote peroxisome proliferation. Whilst oxidative metabolism by CYPs normally results in detoxification, metabolism by, for instance, CYP1 may result in the formation of epoxides.

The COMPACT approach has been built up over a number of years, and is based upon the premise that there are certain, basic, structural requirements of a molecule that make it susceptible to oxidative metabolism. Firstly molecules capable of binding to CYPs are considered to be planar in nature. There is no direct physico-chemical assessment of planarity, although it may be estimated following molecular modeling and 3-D geometry optimization. Thus, molecular planarity is considered to be a function of molecular cross-sectional area (a) and molecular depth (d). It may be defined thus:

$$\text{Molecular Planarity} = \frac{a}{d^2}$$

The second structural requirement for a molecule to be metabolized is that it is capable of chemical oxidation. This can be conveniently calculated from molecular orbital theory as the difference between the energies of the highest occupied and lowest unoccupied molecular orbitals, the so-called electronic activation energy (ΔE). A direct plot of the descriptors for molecular planarity and electronic activation energy for a series of compounds reveals that they are split into categories varying according to the particular enzyme by which they are metabolized [38]. This has been rationalized into the "COMPACT radius," which discriminates between those compounds that are substrates for CYP1 (and so may be carcinogenic) and substrates for other CYPs (which are likely to be non-carcinogenic). Specificity for CYP1 is predicted for compounds that have a value of greater than 15.5 for the COMPACT radius which is defined as:

$$\text{COMPACT radius} = \sqrt{(\Delta E - 9.5)^2 + \left(\frac{a}{d^2} - 7.8\right)^2}$$

The COMPACT radius thus provides a "rule" which may be used to predict toxicity following metabolism, the rule being based upon easily-calculable physico-chemical parameters. It is possible that extensions of this approach will provide

predictive methodologies for the substrates of other enzymes. The prediction rates
from the COMPACT system were found to be considerably improved when combined with the HazardExpert system [39], demonstrating the utility of the battery
approach to prediction.

The COMPACT philosophy has been extended to include to molecular (protein) modeling of the enzymes themselves. This has successfully utilized homology modeling following the determination of the sequences, and isolation of the
enzymes, for rat and man [40].

A further method to enable prediction of potential metabolites is the META
system, which is part of the suite of programs developed by Klopman et al. [41,
42]. The software utilized is similar to that of the other CASE programs (see
Tab. 1 for internet site). Rather than developing the rule-base from a statistical
approach, rules have been taken from standard biochemical sources. Examples of
such rules are provided in Table 3.

Other expert systems to predict toxicity include TOXsys and ONCOis from
SciVision (see Tab. 1 for internet details).

Table 3. Examples of rules from the META metabolism knowledge base (information extracted from
Talafous et al. [42])

Reaction Type	Example of META transformation		Substrate	Enzyme System
	From	To		
aliphatic hydroxylation	$CH_3–CH_2$	$CH_3–CH–OH$	hexane	CYP
deamination	$NH_2–CH–$	$NH=C–$	amphetamine	monoamine oxidase
oxidative dehalogenation	X–C–X *	X–C=O	halothane	CYP
sulfoxidation	S–S–	SO–S–	disulfiram	flavin monooxygenase
nitro reduction	$NO_2–C=$	NO–C=	chloramphenicol	flavin reductase

* X is any halogen

13.3.3   Utility of QSARs and expert systems

A whole host of mathematical models and computational techniques are presented herein to predict metabolism and toxicity. The critical question remains, are
they of any practical use to make predictions of toxicity for drug substances? To
answer this question adequately a number of considerations must be made:
• The particular toxicity endpoint required.
• Whether or not a model is available and whether the training set, or knowledge
  base on which the model is based, is developed sufficiently for the drug in question.
• The nature of the prediction required, e.g., a quantitative or qualitative assessment of toxicity, and whether an estimate of confidence, and the level of confi-

dence, are required.

All these factors must be considered before one makes an attempt to determine whether toxicity prediction is even viable.

A variety of expert systems considered together in a battery-type approach, as opposed to individual QSARs, are more likely to provide the most usable methods to predict toxicity. It can be assumed that, in isolation, individual class or mechanism based QSARs are not going to be suitable to predict the toxicity of heterogeneous drug compounds. The performance of the expert systems is well reviewed, system by system, by Dearden et al. [27]. Basically, all the systems have some predictive value (i.e., they provide estimates of toxicity that are better than random guesses), but no system is infallible. It is not possible to draw any more definitive conclusions regarding comparative capabilities as all the systems were tested using different sets of compounds. To assess the predictivity of toxicological systems a number of approaches can be taken. One can assess how the system predicts the compounds on which the model is based (i.e., is it able to regurgitate the knowledge of the training set?). Various cross-validation or boot-strapping procedures are available which remove compounds in turn and then re-calculate the model. Most stringently, one can check if the model is able to make predictions for a test set of compounds which were not used in the development of the models. This latter approach has been attempted only twice for a large number of toxicity prediction systems. These were the two trials operated by the United States National Toxicological Program (U.S. N.T.P.).

In each trial the N.T.P. solicited predictions for a set of compounds that had yet to be tested. These, to date, remain the only true trials of predictivity, and much has been written and verbalized, both openly and privately, about the use of this type of trial and the scientific meaning of the results. The reader is referred to the excellent review by Benigni [43] and the discussion of this exercise in Dearden et al. [27]) for a full report and interpretation of the results of these trials. It is indicative and instructive, however, that human expert analysis of the data, in conjunction with some biological variables such as Ames Test results, give the best prediction of carcinogenicity. The point should be stressed that such predictions were made incorporating biological information, whereas other users who were denied that information made poorer predictions. The usefulness of any biological data, whether limited or from *in vitro* systems, should not therefore be ignored.

Other aspects regarding system utility relate to what may be termed "user practicalities." Figures are not usually quoted for the speed at which these systems make predictions. Computationally, predictions can be thought of as being virtually "instantaneous," although the exact speed will be obviously dictated by the power and sophistication of the hardware platform and its processors. To all intents and purposes, the time taken for each prediction is negligible, and will become a significant issue only if large combinational chemistry libraries are analyzed. It should be appreciated however that the speed of prediction is mostly meaningless, as it is the interpretation of the prediction that it is relevant to consider. For instance, the TOPKAT system allows the user to obtain a measure of confidence in the estimate, the so-called "optimum prediction space." In addition, the user can search the database on which the prediction was made, to gain an idea, if only subjectively, of how the prediction was made. Similarly the DEREK software, and other rule-based systems, contain a large amount of material which the user can search to obtain information regarding the prediction. All such activities will lengthen considerably the time taken to make a prediction. As such they may compromise the utility of these systems to make adequate predictions for large combinational libraries. Another practical aspect to predictive toxicology is the ultimate compatibility of the hardware and software platforms. This is little addressed, especially in the area of combinatorial chemistry library design (see Chapters 7). Approaches such as the integration of TOPKAT and Q-TOX into the TSAR molecular spreadsheet are, however, encouraging, though it should be obvious that it is only strategic alliances between companies such as combinatorial chemistry and expert system providers that will drive this area of science forward.

13.3.4 Problems and drawbacks of using computer-aided toxicity prediction methods

The drawback with using QSARs and expert systems to predict toxicity is simple to define, yet much more complex to understand and to fix. The drawback is simple: these techniques cannot make adequate predictions for all compounds and for all endpoints! There are many reasons for this:

- Many models are poorly developed in many chemical areas. This is due to there being a paucity of available toxicity data either to build the models, or for their validation. Not only are more data required for modeling, but those data need to be of a high standard (e.g., measured by the same protocol in one laboratory) to provide reliable predictive models. An example of the gaps that are present in the training sets of the systems is that it is only OncoLogic which is able to make predictions for constituents such as fibers, polymers and inorganic-containing compounds. The problem of gaps in the data will be exacerbated by the novel chemistries that are being created by combinatorial chemistry. It is unlikely that a molecular fragment rule-based approach will be able to predict reliably the toxicity of completely novel compounds.
- A considerable amount of expertise is required to interpret and validate the results. The basic premise of an expert system is that it presents the knowledge of an expert for use by a non-expert. It is the contention of the author that users of these systems should not, however, be non-experts. Users require an adequate level of toxicological training and expertise.
- Rule-based expert systems for predicting toxicity may be over-predictive. An example of this was the prediction of skin sensitization by the DEREK software, which predicted a number of non-sensitizers to be sensitizers as these compounds contained a structural alert. Reasons for the over-prediction include lack of knowledge concerning the effect of modulating factors on particular functional groups, and lack of permeability assessment [44]. Whilst the latter

point may be, at least partially, addressed in the StAR and HazardExpert systems, more work is required to predict membrane permeability. Similarly, rule-based systems for the prediction of metabolism will simply provide an idea of the metabolites formed, but not a quantitative estimation of the quantity of each that may be formed.

- There is clearly an issue with the role of mechanisms of action in making toxicological prediction. Systems such as DEREK, HazardExpert and OncoLogic have rule bases developed specifically from a mechanistic viewpoint. Other systems such as TOPKAT and CASE are less, if at all, mechanistically based. The problems of this lack of mechanistic basis to the prediction have never, however, been adequately addressed or quantified.
- The commercial environments in which the systems are placed, effectively as competitors, does little or nothing to assist in the recognition of strengths or weaknesses of each of the systems. It is likely, for instance, that an integrated battery approach to the use of these systems (see next section) will reap the most rewards.
- In many systems there is only limited ability to include proprietary data into the rule-base or predictive system. Some manufacturers such as CASE and TOPKAT will model proprietary data. In other systems, there are opportunities to influence the rule-base either by contributing openly to the rule base, or by the development of proprietary rule bases. Generally, though, these systems require mechanisms to allow users more freely to expand and contribute to the rule-bases.

13.4 Conclusions and perspectives

As described in this chapter, there are a variety of methods and techniques available with which to make predictions of toxicity and metabolism. Before their use a decision is required as to where they will be useful. There are a number of potential uses in the drug development process, including at the library design stage (see also Chapter 7), or in the optimization of a lead compound. As one progresses through the development of a drug the time and expense that may be expended on computer-aided prediction will increase. Concomitantly so will the expected accuracy of the model, and thus its likely complexity. A generalized screen is presented in Figure 1. In this, libraries could be screened to remove the most toxic compounds using basic structure-activity rules and high-throughput screening techniques (see Chapter 4). At this stage it is probable that only "gross structure-based rules" would be of any use, for instance those that will identify toxic compounds with a great degree of certainty (e.g., presence of an aliphatic epoxide). This would allow a large number of compounds to proceed to screening, which will, of course, still require toxicological assessment. Likely hits could be screened in more detail to rank them according to 'desirability'. The top hits could then be more fully assessed with a fuller battery of predictive methods, including molecular modeling and *in vitro* tests. Clearly the key to this whole approach is

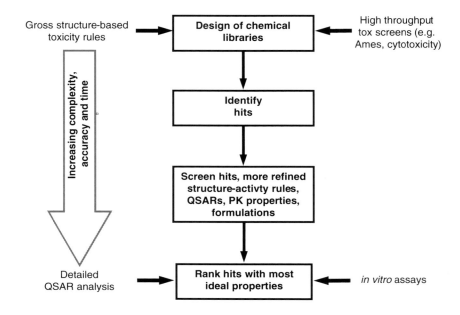

Figure 1. Scheme for the integration of computer-aided predictions into the drug discovery process.

one of integration, not only of the predictive techniques, but also of predictive and *in vitro* modeling.

13.5 References

1 Gorrod JW (ed) (1981) *Testing for toxicity*. Taylor and Francis, London
2 Griffin JP (1998) The evolution of human medicines control from a national to an international perspective. *Adverse Drug React Toxicol Res* 17: 19–50
3 Hodgson E (1997) Toxicity testing and risk assessment. In: E Hodgson, PE Levi (eds): *A textbook of
   modern toxicology*. 2nd Edition. Appleton and Lange, Stamford, Connecticut, 285–338
4 Kroes R (1995) toxicity testing and human health. In: CJ van Leeuwen, JLM Hermens (eds): *Risk
   assessment of chemicals: an introduction*. Kluwer, Dordrecht, The Netherlands, 147–174
5 van Deun K, van Cauteren H, Vandenberghe J et al (1997) Review of alternative methods of carcinogenicity testing and evaluation of human pharmaceuticals. *Adverse Drug React Toxicol Res* 16:
   215–233
6 Rockett JC, Dix DJ (1999) Application of DNA arrays to toxicology. *Environ Health Persp* 107:
   681–685
7 Karcher W, Devillers J (eds) (1990) *Practical applications of quantitative structure-activity relationships (QSAR) in environmental chemistry and toxicology*. Kluwer, Dordrecht
8 Martin YC (1978) *Quantitative drug design*. Marcel Dekker, New York
9 Benfenati E, Gini G (1997) Computational predictive programs (expert systems) in toxicology. *Toxicol*
   119: 213–225
10 Benigni R, Richard AM (1998) Quantitative structure-based modeling applied to characterization and
   prediction of chemical toxicity. *Meth Enzymol* 14: 264–276
11 Cronin MTD (1998) Computer-aided prediction of drug toxicity in high-throughput screening. *Pharm
   Pharmacol Commun* 4: 157–163
12 Cronin MTD, Dearden JC (1995) QSAR in toxicology 1. Prediction of aquatic toxicity. *Quant Struct-*

Act Relat 14: 1–5

13 Cronin MTD, Dearden JC (1995) QSAR in toxicology 2. Prediction of acute mammalian toxicity and interspecies relationships. *Quant Struct-Act Relat* 14: 117–120

14 Cronin MTD, Dearden JC (1995) QSAR in toxicology 3. Prediction of chronic toxicities. *Quant Struct-Act Relat* 14: 329–334

15 Cronin MTD, Bowers GS, Sinks GD et al (2000) Structure-toxicity relationships for aliphatic compounds encompassing a variety of mechanisms of toxic action to *Vibrio fischeri. SAR QSAR Environ Res* 11: 301–312

16 Johnson SR, Jurs PC (1997) Prediction of acute mammalian toxicity from molecular structure for a diverse set of substituted anilines using regression analysis and computational neural networks. In: H van de Waterbeemd, Testa B, Folkers G (eds): *Computer-assisted lead find and optimization: current tools for medicinal chemistry.* Wiley-VCH, Basel, 29–48

17 Guilian W, Naibin B (1998) Structure-activity relationships for rat and mouse LD_{50} of miscellaneous alcohols. *Chemosphere* 36: 1475–1483

18 Hatch FT, Colvin ME (1997) Quantitative structure-activity (QSAR) relationships of mutagenic aromatic and heterocyclic amines. *Mut Res* 376: 87–96

19 Hansch C, Telzer BR, Zhang L (1995) Comparative QSAR in toxicology: examples from teratology and cancer chemotherapy of aniline mustards. *Crit Rev Toxicol* 25: 67–89

20 Bartlett A, Dearden JC, Sibley PR (1995) Quantitative structure-activity relationships in the prediction of penicillin immunotoxicity. *Quant Struct-Act Relat* 14: 258–263

21 Cronin MTD, Dearden JC, Moss GP et al (1999) Investigation of the mechanism of flux across human skin *in vitro* by quantitative structure-permeability relationships. *Eur J Pharm Sci* 7: 325–330

22 Lipinski CA, Lombardo F, Dominy BW (1997) Experimental and computational approaches to estimate solubility and permeability in drug discovery and development settings. *Adv drug Del Rev* 23: 3–25

23 Hansch C, Zhang L (1993) Quantitative structure-activity relationships of cytochrome P-450. *Drug Metab Rev* 25: 1–48

24 Buchwald P, Bodor N (1999) Quantitative structure-metabolism relationships: steric and nonsteric effects in the enzymatic hydrolysis of non-congener carboxylic esters. *J Med Chem* 42: 5160–5168

25 Ghauri FY, Blackledge CA, Glen RC (1992) Quantitative structure metabolism relationships for substituted benzoic-acids in the rat – computational chemistry, NMR-spectroscopy and pattern-recognition studies. *Biochem Pharmacol* 44: 1935–1946

26 Cronin MTD, Dearden JC (1995) QSAR in toxicology 4. Prediction of non-lethal mammalian toxicological endpoints, and expert systems for toxicity prediction. *Quant Struct-Act Relat* 14: 518–523

27 Dearden JC, Barratt MD, Benigni R et al (1997) The development and validation of expert systems for predicting toxicity *ATLA* 25: 223–252

28 Kaiser KLE, Dearden JC, Klein W et al (1999) A note of caution to users of ECOSAR. *Water Qual Res J Canada* 34: 179–182

29 Enslein K (1988) An overview of structure activity relationships as an alternative to testing in animals for carcinogenicity, mutagenicity, dermal and eye irritation and acute oral toxicity. *Toxicol Indust Health* 4: 479–498

30 Klopman G (1984) Artifical intelligence approach to structure-activity studies. Computer automated structure evaluation of biological activity of organic molecules. *J Am Chem Soc* 106: 7315–7320

31 Klopman G (1992) Multi-CASE: a hierarchical computer automated structure evaluation program. *Quant Struct-Act Relat* 11: 176–184

32 Matthews EJ, Contrera JF (1998) A new highly specific method for predicting the carcinogenic potential of pharmaceuticals in rodents using enhanced MCASE QSAR-ES software. *Regul Toxicol Pharmcol* 28: 242–264

33 Greene N, Judson PN, Langowski JJ (1999) Knowledge-based expert systems for toxicity and metabolism prediction: DEREK, StAR and METEOR. *SAR QSAR Environ Res* 10: 299–314

34 Sanderson DM, Earnshaw CG (1991) Computer prediction of possible toxic action from chemical structure; the DEREK system. *Human Exp Toxicol* 10: 261–273

35 Barratt MD, Basketter DA, Chamberlain M et al (1994) An expert system rulebase for identifying contact allergens. *Toxicol in vitro* 8: 1053–1060

36 Woo Y-T, Lai D, Argus M et al (1995) Development of structure-activity relationship rules for predicting carcinogenic potential of chemicals. *Toxicol Lett* 79: 219–228

37 Lewis DFV, Ioannides C, Parke DV (1998) An improved and updated version of the compact procedure for the evaluation of P450-mediated chemical activation. *Drug Metab Rev* 30: 709–737

38 Parke DV, Ioannides C, Lewis DFV (1990) The Safety evaluation of drugs and chemicals by the use of computer optimized molecular parametric analysis of chemical toxicity (COMPACT). *ATLA* 18: 91–102

39 Brown SJ, Raja AA, Lewis DFV (1994) A comparison between COMPACT and Hazard Expert evaluations for 80 chemicals tested by the NTP/NCI rodent bioassay. *ATLA* 22: 482–500

40 Lewis DFV, Lake BG, George SG et al (1999) Molecular modelling of CYPI family enzymes CYP1A1, CYP1A2, CYP1A6 and CYP1B1 based on sequence homology with CYP102. *Toxicol* 139: 53–79

41 Klopman G, Diumayuga M, Galafgous J (1994) META. 1. A program for the evaluation of metabolic transformation of chemicals. *J Chem Inf Comput Sci* 34: 1320–1325

42 Talafous J, Sayre LM, Mieyal JJ et al (1994) META. 2. A dictionary model of mammalian xenobiotic metabolism. *J Chem Inf Comput Sci* 34: 1326–1333

43 Benigni R (1997) The first US National Toxicology Program exercise on the prediction of rodent carcinogenicity: definitive results. *Mut Res* 387: 35–45

44 Barratt MD, Basketter DA (1994) Structure-activity relationships for skin sensitization: an expert system. In: Rougier A, Goldberg AM, Maibach HI (eds): *Alternative methods in toxicology. Vol 10. In vitro skin toxicology. Irritation, phototoxicity, sensitisation.* Mary Ann Liebert, New York, 293–301

Index